Numerical Methods in Engineering with Python

Numerical Methods in Engineering with Python is a text for engineering students and a reference for practicing engineers, especially those who wish to explore the power and efficiency of Python. The choice of numerical methods was based on their relevance to engineering problems. Every method is discussed thoroughly and illustrated with problems involving both hand computation and programming. Computer code accompanies each method and is available on the book web site. This code is made simple and easy to understand by avoiding complex book-keeping schemes, while maintaining the essential features of the method. Python was chosen as the example language because it is elegant, easy to learn and debug, and its facilities for handling arrays are unsurpassed. Moreover, it is an open-source software package that can be downloaded freely on the web. Python is a great language for teaching scientific computation.

Jaan Kiusalaas is a Professor Emeritus in the Department of Engineering Science and Mechanics at the Pennsylvania State University. He has taught computer methods, including finite element and boundary element methods, for over 30 years. He is also the co-author of four other books—*Engineering Mechanics: Statics, Engineering Mechanics: Dynamics, Mechanics of Materials*, and an alternate version of this work with MATLAB® code.

NUMERICAL METHODS IN ENGINEERING WITH
Python

Jaan Kiusalaas
The Pennsylvania State University

CAMBRIDGE
UNIVERSITY PRESS

CAMBRIDGE UNIVERSITY PRESS
Cambridge, New York, Melbourne, Madrid, Cape Town, Singapore, São Paulo

Cambridge University Press
40 West 20th Street, New York, NY 10011-4211, USA

www.cambridge.org
Information on this title: www.cambridge.org/9780521852876

© Jaan Kiusalaas 2005

This book is in copyright. Subject to statutory exception
and to the provisions of relevant collective licensing agreements,
no reproduction of any part may take place without
the written permission of Cambridge University Press.

First published 2005

Printed in the United States of America

A catalogue record for this publication is available from the British Library.

Library of Congress Cataloging in Publication Data

Kiusalaas, Jaan.
Numerical methods in engineering with Python / Jaan Kiusalaas.
   p. cm.
Includes bibliographical references and index.
ISBN-13: 978-0-521-85287-6 (hardback)
ISBN-10: 0-521-85287-0 (hardback)
1. Engineering mathematics – Data processing.   2. Python (Computer program language)   3. Numerical analysis – Data processing.   I. Title.
TA345.K584   2005
620′.001′518 – dc22   2005011265

ISBN-13   978-0-521-85287-6 hardback
ISBN-10   0-521-85287-0 hardback

Cambridge University Press has no responsibility for
the persistence or accuracy of URLs for external or
third-party Internet Web sites referred to in this book
and does not guarantee that any content on such
Web sites is, or will remain, accurate or appropriate.

Contents

Preface vii

1. Introduction to Python 1
2. Systems of Linear Algebraic Equations 27
3. Interpolation and Curve Fitting 103
4. Roots of Equations 142
5. Numerical Differentiation 181
6. Numerical Integration 198
7. Initial Value Problems 248
8. Two-Point Boundary Value Problems 295
9. Symmetric Matrix Eigenvalue Problems 324
10. Introduction to Optimization 381

Appendices 409

Index 419

Preface

This book is targeted primarily toward engineers and engineering students of advanced standing (sophomores, seniors and graduate students). Familiarity with a computer language is required; knowledge of basic engineering mechanics is useful, but not essential.

The text attempts to place emphasis on numerical methods, not programming. Most engineers are not programmers, but problem solvers. They want to know what methods can be applied to a given problem, what are their strengths and pitfalls and how to implement them. Engineers are not expected to write computer code for basic tasks from scratch; they are more likely to utilize functions and subroutines that have been already written and tested. Thus programming by engineers is largely confined to assembling existing pieces of code into a coherent package that solves the problem at hand.

The "piece" of code is usually a function that implements a specific task. For the user the details of the code are unimportant. What matters is the interface (what goes in and what comes out) and an understanding of the method on which the algorithm is based. Since no numerical algorithm is infallible, the importance of understanding the underlying method cannot be overemphasized; it is, in fact, the rationale behind learning numerical methods.

This book attempts to conform to the views outlined above. Each numerical method is explained in detail and its shortcomings are pointed out. The examples that follow individual topics fall into two categories: hand computations that illustrate the inner workings of the method and small programs that show how the computer code is utilized in solving a problem. Problems that require programming are marked with ■.

The material consists of the usual topics covered in an engineering course on numerical methods: solution of equations, interpolation and data fitting, numerical differentiation and integration, solution of ordinary differential equations and eigenvalue problems. The choice of methods within each topic is tilted toward relevance to engineering problems. For example, there is an extensive discussion of symmetric, sparsely populated coefficient matrices in the solution of simultaneous equations. In the same vein, the solution of eigenvalue problems concentrates on methods that efficiently extract specific eigenvalues from banded matrices.

An important criterion used in the selection of methods was clarity. Algorithms requiring overly complex bookkeeping were rejected regardless of their efficiency and robustness. This decision, which was taken with great reluctance, is in keeping with the intent to avoid emphasis on programming.

The selection of algorithms was also influenced by current practice. This disqualified several well-known historical methods that have been overtaken by more recent developments. For example, the secant method for finding roots of equations was omitted as having no advantages over Brent's method. For the same reason, the multistep methods used to solve differential equations (e.g., Milne and Adams methods) were left out in favor of the adaptive Runge–Kutta and Bulirsch–Stoer methods.

Notably absent is a chapter on partial differential equations. It was felt that this topic is best treated by finite element or boundary element methods, which are outside the scope of this book. The finite difference model, which is commonly introduced in numerical methods texts, is just too impractical in handling multidimensional boundary value problems.

As usual, the book contains more material than can be covered in a three-credit course. The topics that can be skipped without loss of continuity are tagged with an asterisk (*).

The programs listed in this book were tested with Python 2.2.2 and 2.3.4 under Windows XP and Red Hat Linux. The source code can be downloaded from the book's website at

www.cambridge.org/0521852870

The author wishes to express his gratitude to the anonymous reviewers and Professor Andrew Pytel for their suggestions for improving the manuscript. Credit is also due to the authors of *Numerical Recipes* (Cambridge University Press) whose presentation of numerical methods was inspirational in writing this book.

1 Introduction to Python

1.1 General Information

Quick Overview

This chapter is not a comprehensive manual of Python. Its sole aim is to provide sufficient information to give you a good start if you are unfamiliar with Python. If you know another computer language, and presumably you do, it is not difficult to pick up the rest as you go.

Python is an object-oriented language that was developed in late 1980s as a scripting language (the name is derived from the British television show Monty Python's Flying Circus). Although Python is not as well known in engineering circles as some other languages, it has a considerable following in the programming community—in fact, Python is considerably more widespread than Fortran. Python may be viewed as an emerging language, since it is still being developed and refined. In the current state, it is an excellent language for developing engineering applications—it possesses a simple elegance that other programming languages cannot match.

Python programs are not compiled into machine code, but are run by an *interpreter*[1]. The great advantage of an interpreted language is that programs can be tested and debugged quickly, allowing the user to concentrate more on the principles behind the program and less on programming itself. Since there is no need to compile, link and execute after each correction, Python programs can be developed in a much shorter time than equivalent Fortran or C programs. On the negative side, interpreted programs do not produce stand-alone applications. Thus a Python program can be run only on computers that have the Python interpreter installed.

---

[1] The Python interpreter also compiles *byte code*, which helps to speed up execution somewhat.

Introduction to Python

Python has other advantages over mainstream languages that are important in a learning environment:

- Python is open-source software, which means that it is *free*; it is included in most Linux distributions.
- Python is available for all major operating systems (Linux, Unix, Windows, Mac OS etc.). A program written on one system runs without modification on all systems.
- Python is easier to learn and produces more readable code than other languages.
- Python and its extensions are easy to install.

Development of Python was clearly influenced by Java and C++, but there is also a remarkable similarity to MATLAB® (another interpreted language, very popular in scientific computing). Python implements the usual concepts of object-oriented languages such as classes, methods, inheritance etc. We will forego these concepts and use Python strictly as a procedural language.

To get an idea of the similarities between MATLAB and Python, let us look at the codes written in the two languages for solution of simultaneous equations $\mathbf{Ax} = \mathbf{b}$ by Gauss elimination. Here is the function written in MATLAB:

```
function [x,det] = gaussElimin(a,b)
n = length(b);
for k = 1:n-1
    for i = k+1:n
        if a(i,k) ~= 0
            lam = a(i,k)/a(k,k);
            a(i,k+1:n) = a(i,k+1:n) - lam*a(k,k+1:n);
            b(i)= b(i) - lam*b(k);
        end
    end
end
det = prod(diag(a));
for k = n:-1:1
    b(k) = (b(k) - a(k,k+1:n)*b(k+1:n))/a(k,k);
end
x = b;
```

The equivalent Python function is:

```
from numarray import dot
def gaussElimin(a,b):
    n = len(b)
```

```
        for k in range(0,n-1):
            for i in range(k+1,n):
                if a[i,k] != 0.0:
                    lam = a [i,k]/a[k,k]
                    a[i,k+1:n] = a[i,k+1:n] - lam*a[k,k+1:n]
                    b[i] = b[i] - lam*b[k]
        for k in range(n-1,-1,-1):
            b[k] = (b[k] - dot(a[k,k+1:n],b[k+1:n]))/a[k,k]
        return b
```

The command from numarray import dot instructs the interpreter to load the function dot (which computes the dot product of two vectors) from the module numarray. The colon (:) operator, known as the *slicing operator* in Python, works the same way it does in MATLAB and Fortran90—it defines a section of an array.

The statement for k = 1:n-1 in MATLAB creates a loop that is executed with $k = 1, 2, \ldots, n - 1$. The same loop appears in Python as for k in range(n-1). Here the function range(n-1) creates the list $[0, 1, \ldots, n-2]$; k then loops over the elements of the list. The differences in the ranges of k reflect the native offsets used for arrays. In Python all sequences have *zero offset*, meaning that the index of the first element of the sequence is always 0. In contrast, the native offset in MATLAB is 1.

Also note that Python has no end statements to terminate blocks of code (loops, conditionals, subroutines etc.). The body of a block is defined by its *indentation*; hence indentation is an integral part of Python syntax.

Like MATLAB, Python is *case sensitive*. Thus the names n and N would represent different objects.

Obtaining Python

Python interpreter can be downloaded from the Python Language Website www.python.org. It normally comes with a nice code editor called *Idle* that allows you to run programs directly from the editor. For scientific programming we also need the *Numarray* module which contains various tools for array operations. It is obtainable from the Numarray Home Page http://www.stsci.edu/resources/ software_hardware/numarray. Both sites also provide documentation for downloading. If you use Linux or Mac OS, it is very likely that Python is already installed on your machine (but you must still download Numarray).

You should acquire other printed material to supplement the on-line documentation. A commendable teaching guide is *Python* by Chris Fehly, Peachpit Press, CA (2002). As a reference, *Python Essential Reference* by David M. Beazley, New Riders

Publishing (2001) is recommended. By the time you read this, newer editions may be available.

1.2 Core Python

Variables

In most computer languages the name of a variable represents a value of a given type stored in a fixed memory location. The value may be changed, but not the type. This it not so in Python, where variables are *typed dynamically*. The following interactive session with the Python interpreter illustrates this (>>> is the Python prompt):

```
>>> b = 2          # b is integer type
>>> print b
2
>>> b = b * 2.0    # Now b is float type
>>> print b
4.0
```

The assignment b = 2 creates an association between the name b and the *integer* value 2. The next statement evaluates the expression b * 2.0 and associates the result with b; the original association with the integer 2 is destroyed. Now b refers to the *floating* point value 4.0.

The pound sign (#) denotes the beginning of a *comment*—all characters between # and the end of the line are ignored by the interpreter.

Strings

A string is a sequence of characters enclosed in single or double quotes. Strings are *concatenated* with the plus (+) operator, whereas *slicing* (:) is used to extract a portion of the string. Here is an example:

```
>>> string1 = 'Press return to exit'
>>> string2 = 'the program'
>>> print string1 + ' ' + string2   # Concatenation
Press return to exit the program
>>> print string1[0:12]             # Slicing
Press return
```

1.2 Core Python

A string is an *immutable* object—its individual characters cannot be modified with an assignment statement and it has a fixed length. An attempt to violate immutability will result in TypeError, as shown below.

```
>>> s = 'Press return to exit'
>>> s[0] = 'p'
Traceback (most recent call last):
  File ''<pyshell#1>'', line 1, in ?
    s[0] = 'p'
TypeError: object doesn't support item assignment
```

Tuples

A *tuple* is a sequence of *arbitrary objects* separated by commas and enclosed in parentheses. If the tuple contains a single object, the parentheses may be omitted. Tuples support the same operations as strings; they are also immutable. Here is an example where the tuple rec contains another tuple (6,23,68):

```
>>> rec = ('Smith','John',(6,23,68))   # This is a tuple
>>> lastName,firstName,birthdate = rec  # Unpacking the tuple
>>> print firstName
John
>>> birthYear = birthdate[2]
>>> print birthYear
68
>>> name = rec[1] + ' ' + rec[0]
>>> print name
John Smith
>>> print rec[0:2]
('Smith', 'John')
```

Lists

A list is similar to a tuple, but it is *mutable,* so that its elements and length can be changed. A list is identified by enclosing it in brackets. Here is a sampling of operations that can be performed on lists:

```
>>> a = [1.0, 2.0, 3.0]      # Create a list
>>> a.append(4.0)            # Append 4.0 to list
>>> print a
[1.0, 2.0, 3.0, 4.0]
```

```
>>> a.insert(0,0.0)          # Insert 0.0 in position 0
>>> print a
[0.0, 1.0, 2.0, 3.0, 4.0]
>>> print len(a)             # Determine length of list
5
>>> a[2:4] = [1.0, 1.0] # Modify selected elements
>>> print a
[0.0, 1.0, 1.0, 1.0, 1.0, 4.0]
```

If *a* is a mutable object, such as a list, the assignment statement b = a does not result in a new object *b*, but simply creates a new reference to *a*. Thus any changes made to *b* will be reflected in *a*. To create an independent copy of a list *a*, use the statement c = a[:], as illustrated below.

```
>>> a = [1.0, 2.0, 3.0]
>>> b = a                # 'b' is an alias of 'a'
>>> b[0] = 5.0           # Change 'b'
>>> print a
[5.0, 2.0, 3.0]          # The change is reflected in 'a'
>>> c = a[:]             # 'c' is an independent copy of 'a'
>>> c[0] = 1.0           # Change 'c'
>>> print a
[5.0, 2.0, 3.0]          # 'a' is not affected by the change
```

Matrices can represented as nested lists with each row being an element of the list. Here is a 3 × 3 matrix *a* in the form of a list:

```
>>> a = [[1, 2, 3], \
         [4, 5, 6], \
         [7, 8, 9]]
>>> print a[1]           # Print second row (element 1)
[4, 5, 6]
>>> print a[1][2]        # Print third element of second row
6
```

The backslash (\) is Python's *continuation character*. Recall that Python sequences have zero offset, so that a[0] represents the first row, a[1] the second row, etc. With very few exceptions we do not use lists for numerical arrays. It is much more convenient

1.2 Core Python

to employ *array objects* provided by the `numarray` module, (an extension of Python language). Array objects will be discussed later.

Arithmetic Operators

Python supports the usual arithmetic operators:

+	Addition
−	Subtraction
*	Multiplication
/	Division
**	Exponentiation
%	Modular division

Some of these operators are also defined for strings and sequences as illustrated below.

```
>>> s = 'Hello '
>>> t = 'to you'
>>> a = [1, 2, 3]
>>> print 3*s                # Repetition
Hello Hello Hello
>>> print 3*a                # Repetition
[1, 2, 3, 1, 2, 3, 1, 2, 3]
>>> print a + [4, 5]         # Append elements
[1, 2, 3, 4, 5]
>>> print s + t              # Concatenation
Hello to you
>>> print 3 + s              # This addition makes no sense
Traceback (most recent call last):
  File ''<pyshell#9>'', line 1, in ?
    print n + s
TypeError: unsupported operand types for +: 'int' and 'str'
```

Python 2.0 and later versions also have *augmented assignment operators,* such as $a\mathrel{+}=b$, that are familiar to the users of C. The augmented operators and the equivalent arithmetic expressions are shown in the following table.

a += b	a = a + b
a -= b	a = a - b
a *= b	a = a*b
a /= b	a = a/b
a **= b	a = a**b
a %= b	a = a%b

Comparison Operators

The comparison (relational) operators return 1 for true and 0 for false. These operators are

<	Less than
>	Greater than
<=	Less than or equal to
>=	Greater than or equal to
==	Equal to
!=	Not equal to

Numbers of different type (integer, floating point etc.) are converted to a common type before the comparison is made. Otherwise, objects of different type are considered to be unequal. Here are a few examples:

```
>>> a = 2           # Integer
>>> b = 1.99        # Floating point
>>> c = '2'         # String
>>> print a > b
1
>>> print a == c
0
>>> print (a > b) and (a != c)
1
>>> print (a > b) or (a == b)
1
```

Conditionals

The `if` construct

> `if` *condition*:
>   *block*

executes a block of statements (which must be indented) if the condition returns true. If the condition returns false, the block skipped. The `if` conditional can be followed by any number of `elif` (short for "else if") constructs

> `elif` *condition*:
>   *block*

which work in the same manner. The `else` clause

> `else`:
>   *block*

can be used to define the block of statements which are to be executed if none of the `if`-`elif` clauses are true. The function sign_of_a below illustrates the use of the conditionals.

```
def sign_of_a(a):
    if a < 0.0:
        sign = 'negative'
    elif a > 0.0:
        sign = 'positive'
    else:
        sign = 'zero'
    return sign

a = 1.5
print 'a is ' + sign_of_a(a)
```

Running the program results in the output

```
a is positive
```

Loops

The `while` construct

> `while` *condition*:
>   *block*

executes a block of (indented) statements if the condition is true. After execution of the block, the condition is evaluated again. If it is still true, the block is executed again. This process is continued until the condition becomes false. The `else` clause

> `else:`
>   *block*

can be used to define the block of statements which are to be executed if condition is false. Here is an example that creates the list $[1, 1/2, 1/3, \ldots]$:

```
nMax = 5
n = 1
a = []                  # Create empty list
while n < nMax:
    a.append(1.0/n)     # Append element to list
    n = n + 1
print a
```

The output of the program is

```
[1.0, 0.5, 0.33333333333333331, 0.25]
```

We met the `for` statement before in Art. 1.1. This statement requires a target and a sequence (usually a list) over which the target loops. The form of the construct is

> `for` *target* `in` *sequence*:
>   *block*

You may add an `else` clause which is executed after the `for` loop has finished. The previous program could be written with the `for` construct as

1.2 Core Python

```
nMax = 5
a = []
for n in range(1,nMax):
    a.append(1.0/n)
print a
```

Here *n* is the target and the list [1,2,...,nMax-1], created by calling the range function, is the sequence.

Any loop can be terminated by the break statement. If there is an else cause associated with the loop, it is not executed. The following program, which searches for a name in a list, illustrates the use of break and else in conjunction with a for loop:

```
list = ['Jack', 'Jill', 'Tim', 'Dave']
name = eval(raw_input('Type a name: '))   # Python input prompt
for i in range(len(list)):
    if list[i] == name:
        print name,'is number',i + 1,'on the list'
        break
else:
    print name,'is not on the list'
```

Here are the results of two searches:

```
Type a name: 'Tim'
Tim is number 3 on the list

Type a name: 'June'
June is not on the list
```

Type Conversion

If an arithmetic operation involves numbers of mixed types, the numbers are automatically converted to a common type before the operation is carried out. Type conversions can also achieved by the following functions:

Introduction to Python

`int(a)`	Converts *a* to integer
`long(a)`	Converts *a* to long integer
`float(a)`	Converts *a* to floating point
`complex(a)`	Converts to complex $a + 0j$
`complex(a,b)`	Converts to complex $a + bj$

The above functions also work for converting strings to numbers as long as the literal in the string represents a valid number. Conversion from float to an integer is carried out by truncation, not by rounding off. Here are a few examples:

```
>>> a = 5
>>> b = -3.6
>>> d = '4.0'
>>> print a + b
1.4
>>> print int(b)
-3
>>> print complex(a,b)
(5-3.6j)
>>> print float(d)
4.0
>>> print int(d)   # This fails: d is not Int type
Traceback (most recent call last):
  File ''<pyshell#7>'', line 1, in ?
    print int(d)
ValueError: invalid literal for int(): 4.0
```

Mathematical Functions

Core Python supports only a few mathematical functions. They are:

`abs(a)`	Absolute value of a
`max(sequence)`	Largest element of *sequence*
`min(sequence)`	Smallest element of *sequence*
`round(a,n)`	Round a to n decimal places
`cmp(a,b)`	Returns $\begin{cases} -1 \text{ if } a < b \\ 0 \text{ if } a = b \\ 1 \text{ if } a > b \end{cases}$

1.2 Core Python

The majority of mathematical functions are available in the `math` module.

Reading Input

The intrinsic function for accepting user input is

$$\texttt{raw\_input}(prompt)$$

It displays the prompt and then reads a line of input which is converted to a *string*. To convert the string into a numerical value use the function

$$\texttt{eval}(string)$$

The following program illustrates the use of these functions:

```
a = raw_input('Input a: ')
print a, type(a)            # Print a and its type
b = eval(a)
print b,type(b)             # Print b and its type
```

The function `type(a)` returns the type of the object *a*; it is a very useful tool in debugging. The program was run twice with the following results:

```
Input a: 10.0
10.0 <type 'str'>
10.0 <type 'float'>

Input a: 11**2
11**2 <type 'str'>
121 <type 'int'>
```

A convenient way to input a number and assign it to the variable *a* is

$$\texttt{a = eval(raw\_input}(prompt))$$

Printing Output

Output can be displayed with the print statement:

> print *object1, object2, ...*

which converts *object1, object2* etc. to strings and prints them on the same line, separated by spaces. The *newline* character '\n' can be uses to force a new line. For example,

```
>>> a = 1234.56789
>>> b = [2, 4, 6, 8]
>>> print a,b
1234.56789 [2, 4, 6, 8]
>>> print 'a =',a, '\nb =',b
a = 1234.56789
b = [2, 4, 6, 8]
```

The *modulo operator* (%) can be used to format a tuple. The form of the conversion statement is

> '*%format1 %format2 ···*' % *tuple*

where *format1, format2* ··· are the format specifications for each object in the tuple. Typically used format specifications are

wd	Integer
$w.d\mathtt{f}$	Floating point notation
$w.d\mathtt{e}$	Exponential notation

where w is the width of the field and d is the number of digits after the decimal point. The output is right-justified in the specified field and padded with blank spaces (there are provisions for changing the justification and padding). Here are a couple of examples:

```
>>> a = 1234.56789
>>> n = 9876
>>> print '%7.2f' % a
1234.57
>>> print 'n = %6d' % n    # Pad with 2 spaces
n =   9876
```

```
>>> print 'n = %06d' %n   # Pad with 2 zeroes
n = 009876
>>> print '%12.4e %6d' % (a,n)
  1.2346e+003   9876
```

Error Control

When an error occurs during execution of a program an *exception* is raised and the program stops. Exceptions can be caught with `try` and `except` statements:

> try:
>   *do something*
> except *error*:
>   *do something else*

where *error* is the name of a built-in Python exception. If the exception *error* is not raised, the `try` block is executed; otherwise the execution passes to the `except` block. All exceptions can be caught by omitting *error* from the `except` statement.

Here is a statement that raises the exception `ZeroDivisionError`:

```
>>> c = 12.0/0.0
Traceback (most recent call last):
  File ''<pyshell#0>'', line 1, in ?
    c = 12.0/0.0
ZeroDivisionError: float division
```

This error can be caught by

```
try:
    c = 12.0/0.0
except ZeroDivisionError:
    print 'Division by zero'
```

1.3 Functions and Modules

Functions

The structure of a Python function is

> def *func_name*(*param1, param2,...*):
>   *statements*
>   return *return_values*

where *param1, param2,...* are the parameters. A parameter can be any Python object, including a function. Parameters may be given default values, in which case the parameter in the function call is optional. If the `return` statement or *return_values* are omitted, the function returns the null object.

The following example computes the first two derivatives of arctan(x) by finite differences:

```
from math import arctan
def  finite_diff(f,x,h=0.0001):    # h has a default value
    df =(f(x+h) - f(x-h))/(2.0*h)
    ddf =(f(x+h) - 2.0*f(x) + f(x-h))/h**2
    return df,ddf
x = 0.5
df,ddf = finite_diff(arctan,x)    # Uses default value of h
print 'First derivative  =',df
print 'Second derivative =',ddf
```

Note that `arctan` is passed to `finite_diff` as a parameter. The output from the program is

```
First derivative  = 0.799999999573
Second derivative = -0.639999991892
```

If a mutable object, such as a list, is passed to a function where it is modified, the changes will also appear in the calling program. Here is an example:

```
def squares(a):
    for i in range(len(a)):
        a[i] = a[i]**2

a = [1, 2, 3, 4]
squares(a)
print a
```

The output is

```
[1, 4, 9, 16]
```

Modules

It is sound practice to store useful functions in modules. A module is simply a file where the functions reside; the name of the module is the name of the file. A module can be loaded into a program by the statement

```
from module_name import *
```

Python comes with a large number of modules containing functions and methods for various tasks. Two of the modules are described briefly in the next section. Additional modules, including graphics packages, are available for downloading on the Web.

1.4 Mathematics Modules

`math` Module

Most mathematical functions are not built into core Python, but are available by loading the `math` module. There are three ways of accessing the functions in a module. The statement

```
from math import *
```

loads *all* the function definitions in the `math` module into the current function or module. The use of this method is discouraged because it is not only wasteful, but can also lead to conflicts with definitions loaded from other modules.

You can load selected definitions by

```
from math import func1, func2,...
```

as illustrated below.

```
>>> from math import log,sin
>>> print log(sin(0.5))
-0.735166686385
```

Introduction to Python

The third method, which is used by the majority of programmers, is to make the module available by

```
import math
```

The functions in the module can then be accessed by using the module name as a prefix:

```
>>> import math
>>> print math.log(math.sin(0.5))
-0.735166686385
```

The contents of a module can be printed by calling dir(*module*). Here is how to obtain a list of the functions in the math module:

```
>>> import math
>>> dir(math)
['__doc__', '__name__', 'acos', 'asin', 'atan',
 'atan2', 'ceil', 'cos', 'cosh', 'e', 'exp', 'fabs',
 'floor', 'fmod', 'frexp', 'hypot', 'ldexp', 'log',
 'log10', 'modf', 'pi', 'pow', 'sin', 'sinh', 'sqrt',
 'tan', 'tanh']
```

Most of these functions are familiar to programmers. Note that the module includes two constants: π and e.

cmath Module

The cmath module provides many of the functions found in the math module, but these accept complex numbers. The functions in the module are:

```
['__doc__', '__name__', 'acos', 'acosh', 'asin', 'asinh',
 'atan', 'atanh', 'cos', 'cosh', 'e', 'exp', 'log',
 'log10', 'pi', 'sin', 'sinh', 'sqrt', 'tan', 'tanh']
```

Here are examples of complex arithmetic:

```
>>> from cmath import sin
>>> x = 3.0 -4.5j
>>> y = 1.2 + 0.8j
>>> z = 0.8
```

```
>>> print x/y
(-2.56205313375e-016-3.75j)
>>> print sin(x)
(6.35239299817+44.5526433649j)
>>> print sin(z)
(0.7173560909+0j)
```

1.5 numarray Module

General Information

The numarray module[2] is not a part of the standard Python release. As pointed out before, it must be obtained separately and installed (the installation is very easy). The module introduces *array objects* which are similar to lists, but can be manipulated by numerous functions contained in the module. The size of the array is immutable and no empty elements are allowed.

The complete set of functions in numarray is too long to be printed in its entirety. The list below is limited to the most commonly used functions.

```
['Complex', 'Complex32', 'Complex64', 'Float',
'Float32', 'Float64', 'abs', 'arccos',
'arccosh', 'arcsin', 'arcsinh', 'arctan',
'arctan2', 'arctanh', 'argmax', 'argmin',
'cos', 'cosh', 'diagonal', 'dot', 'e', 'exp',
'floor', 'identity', 'innerproduct', 'log',
'log10', 'matrixmultiply', 'maximum', 'minimum',
'numarray', 'ones', 'pi', 'product', 'sin', 'sinh',
'size', 'sqrt', 'sum', 'tan', 'tanh', 'trace',
'transpose', 'zeros']
```

Creating an Array

Arrays can be created in several ways. One of them is to use the array function to turn a list into an array:

$$\text{array}(list, \text{type} = type\_specification)$$

---

[2] *Numarray* is based on an older Python array module called *Numeric*. Their interfaces and capabilities are very similar and they are largely compatible. Although *Numeric* is still available, it is no longer supported.

Here are two examples of creating a 2 × 2 array with floating-point elements:

```
>>> from numarray import array,Float
>>> a = array([[2.0, -1.0],[-1.0, 3.0]])
>>> print a
[[ 2. -1.]
 [-1.  3.]]
>>> b = array([[2, -1],[-1, 3]],type = Float)
>>> print b
[[ 2. -1.]
 [-1.  3.]]
```

Other available functions are

$$\texttt{zeros}((dim1, dim2), \texttt{type} = type\_specification)$$

which creates a *dim1* × *dim2* array and fills it with zeroes, and

$$\texttt{ones}((dim1, dim2), \texttt{type} = type\_specification)$$

which fills the array with ones. The default `type` in both cases is `Int`.

Finally, there is the function

$$\texttt{arange}(from, to, increment)$$

which works just like the `range` function, but returns an array rather than a list. Here are examples of creating arrays:

```
>>> from numarray import arange,zeros,ones,Float
>>> a = arange(2,10,2)
>>> print a
[2 4 6 8]
>>> b = arange(2.0,10.0,2.0)
>>> print b
[ 2.  4.  6.  8.]
>>> z = zeros((4))
>>> print z
[0 0 0 0]
```

1.5 numarray Module

```
>>> y = ones((3,3),type= Float)
>>> print y
[[ 1.  1.  1.]
 [ 1.  1.  1.]
 [ 1.  1.  1.]]
```

Accessing and Changing Array Elements

If a is a rank-2 array, then a[i, j] accesses the element in row i and column j, whereas a[i] refers to row i. The elements of an array can be changed by assignment as shown below.

```
>>> from numarray import *
>>> a = zeros((3,3),type=Float)
>>> a[0] = [2.0, 3.1, 1.8]    # Change a row
>>> a[1,1] = 5.2              # Change an element
>>> a[2,0:2] = [8.0, -3.3]    # Change part of a row
>>> print a
[[ 2.   3.1  1.8]
 [ 0.   5.2  0. ]
 [ 8.  -3.3  0. ]]
```

Operations on Arrays

Arithmetic operators work differently on arrays than they do on tuples and lists—the operation is *broadcast* to all the elements of the array; that is, the operation is applied to each element in the array. Here are examples:

```
>>> from numarray import array
>>> a = array([0.0, 4.0, 9.0, 16.0])
>>> print a/16.0
[ 0.      0.25    0.5625  1.    ]
>>> print a - 4.0
[ -4.   0.   5.  12.]
```

The mathematical functions available in numarray are also broadcast, as illustrated below

```
>>> from numarray import array,sqrt,sin
>>> a = array([1.0, 4.0, 9.0, 16.0])
```

```
>>> print sqrt(a)
[ 1.  2.  3.  4.]
>>> print sin(a)
[ 0.84147098 -0.7568025   0.41211849 -0.28790332]
```

Functions imported from the math module will work on the individual elements, of course, but not on the array itself. Here is an example:

```
>>> from numarray import array
>>> from math import sqrt
>>> a = array([1.0, 4.0, 9.0, 16.0])
>>> print sqrt(a[1])
2.0
>>> print sqrt(a)
Traceback (most recent call last):
    ⋮
TypeError: Only rank-0 arrays can be cast to floats.
```

Array Functions

There are numerous array functions in numarray that perform matrix operations and other useful tasks. Here are a few examples:

```
>>> from numarray import *
>>> a = array([[ 4.0, -2.0,  1.0], \
               [-2.0,  4.0, -2.0], \
               [ 1.0, -2.0,  3.0]])
>>> b = array([1.0, 4.0, 2.0])
>>> print dot(b,b)              # Dot product
21.0
>>> print matrixmultiply(a,b)   # Matrix multiplication
[ -2.  10.  -1.]
>>> print diagonal(a)           # Principal diagonal
[ 4.  4.  3.]
>>> print diagonal(a,1)         # First subdiagonal
[-2. -2.]
>>> print trace(a)              # Sum of diagonal elements
11.0
```

```
>>> print argmax(b)           # Index of largest element
1
>>> print identity(3)         # Identity matrix
[[1 0 0]
 [0 1 0]
 [0 0 1]]
```

Copying Arrays

We explained before that if *a* is a mutable object, such as a list, the assignment statement b = a does not result in a new object *b*, but simply creates a new reference to *a*, called a *deep copy*. This also applies to arrays. To make an independent copy of an array *a*, use the copy method in the numarray module:

```
b = a.copy()
```

1.6 Scoping of Variables

Namespace is a dictionary that contains the names of the variables and their values. The namespaces are automatically created and updated as a program runs. There are three levels of namespaces in Python:

- Local namespace, which is created when a function is called. It contains the variables passed to the function as arguments and the variables created within the function. The namespace is deleted when the function terminates. If a variable is created inside a function, its scope is the function's local namespace. It is not visible outside the function.
- A global namespace is created when a module is loaded. Each module has its own namespace. Variables assigned in a global namespace are visible to any function within the module.
- Built-in namespace is created when the interpreter starts. It contains the functions that come with the Python interpreter. These functions can be accessed by any program unit.

When a name is encountered during execution of a function, the interpreter tries to resolve it by searching the following in the order shown: (1) local namespace, (2) global namespace, and (3) built-in namespace. If the name cannot be resolved, Python raises a NameError exception.

Since the variables residing in a global namespace are visible to functions within the module, it is not necessary to pass them to the functions as arguments (although is good programming practice to do so), as the following program illustrates.

```
def divide():
    c = a/b
    print 'a/b =',c


a = 100.0
b = 5.0
divide()
>>>
a/b = 20.0
```

Note that the variable c is created inside the function divide and is thus not accessible to statements outside the function. Hence an attempt to move the print statement out of the function fails:

```
def divide():
    c = a/b

a = 100.0
b = 5.0
divide()
print 'a/b =',c

>>>
Traceback (most recent call last):
  File ''C:\Python22\scope.py'', line 8, in ?
    print c
NameError: name 'c' is not defined
```

1.7 Writing and Running Programs

When the Python editor *Idle* is opened, the user is faced with the prompt >>>, indicating that the editor is in interactive mode. Any statement typed into the editor is immediately processed upon pressing the enter key. The interactive mode is a good way to learn the language by experimentation and to try out new programming ideas.

Opening a new window places Idle in the batch mode, which allows typing and saving of programs. One can also use a text editor to enter program lines, but Idle has Python-specific features, such as color coding of keywords and automatic indentation, that make work easier. Before a program can be run, it must be saved as a Python file with the .py extension, e.g., myprog.py. The program can then be executed by typing

1.7 Writing and Running Programs

`python myprog.py`; in Windows, double-clicking on the program icon will also work. But beware: the program window closes immediately after execution, before you get a chance to read the output. To prevent this from happening, conclude the program with the line

```
raw_input('press return')
```

Double-clicking the program icon also works in Unix and Linux if the first line of the program specifies the path to the Python interpreter (or a shell script that provides a link to Python). The path name must be preceded by the symbols #!. On my computer the path is `/usr/bin/python`, so that all my programs start with the line

```
#!/usr/bin/python
```

On multiuser systems the path is usually `/usr/local/bin/python`.

When a module is loaded into a program for the first time with the `import` statement, it is compiled into bytecode and written in a file with the extension `.pyc`. The next time the program is run, the interpreter loads the bytecode rather than the original Python file. If in the meantime changes have been made to the module, the module is automatically recompiled. A program can also be run from Idle using *edit/run script* menu, but automatic recompilation of modules will not take place, unless the existing bytecode file is deleted and the program window is closed and reopened.

Python's error messages can sometimes be confusing, as seen in the following example:

```
from numarray import array
a = array([1.0, 2.0, 3.0]
print a
raw_input('press return')
```

The output is

```
File ''C:\Python22\test_module.py'', line 3
    print a
        ^
SyntaxError: invalid syntax
```

What could possibly be wrong with the line `print a`? The answer is nothing. The problem is actually in the preceding line, where the closing parenthesis is missing, making the statement incomplete. Consequently, the interpreter views the third line as continuation of the second line, so that it tries to interpret the statement

```
a = array([1.0, 2.0, 3.0]print a
```

The lesson is this: when faced with a `SyntaxError`, look at the line preceding the alleged offender. It can save a lot of frustration.

It is a good idea to document your modules by adding a *docstring* the beginning of each module. The docstring, which is enclosed in triple quotes, should explain what the module does. Here is an example that documents the module `error` (we use this module in several of our programs):

```
## module error
''' err(string).
    Prints 'string' and terminates program.
'''
import sys
def err(string):
    print string
    raw_input('Press return to exit')
    sys.exit()
```

The docstring of a module can be printed with the statement

> print *module_name*.__doc__

For example, the docstring of `error` is displayed by

```
>>> import error
>>> print error.__doc__
 err(string).
    Prints 'string' and terminates program.
```

2 Systems of Linear Algebraic Equations

> Solve the simultaneous equations $\mathbf{Ax} = \mathbf{b}$

2.1 Introduction

In this chapter we look at the solution of n linear, algebraic equations in n unknowns. It is by far the longest and arguably the most important topic in the book. There is a good reason for this—it is almost impossible to carry out numerical analysis of any sort without encountering simultaneous equations. Moreover, equation sets arising from physical problems are often very large, consuming a lot of computational resources. It usually possible to reduce the storage requirements and the run time by exploiting special properties of the coefficient matrix, such as sparseness (most elements of a sparse matrix are zero). Hence there are many algorithms dedicated to the solution of large sets of equations, each one being tailored to a particular form of the coefficient matrix (symmetric, banded, sparse etc.). A well-known collection of these routines is LAPACK—Linear Algebra PACKage, originally written in Fortran77[3].

We cannot possibly discuss all the special algorithms in the limited space available. The best we can do is to present the basic methods of solution, supplemented by a few useful algorithms for banded and sparse coefficient matrices.

Notation

A system of algebraic equations has the form

$$A_{11}x_1 + A_{12}x_2 + \cdots + A_{1n}x_n = b_1$$

---

[3] LAPACK is the successor of LINPACK, a 1970s and 80s collection of Fortran subroutines.

$$A_{21}x_1 + A_{22}x_2 + \cdots + A_{2n}x_n = b_2 \tag{2.1}$$

$$\vdots$$

$$A_{n1}x_1 + A_{n2}x_2 + \cdots + A_{nn}x_n = b_n$$

where the coefficients A_{ij} and the constants b_j are known, and x_i represent the unknowns. In matrix notation the equations are written as

$$\begin{bmatrix} A_{11} & A_{12} & \cdots & A_{1n} \\ A_{21} & A_{22} & \cdots & A_{2n} \\ \vdots & \vdots & \ddots & \vdots \\ A_{n1} & A_{n2} & \cdots & A_{nn} \end{bmatrix} \begin{bmatrix} x_1 \\ x_2 \\ \vdots \\ x_n \end{bmatrix} = \begin{bmatrix} b_1 \\ b_2 \\ \vdots \\ b_n \end{bmatrix} \tag{2.2}$$

or, simply

$$\mathbf{Ax} = \mathbf{b} \tag{2.3}$$

A particularly useful representation of the equations for computational purposes is the *augmented coefficient matrix* obtained by adjoining the constant vector **b** to the coefficient matrix **A** in the following fashion:

$$\begin{bmatrix} \mathbf{A} \mid \mathbf{b} \end{bmatrix} = \begin{bmatrix} A_{11} & A_{12} & \cdots & A_{1n} & b_1 \\ A_{21} & A_{22} & \cdots & A_{2n} & b_2 \\ \vdots & \vdots & \ddots & \vdots & \vdots \\ A_{n1} & A_{n2} & \cdots & A_{nn} & b_n \end{bmatrix} \tag{2.4}$$

Uniqueness of Solution

A system of n linear equations in n unknowns has a unique solution, provided that the determinant of the coefficient matrix is *nonsingular*; that is, $|\mathbf{A}| \neq 0$. The rows and columns of a nonsingular matrix are *linearly independent* in the sense that no row (or column) is a linear combination of other rows (or columns).

If the coefficient matrix is *singular*, the equations may have an infinite number of solutions, or no solutions at all, depending on the constant vector. As an illustration, take the equations

$$2x + y = 3 \qquad 4x + 2y = 6$$

Since the second equation can be obtained by multiplying the first equation by two, any combination of x and y that satisfies the first equation is also a solution of the second equation. The number of such combinations is infinite. On the other hand, the equations

$$2x + y = 3 \qquad 4x + 2y = 0$$

have no solution because the second equation, being equivalent to $2x + y = 0$, contradicts the first one. Therefore, any solution that satisfies one equation cannot satisfy the other one.

Ill-Conditioning

An obvious question is: what happens when the coefficient matrix is almost singular; i.e., if $|\mathbf{A}|$ is very small? In order to determine whether the determinant of the coefficient matrix is "small," we need a reference against which the determinant can be measured. This reference is called the *norm* of the matrix and is denoted by $\|\mathbf{A}\|$. We can then say that the determinant is small if

$$|\mathbf{A}| << \|\mathbf{A}\|$$

Several norms of a matrix have been defined in existing literature, such as

$$\|\mathbf{A}\| = \sqrt{\sum_{i=1}^{n}\sum_{j=1}^{n} A_{ij}^2} \qquad \|\mathbf{A}\| = \max_{1 \le i \le n} \sum_{j=1}^{n} |A_{ij}| \qquad (2.5a)$$

A formal measure of conditioning is the *matrix condition number*, defined as

$$\text{cond}(\mathbf{A}) = \|\mathbf{A}\|\, \|\mathbf{A}^{-1}\| \qquad (2.5b)$$

If this number is close to unity, the matrix is well-conditioned. The condition number increases with the degree of ill-conditioning, reaching infinity for a singular matrix. Note that the condition number is not unique, but depends on the choice of the matrix norm. Unfortunately, the condition number is expensive to compute for large matrices. In most cases it is sufficient to gauge conditioning by comparing the determinant with the magnitudes of the elements in the matrix.

If the equations are ill-conditioned, small changes in the coefficient matrix result in large changes in the solution. As an illustration, consider the equations

$$2x + y = 3 \qquad 2x + 1.001y = 0$$

that have the solution $x = 1501.5$, $y = -3000$. Since $|\mathbf{A}| = 2(1.001) - 2(1) = 0.002$ is much smaller than the coefficients, the equations are ill-conditioned. The effect of ill-conditioning can be verified by changing the second equation to $2x + 1.002y = 0$ and re-solving the equations. The result is $x = 751.5$, $y = -1500$. Note that a 0.1% change in the coefficient of y produced a 100% change in the solution!

Numerical solutions of ill-conditioned equations are not to be trusted. The reason is that the inevitable roundoff errors during the solution process are equivalent to introducing small changes into the coefficient matrix. This in turn introduces large errors into the solution, the magnitude of which depends on the severity of ill-conditioning. In suspect cases the determinant of the coefficient matrix should be computed so that the degree of ill-conditioning can be estimated. This can be done during or after the solution with only a small computational effort.

Linear Systems

Linear, algebraic equations occur in almost all branches of numerical analysis. But their most visible application in engineering is in the analysis of linear systems (any system whose response is proportional to the input is deemed to be linear). Linear systems include structures, elastic solids, heat flow, seepage of fluids, electromagnetic fields and electric circuits, i.e., most topics taught in an engineering curriculum.

If the system is discrete, such as a truss or an electric circuit, then its analysis leads directly to linear algebraic equations. In the case of a statically determinate truss, for example, the equations arise when the equilibrium conditions of the joints are written down. The unknowns $x_1, x_2, \ldots, x_n$ represent the forces in the members and the support reactions, and the constants $b_1, b_2, \ldots, b_n$ are the prescribed external loads.

The behavior of continuous systems is described by differential equations, rather than algebraic equations. However, because numerical analysis can deal only with discrete variables, it is first necessary to approximate a differential equation with a system of algebraic equations. The well-known finite difference, finite element and boundary element methods of analysis work in this manner. They use different approximations to achieve the "discretization," but in each case the final task is the same: solve a system (often a very large system) of linear, algebraic equations.

In summary, the modeling of linear systems invariably gives rise to equations of the form $\mathbf{Ax} = \mathbf{b}$, where $\mathbf{b}$ is the input and $\mathbf{x}$ represents the response of the system. The coefficient matrix $\mathbf{A}$, which reflects the characteristics of the system, is independent of the input. In other words, if the input is changed, the equations have to be solved again with a different $\mathbf{b}$, but the same $\mathbf{A}$. Therefore, it is desirable to have an equation solving algorithm that can handle any number of constant vectors with minimal computational effort.

Methods of Solution

There are two classes of methods for solving systems of linear, algebraic equations: direct and iterative methods. The common characteristic of *direct methods* is that they transform the original equations into *equivalent equations* (equations that have the same solution) that can be solved more easily. The transformation is carried out by applying the three operations listed below. These so-called *elementary operations* do not change the solution, but they may affect the determinant of the coefficient matrix as indicated in parenthesis.

- Exchanging two equations (changes sign of $|\mathbf{A}|$).
- Multiplying an equation by a nonzero constant (multiplies $|\mathbf{A}|$ by the same constant).
- Multiplying an equation by a nonzero constant and then subtracting it from another equation (leaves $|\mathbf{A}|$ unchanged).

Iterative, or *indirect methods*, start with a guess of the solution $\mathbf{x}$, and then repeatedly refine the solution until a certain convergence criterion is reached. Iterative methods are generally less efficient than their direct counterparts due to the large number of iterations required. But they do have significant computational advantages if the coefficient matrix is very large and sparsely populated (most coefficients are zero).

Overview of Direct Methods

Table 2.1 lists three popular direct methods, each of which uses elementary operations to produce its own final form of easy-to-solve equations.

Method	Initial form	Final form
Gauss elimination	$\mathbf{Ax} = \mathbf{b}$	$\mathbf{Ux} = \mathbf{c}$
LU decomposition	$\mathbf{Ax} = \mathbf{b}$	$\mathbf{LUx} = \mathbf{b}$
Gauss–Jordan elimination	$\mathbf{Ax} = \mathbf{b}$	$\mathbf{Ix} = \mathbf{c}$

Table 2.1

In the above table $\mathbf{U}$ represents an upper triangular matrix, $\mathbf{L}$ is a lower triangular matrix and $\mathbf{I}$ denotes the identity matrix. A square matrix is called *triangular* if it contains only zero elements on one side of the leading diagonal. Thus a 3×3 upper triangular matrix has the form

$$\mathbf{U} = \begin{bmatrix} U_{11} & U_{12} & U_{13} \\ 0 & U_{22} & U_{23} \\ 0 & 0 & U_{33} \end{bmatrix}$$

Systems of Linear Algebraic Equations

and a 3 × 3 lower triangular matrix appears as

$$\mathbf{L} = \begin{bmatrix} L_{11} & 0 & 0 \\ L_{21} & L_{22} & 0 \\ L_{31} & L_{32} & L_{33} \end{bmatrix}$$

Triangular matrices play an important role in linear algebra, since they simplify many computations. For example, consider the equations $\mathbf{Lx} = \mathbf{c}$, or

$$L_{11}x_1 = c_1$$
$$L_{21}x_1 + L_{22}x_2 = c_2$$
$$L_{31}x_1 + L_{32}x_2 + L_{33}x_3 = c_3$$
$$\vdots$$

If we solve the equations forward, starting with the first equation, the computations are very easy, since each equation contains only one unknown at a time. The solution would thus proceed as follows:

$$x_1 = c_1/L_{11}$$
$$x_2 = (c_2 - L_{21}x_1)/L_{22}$$
$$x_3 = (c_3 - L_{31}x_1 - L_{32}x_2)/L_{33}$$
$$\vdots$$

This procedure is known as *forward substitution*. In a similar way, $\mathbf{Ux} = \mathbf{c}$, encountered in Gauss elimination, can easily be solved by *back substitution*, which starts with the last equation and proceeds backward through the equations.

The equations $\mathbf{LUx} = \mathbf{b}$, which are associated with LU decomposition, can also be solved quickly if we replace them with two sets of equivalent equations: $\mathbf{Ly} = \mathbf{b}$ and $\mathbf{Ux} = \mathbf{y}$. Now $\mathbf{Ly} = \mathbf{b}$ can be solved for $\mathbf{y}$ by forward substitution, followed by the solution of $\mathbf{Ux} = \mathbf{y}$ by means of back substitution.

The equations $\mathbf{Ix} = \mathbf{c}$, which are produced by Gauss–Jordan elimination, are equivalent to $\mathbf{x} = \mathbf{c}$ (recall the identity $\mathbf{Ix} = \mathbf{x}$), so that $\mathbf{c}$ is already the solution.

EXAMPLE 2.1
Determine whether the following matrix is singular:

$$\mathbf{A} = \begin{bmatrix} 2.1 & -0.6 & 1.1 \\ 3.2 & 4.7 & -0.8 \\ 3.1 & -6.5 & 4.1 \end{bmatrix}$$

Solution Laplace's development of the determinant (see Appendix A2) about the first row of **A** yields

$$|A| = 2.1 \begin{vmatrix} 4.7 & -0.8 \\ -6.5 & 4.1 \end{vmatrix} - (-0.6) \begin{vmatrix} 3.2 & -0.8 \\ 3.1 & 4.1 \end{vmatrix} + 1.1 \begin{vmatrix} 3.2 & 4.7 \\ 3.1 & -6.5 \end{vmatrix}$$
$$= 2.1(14.07) + 0.6(15.60) + 1.1(-35.37) = 0$$

Since the determinant is zero, the matrix is singular. It can be verified that the singularity is due to the following row dependency: (row 3) = (3 × row 1) − (row 2).

EXAMPLE 2.2
Solve the equations $\mathbf{Ax} = \mathbf{b}$, where

$$\mathbf{A} = \begin{bmatrix} 8 & -6 & 2 \\ -4 & 11 & -7 \\ 4 & -7 & 6 \end{bmatrix} \quad \mathbf{b} = \begin{bmatrix} 28 \\ -40 \\ 33 \end{bmatrix}$$

knowing that the LU decomposition of the coefficient matrix is (you should verify this)

$$\mathbf{A} = \mathbf{LU} = \begin{bmatrix} 2 & 0 & 0 \\ -1 & 2 & 0 \\ 1 & -1 & 1 \end{bmatrix} \begin{bmatrix} 4 & -3 & 1 \\ 0 & 4 & -3 \\ 0 & 0 & 2 \end{bmatrix}$$

Solution We first solve the equations $\mathbf{Ly} = \mathbf{b}$ by forward substitution:

$$2y_1 = 28 \quad y_1 = 28/2 = 14$$
$$-y_1 + 2y_2 = -40 \quad y_2 = (-40 + y_1)/2 = (-40 + 14)/2 = -13$$
$$y_1 - y_2 + y_3 = 33 \quad y_3 = 33 - y_1 + y_2 = 33 - 14 - 13 = 6$$

The solution **x** is then obtained from $\mathbf{Ux} = \mathbf{y}$ by back substitution:

$$2x_3 = y_3 \quad x_3 = y_3/2 = 6/2 = 3$$
$$4x_2 - 3x_3 = y_2 \quad x_2 = (y_2 + 3x_3)/4 = [-13 + 3(3)]/4 = -1$$
$$4x_1 - 3x_2 + x_3 = y_1 \quad x_1 = (y_1 + 3x_2 - x_3)/4 = [14 + 3(-1) - 3]/4 = 2$$

Hence the solution is $\mathbf{x} = \begin{bmatrix} 2 & -1 & 3 \end{bmatrix}^T$

2.2 Gauss Elimination Method

Introduction

Gauss elimination is the most familiar method for solving simultaneous equations. It consists of two parts: the elimination phase and the solution phase. As indicated in Table 2.1, the function of the elimination phase is to transform the equations into the form $\mathbf{Ux} = \mathbf{c}$. The equations are then solved by back substitution. In order to illustrate the procedure, let us solve the equations

$$4x_1 - 2x_2 + x_3 = 11 \tag{a}$$

$$-2x_1 + 4x_2 - 2x_3 = -16 \tag{b}$$

$$x_1 - 2x_2 + 4x_3 = 17 \tag{c}$$

Elimination phase The elimination phase utilizes only one of the elementary operations listed in Table 2.1—multiplying one equation (say, equation j) by a constant λ and subtracting it from another equation (equation i). The symbolic representation of this operation is

$$\text{Eq. } (i) \leftarrow \text{Eq. } (i) - \lambda \times \text{Eq. } (j) \tag{2.6}$$

The equation being subtracted, namely Eq. (j), is called the *pivot equation*.

We start the elimination by taking Eq. (a) to be the pivot equation and choosing the multipliers λ so as to eliminate x_1 from Eqs. (b) and (c):

$$\text{Eq. (b)} \leftarrow \text{Eq. (b)} - (-0.5) \times \text{Eq. (a)}$$

$$\text{Eq. (c)} \leftarrow \text{Eq. (c)} - 0.25 \times \text{Eq. (a)}$$

After this transformation, the equations become

$$4x_1 - 2x_2 + x_3 = 11 \tag{a}$$

$$3x_2 - 1.5x_3 = -10.5 \tag{b}$$

$$-1.5x_2 + 3.75x_3 = 14.25 \tag{c}$$

This completes the first pass. Now we pick (b) as the pivot equation and eliminate x_2 from (c):

$$\text{Eq. (c)} \leftarrow \text{Eq. (c)} - (-0.5) \times \text{Eq.(b)}$$

which yields the equations

$$4x_1 - 2x_2 + x_3 = 11 \tag{a}$$

$$3x_2 - 1.5x_3 = -10.5 \tag{b}$$

$$3x_3 = 9 \tag{c}$$

The elimination phase is now complete. The original equations have been replaced by equivalent equations that can be easily solved by back substitution.

As pointed out before, the augmented coefficient matrix is a more convenient instrument for performing the computations. Thus the original equations

2.2 Gauss Elimination Method

would be written as

$$\begin{bmatrix} 4 & -2 & 1 & | & 11 \\ -2 & 4 & -2 & | & -16 \\ 1 & -2 & 4 & | & 17 \end{bmatrix}$$

and the equivalent equations produced by the first and the second passes of Gauss elimination would appear as

$$\begin{bmatrix} 4 & -2 & 1 & | & 11.00 \\ 0 & 3 & -1.5 & | & -10.50 \\ 0 & -1.5 & 3.75 & | & 14.25 \end{bmatrix}$$

$$\begin{bmatrix} 4 & -2 & 1 & | & 11.0 \\ 0 & 3 & -1.5 & | & -10.5 \\ 0 & 0 & 3 & | & 9.0 \end{bmatrix}$$

It is important to note that the elementary row operation in Eq. (2.6) leaves the determinant of the coefficient matrix unchanged. This is rather fortunate, since the determinant of a triangular matrix is very easy to compute—it is the product of the diagonal elements. In other words,

$$|\mathbf{A}| = |\mathbf{U}| = U_{11} \times U_{22} \times \cdots \times U_{nn} \tag{2.7}$$

Back substitution phase The unknowns can now be computed by back substitution in the manner described in the previous article. Solving Eqs. (c), (b) and (a) in that order, we get

$$x_3 = 9/3 = 3$$
$$x_2 = (-10.5 + 1.5x_3)/3 = [-10.5 + 1.5(3)]/3 = -2$$
$$x_1 = (11 + 2x_2 - x_3)/4 = [11 + 2(-2) - 3]/4 = 1$$

Algorithm for Gauss Elimination Method

Elimination phase

Let us look at the equations at some instant during the elimination phase. Assume that the first k rows of $\mathbf{A}$ have already been transformed to upper triangular form. Therefore, the current pivot equation is the kth equation, and all the equations below it are still to be transformed. This situation is depicted by the augmented coefficient matrix shown below. Note that the components of $\mathbf{A}$ are not the coefficients of the original equations (except for the first row), since they have been altered by the elimination procedure.

Systems of Linear Algebraic Equations

The same applies to the components of the constant vector **b**.

$$\left[\begin{array}{ccccccc|c} A_{11} & A_{12} & A_{13} & \cdots & A_{1k} & \cdots & A_{1j} & \cdots & A_{1n} & b_1 \\ 0 & A_{22} & A_{23} & \cdots & A_{2k} & \cdots & A_{2j} & \cdots & A_{2n} & b_2 \\ 0 & 0 & A_{33} & \cdots & A_{3k} & \cdots & A_{3j} & \cdots & A_{3n} & b_3 \\ \vdots & \vdots & \vdots & & \vdots & & \vdots & & \vdots & \vdots \\ 0 & 0 & 0 & \cdots & A_{kk} & \cdots & A_{kj} & \cdots & A_{kn} & b_k \\ \vdots & \vdots & \vdots & & \vdots & & \vdots & & \vdots & \vdots \\ 0 & 0 & 0 & \cdots & A_{ik} & \cdots & A_{ij} & \cdots & A_{in} & b_i \\ \vdots & \vdots & \vdots & & \vdots & & \vdots & & \vdots & \vdots \\ 0 & 0 & 0 & \cdots & A_{nk} & \cdots & A_{nj} & \cdots & A_{nn} & b_n \end{array}\right] \begin{array}{l} \\ \\ \\ \\ \leftarrow \text{pivot row} \\ \\ \leftarrow \text{row being} \\ \text{transformed} \\ \\ \end{array}$$

Let the ith row be a typical row below the pivot equation that is to be transformed, meaning that the element A_{ik} is to be eliminated. We can achieve this by multiplying the pivot row by $\lambda = A_{ik}/A_{kk}$ and subtracting it from the ith row. The corresponding changes in the ith row are

$$A_{ij} \leftarrow A_{ij} - \lambda A_{kj}, \quad j = k, k+1, \ldots, n \tag{2.8a}$$

$$b_i \leftarrow b_i - \lambda b_k \tag{2.8b}$$

To transform the entire coefficient matrix to upper triangular form, k and i in Eqs. (2.8) must have the ranges $k = 1, 2, \ldots, n-1$ (chooses the pivot row), $i = k+1, k+2 \ldots, n$ (chooses the row to be transformed). The algorithm for the elimination phase now almost writes itself:

```
for k in range(0,n-1):
    for i in range(k+1,n):
        if a[i,k] != 0.0:
            lam = a[i,k]/a[k,k]
            a[i,k+1:n] = a[i,k+1:n] - lam*a[k,k+1:n]
            b[i] = b[i] - lam*b[k]
```

In order to avoid unnecessary operations, the above algorithm departs slightly from Eqs. (2.8) in the following ways:

- If A_{ik} happens to be zero, the transformation of row i is skipped.
- The index j in Eq. (2.8a) starts with $k+1$ rather than k. Therefore, A_{ik} is not replaced by zero, but retains its original value. As the solution phase never accesses the lower triangular portion of the coefficient matrix anyway, its contents are irrelevant.

2.2 Gauss Elimination Method

Back Substitution Phase

After Gauss elimination the augmented coefficient matrix has the form

$$\begin{bmatrix} \mathbf{A} \mid \mathbf{b} \end{bmatrix} = \begin{bmatrix} A_{11} & A_{12} & A_{13} & \cdots & A_{1n} & \mid & b_1 \\ 0 & A_{22} & A_{23} & \cdots & A_{2n} & \mid & b_2 \\ 0 & 0 & A_{33} & \cdots & A_{3n} & \mid & b_3 \\ \vdots & \vdots & \vdots & & \vdots & \mid & \vdots \\ 0 & 0 & 0 & \cdots & A_{nn} & \mid & b_n \end{bmatrix}$$

The last equation, $A_{nn}x_n = b_n$, is solved first, yielding

$$x_n = b_n/A_{nn} \tag{2.9}$$

Consider now the stage of back substitution where $x_n, x_{n-1}, \ldots, x_{k+1}$ have been already been computed (in that order), and we are about to determine x_k from the kth equation

$$A_{kk}x_k + A_{k,k+1}x_{k+1} + \cdots + A_{kn}x_n = b_k$$

The solution is

$$x_k = \left(b_k - \sum_{j=k+1}^{n} A_{kj}x_j \right) \frac{1}{A_{kk}}, \quad k = n-1, n-2, \ldots, 1 \tag{2.10}$$

The corresponding algorithm for back substitution is:

```
for k in range(n-1,-1,-1):
    x[k]=(b[k] - dot(a[k,k+1:n],x[k+1:n]))/a[k,k]
```

■ gaussElimin

The function `gaussElimin` combines the elimination and the back substitution phases. During back substitution b is overwritten by the solution vector x, so that b contains the solution upon exit.

```
## module gaussElimin
''' x = gaussElimin(a,b).
    Solves [a]{b} = {x} by Gauss elimination.
'''
from numarray import dot

def gaussElimin(a,b):
    n = len(b)
  # Elimination phase
```

```
        for k in range(0,n-1):
            for i in range(k+1,n):
                if a[i,k] != 0.0:
                    lam = a [i,k]/a[k,k]
                    a[i,k+1:n] = a[i,k+1:n] - lam*a[k,k+1:n]
                    b[i] = b[i] - lam*b[k]
    # Back substitution
        for k in range(n-1,-1,-1):
            b[k] = (b[k] - dot(a[k,k+1:n],b[k+1:n]))/a[k,k]
        return b
```

Multiple Sets of Equations

As mentioned before, it is frequently necessary to solve the equations $\mathbf{Ax} = \mathbf{b}$ for several constant vectors. Let there be m such constant vectors, denoted by $\mathbf{b}_1, \mathbf{b}_2, \ldots, \mathbf{b}_m$ and let the corresponding solution vectors be $\mathbf{x}_1, \mathbf{x}_2, \ldots, \mathbf{x}_m$. We denote multiple sets of equations by $\mathbf{AX} = \mathbf{B}$, where

$$\mathbf{X} = \begin{bmatrix} \mathbf{x}_1 & \mathbf{x}_2 & \cdots & \mathbf{x}_m \end{bmatrix} \quad \mathbf{B} = \begin{bmatrix} \mathbf{b}_1 & \mathbf{b}_2 & \cdots & \mathbf{b}_m \end{bmatrix}$$

are $n \times m$ matrices whose columns consist of solution vectors and constant vectors, respectively.

An economical way to handle such equations during the elimination phase is to include all m constant vectors in the augmented coefficient matrix, so that they are transformed simultaneously with the coefficient matrix. The solutions are then obtained by back substitution in the usual manner, one vector at a time. It would be quite easy to make the corresponding changes in `gaussElimin`. However, the LU decomposition method, described in the next article, is more versatile in handling multiple constant vectors.

EXAMPLE 2.3
Use Gauss elimination to solve the equations $\mathbf{AX} = \mathbf{B}$, where

$$\mathbf{A} = \begin{bmatrix} 6 & -4 & 1 \\ -4 & 6 & -4 \\ 1 & -4 & 6 \end{bmatrix} \quad \mathbf{B} = \begin{bmatrix} -14 & 22 \\ 36 & -18 \\ 6 & 7 \end{bmatrix}$$

Solution The augmented coefficient matrix is

$$\begin{bmatrix} 6 & -4 & 1 & -14 & 22 \\ -4 & 6 & -4 & 36 & -18 \\ 1 & -4 & 6 & 6 & 7 \end{bmatrix}$$

2.2 Gauss Elimination Method

The elimination phase consists of the following two passes:

$$\text{row } 2 \leftarrow \text{row } 2 + (2/3) \times \text{row } 1$$
$$\text{row } 3 \leftarrow \text{row } 3 - (1/6) \times \text{row } 1$$

$$\begin{bmatrix} 6 & -4 & 1 & -14 & 22 \\ 0 & 10/3 & -10/3 & 80/3 & -10/3 \\ 0 & -10/3 & 35/6 & 25/3 & 10/3 \end{bmatrix}$$

and

$$\text{row } 3 \leftarrow \text{row } 3 + \text{row } 2$$

$$\begin{bmatrix} 6 & -4 & 1 & -14 & 22 \\ 0 & 10/3 & -10/3 & 80/3 & -10/3 \\ 0 & 0 & 5/2 & 35 & 0 \end{bmatrix}$$

In the solution phase, we first compute x_1 by back substitution:

$$X_{31} = \frac{35}{5/2} = 14$$

$$X_{21} = \frac{80/3 + (10/3)X_{31}}{10/3} = \frac{80/3 + (10/3)14}{10/3} = 22$$

$$X_{11} = \frac{-14 + 4X_{21} - X_{31}}{6} = \frac{-14 + 4(22) - 14}{6} = 10$$

Thus the first solution vector is

$$\mathbf{x}_1 = \begin{bmatrix} X_{11} & X_{21} & X_{31} \end{bmatrix}^T = \begin{bmatrix} 10 & 22 & 14 \end{bmatrix}^T$$

The second solution vector is computed next, also using back substitution:

$$X_{32} = 0$$

$$X_{22} = \frac{-10/3 + (10/3)X_{32}}{10/3} = \frac{-10/3 + 0}{10/3} = -1$$

$$X_{12} = \frac{22 + 4X_{22} - X_{32}}{6} = \frac{22 + 4(-1) - 0}{6} = 3$$

Therefore,

$$\mathbf{x}_2 = \begin{bmatrix} X_{12} & X_{22} & X_{32} \end{bmatrix}^T = \begin{bmatrix} 3 & -1 & 0 \end{bmatrix}^T$$

EXAMPLE 2.4
An $n \times n$ Vandermode matrix **A** is defined by

$$A_{ij} = v_i^{n-j}, \quad i = 1, 2, \ldots, n, \quad j = 1, 2, \ldots, n$$

where **v** is a vector. Use the function gaussElimin to compute the solution of **Ax** = **b**, where **A** is the 6×6 Vandermode matrix generated from the vector

$$\mathbf{v} = \begin{bmatrix} 1.0 & 1.2 & 1.4 & 1.6 & 1.8 & 2.0 \end{bmatrix}^T$$

and

$$\mathbf{b} = \begin{bmatrix} 0 & 1 & 0 & 1 & 0 & 1 \end{bmatrix}^T$$

Also evaluate the accuracy of the solution (Vandermode matrices tend to be ill-conditioned).

Solution
```
#!/usr/bin/python
## example2_4
from numarray import zeros,Float64,array,product, \
                     diagonal,matrixmultiply
from gaussElimin import *

def vandermode(v):
    n = len(v)
    a = zeros((n,n),type=Float64)
    for j in range(n):
        a[:,j] = v**(n-j-1)
    return a

v = array([1.0, 1.2, 1.4, 1.6, 1.8, 2.0])
b = array([0.0, 1.0, 0.0, 1.0, 0.0, 1.0])
a = vandermode(v)
aOrig = a.copy()       # Save original matrix
bOrig = b.copy()       # and the constant vector
x = gaussElimin(a,b)
det = product(diagonal(a))
print 'x =\n',x
print '\ndet =',det
```

2.3 LU Decomposition Methods

```
print '\nCheck result: [a]{x} - b =\n', \
      matrixmultiply(aOrig,x) - bOrig
raw_input(''\nPress return to exit'')
```

The program produced the following results:

```
x =
[ 416.66666667   -3125.00000004    9250.00000012 -13500.00000017
  9709.33333345   -2751.00000003]

det = -1.13246207999e-006

Check result: [a]{x} - b =
[-4.54747351e-13    4.54747351e-13  -1.36424205e-12    4.54747351e-13
 -3.41060513e-11    9.54969437e-12]
```

As the determinant is quite small relative to the elements of **A** (you may want to print **A** to verify this), we expect detectable roundoff error. Inspection of **x** leads us to suspect that the exact solution is

$$\mathbf{x} = \begin{bmatrix} 1250/3 & -3125 & 9250 & -13500 & 29128/3 & -2751 \end{bmatrix}^T$$

in which case the numerical solution would be accurate to about 10 decimal places. Another way to gauge the accuracy of the solution is to compute **Ax** − **b** (the result should be **0**). The printout indicates that the solution is indeed accurate to at least 10 decimal places.

2.3 LU Decomposition Methods

Introduction

It is possible to show that any square matrix **A** can be expressed as a product of a lower triangular matrix **L** and an upper triangular matrix **U**:

$$\mathbf{A} = \mathbf{LU} \tag{2.11}$$

The process of computing **L** and **U** for a given **A** is known as *LU decomposition* or *LU factorization*. LU decomposition is not unique (the combinations of **L** and **U** for a prescribed **A** are endless), unless certain constraints are placed on **L** or **U**. These constraints distinguish one type of decomposition from another. Three commonly used decompositions are listed in Table 2.2.

Systems of Linear Algebraic Equations

Name	Constraints
Doolittle's decomposition	$L_{ii} = 1, \quad i = 1, 2, \ldots, n$
Crout's decomposition	$U_{ii} = 1, \quad i = 1, 2, \ldots, n$
Choleski's decomposition	$\mathbf{L} = \mathbf{U}^T$

Table 2.2

After decomposing $\mathbf{A}$, it is easy to solve the equations $\mathbf{Ax} = \mathbf{b}$, as pointed out in Art. 2.1. We first rewrite the equations as $\mathbf{LUx} = \mathbf{b}$. Upon using the notation $\mathbf{Ux} = \mathbf{y}$, the equations become

$$\mathbf{Ly} = \mathbf{b}$$

which can be solved for $\mathbf{y}$ by forward substitution. Then

$$\mathbf{Ux} = \mathbf{y}$$

will yield $\mathbf{x}$ by the back substitution process.

The advantage of LU decomposition over the Gauss elimination method is that once $\mathbf{A}$ is decomposed, we can solve $\mathbf{Ax} = \mathbf{b}$ for as many constant vectors $\mathbf{b}$ as we please. The cost of each additional solution is relatively small, since the forward and back substitution operations are much less time consuming than the decomposition process.

Doolittle's Decomposition Method

Decomposition Phase

Doolittle's decomposition is closely related to Gauss elimination. In order to illustrate the relationship, consider a 3×3 matrix $\mathbf{A}$ and assume that there exist triangular matrices

$$\mathbf{L} = \begin{bmatrix} 1 & 0 & 0 \\ L_{21} & 1 & 0 \\ L_{31} & L_{32} & 1 \end{bmatrix} \quad \mathbf{U} = \begin{bmatrix} U_{11} & U_{12} & U_{13} \\ 0 & U_{22} & U_{23} \\ 0 & 0 & U_{33} \end{bmatrix}$$

such that $\mathbf{A} = \mathbf{LU}$. After completing the multiplication on the right hand side, we get

$$\mathbf{A} = \begin{bmatrix} U_{11} & U_{12} & U_{13} \\ U_{11}L_{21} & U_{12}L_{21} + U_{22} & U_{13}L_{21} + U_{23} \\ U_{11}L_{31} & U_{12}L_{31} + U_{22}L_{32} & U_{13}L_{31} + U_{23}L_{32} + U_{33} \end{bmatrix} \quad (2.12)$$

Let us now apply Gauss elimination to Eq. (2.12). The first pass of the elimination procedure consists of choosing the first row as the pivot row and applying the

2.3 LU Decomposition Methods

elementary operations

$$\text{row 2} \leftarrow \text{row 2} - L_{21} \times \text{row 1 (eliminates } A_{21})$$

$$\text{row 3} \leftarrow \text{row 3} - L_{31} \times \text{row 1 (eliminates } A_{31})$$

The result is

$$\mathbf{A}' = \begin{bmatrix} U_{11} & U_{12} & U_{13} \\ 0 & U_{22} & U_{23} \\ 0 & U_{22}L_{32} & U_{23}L_{32} + U_{33} \end{bmatrix}$$

In the next pass we take the second row as the pivot row, and utilize the operation

$$\text{row 3} \leftarrow \text{row 3} - L_{32} \times \text{row 2 (eliminates } A_{32})$$

ending up with

$$\mathbf{A}'' = \mathbf{U} = \begin{bmatrix} U_{11} & U_{12} & U_{13} \\ 0 & U_{22} & U_{23} \\ 0 & 0 & U_{33} \end{bmatrix}$$

The foregoing illustration reveals two important features of Doolittle's decomposition:

- The matrix **U** is identical to the upper triangular matrix that results from Gauss elimination.
- The off-diagonal elements of **L** are the pivot equation multipliers used during Gauss elimination; that is, L_{ij} is the multiplier that eliminated A_{ij}.

It is usual practice to store the multipliers in the lower triangular portion of the coefficient matrix, replacing the coefficients as they are eliminated (L_{ij} replacing A_{ij}). The diagonal elements of **L** do not have to be stored, since it is understood that each of them is unity. The final form of the coefficient matrix would thus be the following mixture of **L** and **U**:

$$[\mathbf{L} \backslash \mathbf{U}] = \begin{bmatrix} U_{11} & U_{12} & U_{13} \\ L_{21} & U_{22} & U_{23} \\ L_{31} & L_{32} & U_{33} \end{bmatrix} \quad (2.13)$$

The algorithm for Doolittle's decomposition is thus identical to the Gauss elimination procedure in `gaussElimin`, except that each multiplier λ is now stored in the lower triangular portion of **A**:

```
for k in range(0,n-1):
    for i in range(k+1,n):
```

Systems of Linear Algebraic Equations

```
if a[i,k] != 0.0:
    lam = a[i,k]/a[k,k]
    a[i,k+1:n] = a[i,k+1:n] - lam*a[k,k+1:n]
    a[i,k] = lam
```

Solution Phase

Consider now the procedure for solving $\mathbf{Ly} = \mathbf{b}$ by forward substitution. The scalar form of the equations is (recall that $L_{ii} = 1$)

$$y_1 = b_1$$
$$L_{21}y_1 + y_2 = b_2$$
$$\vdots$$
$$L_{k1}y_1 + L_{k2}y_2 + \cdots + L_{k,k-1}y_{k-1} + y_k = b_k$$
$$\vdots$$

Solving the kth equation for y_k yields

$$y_k = b_k - \sum_{j=1}^{k-1} L_{kj}y_j, \quad k = 2, 3, \ldots, n \tag{2.14}$$

Therefore, the forward substitution algorithm is

```
y[0] = b[0]
for k in range(1,n):
    y[k] = b[k] - dot(a[k,0:k],y[0:k])
```

The back substitution phase for solving $\mathbf{Ux} = \mathbf{y}$ is identical to that used in the Gauss elimination method.

■ LUdecomp

This module contains both the decomposition and solution phases. The decomposition phase returns the matrix $[\mathbf{L}\backslash\mathbf{U}]$ shown in Eq. (2.13). In the solution phase, the contents of $\mathbf{b}$ are replaced by $\mathbf{y}$ during forward substitution. Similarly, back substitution overwrites $\mathbf{y}$ with the solution $\mathbf{x}$.

```
## module LUdecomp
''' a = LUdecomp(a).
    LU decomposition: [L][U] = [a]. The returned matrix
```

```
    [a] = [L\U] contains [U] in the upper triangle and
    the nondiagonal terms of [L] in the lower triangle.

    x = LUsolve(a,b).
    Solves [L][U]{x} = b, where [a] = [L\U] is the matrix
    returned from LUdecomp.
'''
from numarray import dot

def LUdecomp(a):
    n = len(a)
    for k in range(0,n-1):
        for i in range(k+1,n):
            if a[i,k] != 0.0:
                lam = a [i,k]/a[k,k]
                a[i,k+1:n] = a[i,k+1:n] - lam*a[k,k+1:n]
                a[i,k] = lam
    return a

def LUsolve(a,b):
    n = len(a)
    for k in range(1,n):
        b[k] = b[k] - dot(a[k,0:k],b[0:k])
    for k in range(n-1,-1,-1):
        b[k] = (b[k] - dot(a[k,k+1:n],b[k+1:n]))/a[k,k]
    return b
```

Choleski's Decomposition

Choleski's decomposition $\mathbf{A} = \mathbf{LL}^T$ has two limitations:

- Since $\mathbf{LL}^T$ is always a symmetric matrix, Choleski's decomposition requires A to be *symmetric*.
- The decomposition process involves taking square roots of certain combinations of the elements of A. It can be shown that in order to avoid square roots of negative numbers A must be *positive definite*.

Although the number of multiplications in all the decomposition methods is about the same, Choleski's decomposition is not a particularly popular means of solving simultaneous equations due to the restrictions listed above. We study it here because it is invaluable in certain applications that we encounter later on.

Let us start by looking at Choleski's decomposition

$$\mathbf{A} = \mathbf{L}\mathbf{L}^T \tag{2.15}$$

of a 3×3 matrix:

$$\begin{bmatrix} A_{11} & A_{12} & A_{13} \\ A_{21} & A_{22} & A_{23} \\ A_{31} & A_{32} & A_{33} \end{bmatrix} = \begin{bmatrix} L_{11} & 0 & 0 \\ L_{21} & L_{22} & 0 \\ L_{31} & L_{32} & L_{33} \end{bmatrix} \begin{bmatrix} L_{11} & L_{21} & L_{31} \\ 0 & L_{22} & L_{32} \\ 0 & 0 & L_{33} \end{bmatrix}$$

After completing the matrix multiplication on the right hand side, we get

$$\begin{bmatrix} A_{11} & A_{12} & A_{13} \\ A_{21} & A_{22} & A_{23} \\ A_{31} & A_{32} & A_{33} \end{bmatrix} = \begin{bmatrix} L_{11}^2 & L_{11}L_{21} & L_{11}L_{31} \\ L_{11}L_{21} & L_{21}^2 + L_{22}^2 & L_{21}L_{31} + L_{22}L_{32} \\ L_{11}L_{31} & L_{21}L_{31} + L_{22}L_{32} & L_{31}^2 + L_{32}^2 + L_{33}^2 \end{bmatrix} \tag{2.16}$$

Note that the right-hand-side matrix is symmetric, as pointed out before. Equating the matrices $\mathbf{A}$ and $\mathbf{L}\mathbf{L}^T$ element-by-element, we obtain six equations (due to symmetry only lower or upper triangular elements have to be considered) in the six unknown components of $\mathbf{L}$. By solving these equations in a certain order, it is possible to have only one unknown in each equation.

Consider the lower triangular portion of each matrix in Eq. (2.16) (the upper triangular portion would do as well). By equating the elements in the first column, starting with the first row and proceeding downward, we can compute L_{11}, L_{21} and L_{31} in that order:

$$A_{11} = L_{11}^2 \qquad L_{11} = \sqrt{A_{11}}$$
$$A_{21} = L_{11}L_{21} \qquad L_{21} = A_{21}/L_{11}$$
$$A_{31} = L_{11}L_{31} \qquad L_{31} = A_{31}/L_{11}$$

The second column, starting with second row, yields L_{22} and L_{32}:

$$A_{22} = L_{21}^2 + L_{22}^2 \qquad L_{22} = \sqrt{A_{22} - L_{21}^2}$$
$$A_{32} = L_{21}L_{31} + L_{22}L_{32} \qquad L_{32} = (A_{32} - L_{21}L_{31})/L_{22}$$

Finally the third column, third row gives us L_{33}:

$$A_{33} = L_{31}^2 + L_{32}^2 + L_{33}^2 \qquad L_{33} = \sqrt{A_{33} - L_{31}^2 - L_{32}^2}$$

2.3 LU Decomposition Methods

We can now extrapolate the results for an $n \times n$ matrix. We observe that a typical element in the lower-triangular portion of $\mathbf{LL}^T$ is of the form

$$(\mathbf{LL}^T)_{ij} = L_{i1}L_{j1} + L_{i2}L_{j2} + \cdots + L_{ij}L_{jj} = \sum_{k=1}^{j} L_{ik}L_{jk}, \quad i \geq j$$

Equating this term to the corresponding element of $\mathbf{A}$ yields

$$A_{ij} = \sum_{k=1}^{j} L_{ik}L_{jk}, \quad i = j, j+1, \ldots, n, \quad j = 1, 2, \ldots, n \quad (2.17)$$

The range of indices shown limits the elements to the lower triangular part. For the first column ($j = 1$), we obtain from Eq. (2.17)

$$L_{11} = \sqrt{A_{11}} \qquad L_{i1} = A_{i1}/L_{11}, \quad i = 2, 3, \ldots, n \quad (2.18)$$

Proceeding to other columns, we observe that the unknown in Eq. (2.17) is L_{ij} (the other elements of $\mathbf{L}$ appearing in the equation have already been computed). Taking the term containing L_{ij} outside the summation in Eq. (2.17), we obtain

$$A_{ij} = \sum_{k=1}^{j-1} L_{ik}L_{jk} + L_{ij}L_{jj}$$

If $i = j$ (a diagonal term), the solution is

$$L_{jj} = \sqrt{A_{jj} - \sum_{k=1}^{j-1} L_{jk}^2}, \quad j = 2, 3, \ldots, n \quad (2.19)$$

For a nondiagonal term we get

$$L_{ij} = \left(A_{ij} - \sum_{k=1}^{j-1} L_{ik}L_{jk}\right)/L_{jj}, \quad j = 2, 3, \ldots, n-1, \quad i = j+1, j+2, \ldots, n \quad (2.20)$$

■ choleski(a)

Before presenting the algorithm for Choleski's decomposition, we make a useful observation: A_{ij} appears only in the formula for L_{ij}. Therefore, once L_{ij} has been computed, A_{ij} is no longer needed. This makes it possible to write the elements of $\mathbf{L}$ over the lower triangular portion of $\mathbf{A}$ as they are computed. The elements above the leading diagonal of $\mathbf{A}$ will remain untouched. The function listed below implements Choleski's decomposition. If a negative diagonal term is encountered during decomposition, an error message is printed and the program is terminated.

```
## module choleski
''' L = choleski(a).
```

Choleski decomposition: [L][L]transpose = [a].
'''
from numarray import dot
from math import sqrt
import error

def choleski(a):
    n = len(a)
    for k in range(n):
        try:
            a[k,k] = sqrt(a[k,k] - dot(a[k,0:k],a[k,0:k]))
        except ValueError:
            error.err('Matrix is not positive definite')
        for i in range(k+1,n):
            a[i,k] = (a[i,k] - dot(a[i,0:k],a[k,0:k]))/a[k,k]
    for k in range(1,n): a[0:k,k] = 0.0
    return a
```

We could also write the algorithm for forward and back substitutions that are necessary in the solution of $\mathbf{Ax} = \mathbf{b}$. But since Choleski's decomposition has no advantages over Doolittle's decomposition in the solution of simultaneous equations, we will skip that.

### EXAMPLE 2.5
Use Doolittle's decomposition method to solve the equations $\mathbf{Ax} = \mathbf{b}$, where

$$\mathbf{A} = \begin{bmatrix} 1 & 4 & 1 \\ 1 & 6 & -1 \\ 2 & -1 & 2 \end{bmatrix} \quad \mathbf{b} = \begin{bmatrix} 7 \\ 13 \\ 5 \end{bmatrix}$$

**Solution** We first decompose $\mathbf{A}$ by Gauss elimination. The first pass consists of the elementary operations

$$\text{row } 2 \leftarrow \text{row } 2 - 1 \times \text{row } 1 \text{ (eliminates } A_{21}\text{)}$$

$$\text{row } 3 \leftarrow \text{row } 3 - 2 \times \text{row } 1 \text{ (eliminates } A_{31}\text{)}$$

Storing the multipliers $L_{21} = 1$ and $L_{31} = 2$ in place of the eliminated terms, we obtain

$$\mathbf{A}' = \begin{bmatrix} 1 & 4 & 1 \\ 1 & 2 & -2 \\ 2 & -9 & 0 \end{bmatrix}$$
```

2.3 LU Decomposition Methods

The second pass of Gauss elimination uses the operation

$$\text{row 3} \leftarrow \text{row 3} - (-4.5) \times \text{row 2} \quad (\text{eliminates } A_{32})$$

Storing the multiplier $L_{32} = -4.5$ in place of A_{32}, we get

$$\mathbf{A}'' = [\mathbf{L}\backslash\mathbf{U}] = \begin{bmatrix} 1 & 4 & 1 \\ 1 & 2 & -2 \\ 2 & -4.5 & -9 \end{bmatrix}$$

The decomposition is now complete, with

$$\mathbf{L} = \begin{bmatrix} 1 & 0 & 0 \\ 1 & 1 & 0 \\ 2 & -4.5 & 1 \end{bmatrix} \quad \mathbf{U} = \begin{bmatrix} 1 & 4 & 1 \\ 0 & 2 & -2 \\ 0 & 0 & -9 \end{bmatrix}$$

Solution of $\mathbf{Ly} = \mathbf{b}$ by forward substitution comes next. The augmented coefficient form of the equations is

$$\begin{bmatrix} \mathbf{L} \mid \mathbf{b} \end{bmatrix} = \begin{bmatrix} 1 & 0 & 0 & | & 7 \\ 1 & 1 & 0 & | & 13 \\ 2 & -4.5 & 1 & | & 5 \end{bmatrix}$$

The solution is

$$y_1 = 7$$
$$y_2 = 13 - y_1 = 13 - 7 = 6$$
$$y_3 = 5 - 2y_1 + 4.5y_2 = 5 - 2(7) + 4.5(6) = 18$$

Finally, the equations $\mathbf{Ux} = \mathbf{y}$, or

$$\begin{bmatrix} \mathbf{U} \mid \mathbf{y} \end{bmatrix} = \begin{bmatrix} 1 & 4 & 1 & | & 7 \\ 0 & 2 & -2 & | & 6 \\ 0 & 0 & -9 & | & 18 \end{bmatrix}$$

are solved by back substitution. This yields

$$x_3 = \frac{18}{-9} = -2$$
$$x_2 = \frac{6 + 2x_3}{2} = \frac{6 + 2(-2)}{2} = 1$$
$$x_1 = 7 - 4x_2 - x_3 = 7 - 4(1) - (-2) = 5$$

EXAMPLE 2.6
Compute Choleski's decomposition of the matrix

$$\mathbf{A} = \begin{bmatrix} 4 & -2 & 2 \\ -2 & 2 & -4 \\ 2 & -4 & 11 \end{bmatrix}$$

Solution First we note that **A** is symmetric. Therefore, Choleski's decomposition is applicable, provided that the matrix is also positive definite. An *a priori* test for positive definiteness is not needed, since the decomposition algorithm contains its own test: if the square root of a negative number is encountered, the matrix is not positive definite and the decomposition fails.

Substituting the given matrix for **A** in Eq. (2.16), we obtain

$$\begin{bmatrix} 4 & -2 & 2 \\ -2 & 2 & -4 \\ 2 & -4 & 11 \end{bmatrix} = \begin{bmatrix} L_{11}^2 & L_{11}L_{21} & L_{11}L_{31} \\ L_{11}L_{21} & L_{21}^2 + L_{22}^2 & L_{21}L_{31} + L_{22}L_{32} \\ L_{11}L_{31} & L_{21}L_{31} + L_{22}L_{32} & L_{31}^2 + L_{32}^2 + L_{33}^2 \end{bmatrix}$$

Equating the elements in the lower (or upper) triangular portions yields

$$L_{11} = \sqrt{4} = 2$$

$$L_{21} = -2/L_{11} = -2/2 = -1$$

$$L_{31} = 2/L_{11} = 2/2 = 1$$

$$L_{22} = \sqrt{2 - L_{21}^2} = \sqrt{2 - 1^2} = 1$$

$$L_{32} = \frac{-4 - L_{21}L_{31}}{L_{22}} = \frac{-4 - (-1)(1)}{1} = -3$$

$$L_{33} = \sqrt{11 - L_{31}^2 - L_{32}^2} = \sqrt{11 - (1)^2 - (-3)^2} = 1$$

Therefore,

$$\mathbf{L} = \begin{bmatrix} 2 & 0 & 0 \\ -1 & 1 & 0 \\ 1 & -3 & 1 \end{bmatrix}$$

The result can easily be verified by performing the multiplication $\mathbf{LL}^T$.

EXAMPLE 2.7
Write a program that solves $\mathbf{AX} = \mathbf{B}$ with Doolittle's decomposition method and computes $|\mathbf{A}|$. Utilize the functions LUdecomp and LUsolve. Test the program with

$$\mathbf{A} = \begin{bmatrix} 3 & -1 & 4 \\ -2 & 0 & 5 \\ 7 & 2 & -2 \end{bmatrix} \quad \mathbf{B} = \begin{bmatrix} 6 & -4 \\ 3 & 2 \\ 7 & -5 \end{bmatrix}$$

2.3 LU Decomposition Methods

Solution The program listed below decomposes A and then prompts for the constant vectors. After a constant vector is entered, the corresponding solution is computed and the program prompts for another constant vector. The program is terminated when a SyntaxError is encountered in input (e.g., when the "return" key is pressed).

```
#!/usr/bin/python
## example2_7
from numarray import zeros,array,Float64,product,diagonal
from LUdecomp import *

a = array([[ 3.0, -1.0,  4.0], \
           [-2.0,  0.0,  5.0], \
           [ 7.0,  2.0, -2.0]])
a = LUdecomp(a)
det = product(diagonal(a))
print ''\nDeterminant ='',det
while 1:
    print ''\nInput constant vector (press return to exit):''
    try:
        b = array(eval(raw_input(''==> '')),type=Float64)
    except SyntaxError: break
    x = LUsolve(a,b)
    print ''The solution is:\n'',x
raw_input(''\nPress return to exit'')
```

Running the program produced the following display:

```
Determinant = -77.0

Input constant vector (press return to exit):
==> [6.0, 3.0, 7.0]
The solution is:
[ 1.  1.  1.]

Input constant vector (press return to exit):
==> [-4.0, 2.0, -5.0]
The solution is:
[ -1.00000000e+00   1.00000000e+00   2.30695693e-17]

Input constant vector (press return to exit):
==>
```

Systems of Linear Algebraic Equations

EXAMPLE 2.8

Test the function choleski by decomposing

$$A = \begin{bmatrix} 1.44 & -0.36 & 5.52 & 0.00 \\ -0.36 & 10.33 & -7.78 & 0.00 \\ 5.52 & -7.78 & 28.40 & 9.00 \\ 0.00 & 0.00 & 9.00 & 61.00 \end{bmatrix}$$

Solution

```
#!/usr/bin/python
## example2_8
from numarray import array,matrixmultiply,transpose
from choleski import *

a = array([[ 1.44, -0.36,  5.52,  0.0], \
           [-0.36, 10.33, -7.78,  0.0], \
           [ 5.52, -7.78, 28.40,  9.0], \
           [ 0.0,   0.0,   9.0,  61.0]])
L = choleski(a)
print 'L =\n',L
print '\nCheck: L*L_transpose =\n', \
      matrixmultiply(L,transpose(L))
raw_input(''\nPress return to exit'')
```

The output is:

```
L =
[[ 1.2  0.   0.   0. ]
 [-0.3  3.2  0.   0. ]
 [ 4.6 -2.   1.8  0. ]
 [ 0.   0.   5.   6. ]]

Check: L*L_transpose =
[[  1.44  -0.36   5.52   0.  ]
 [ -0.36  10.33  -7.78   0.  ]
 [  5.52  -7.78  28.4    9.  ]
 [  0.     0.     9.    61.  ]]
```

2.3 LU Decomposition Methods

PROBLEM SET 2.1

1. By evaluating the determinant, classify the following matrices as singular, ill-conditioned, or well-conditioned.

   (a) $\mathbf{A} = \begin{bmatrix} 1 & 2 & 3 \\ 2 & 3 & 4 \\ 3 & 4 & 5 \end{bmatrix}$ (b) $\mathbf{A} = \begin{bmatrix} 2.11 & -0.80 & 1.72 \\ -1.84 & 3.03 & 1.29 \\ -1.57 & 5.25 & 4.30 \end{bmatrix}$

   (c) $\mathbf{A} = \begin{bmatrix} 2 & -1 & 0 \\ -1 & 2 & -1 \\ 0 & -1 & 2 \end{bmatrix}$ (d) $\mathbf{A} = \begin{bmatrix} 4 & 3 & -1 \\ 7 & -2 & 3 \\ 5 & -18 & 13 \end{bmatrix}$

2. Given the LU decomposition $\mathbf{A} = \mathbf{LU}$, determine $\mathbf{A}$ and $|\mathbf{A}|$.

   (a) $\mathbf{L} = \begin{bmatrix} 1 & 0 & 0 \\ 1 & 1 & 0 \\ 1 & 5/3 & 1 \end{bmatrix} \quad \mathbf{U} = \begin{bmatrix} 1 & 2 & 4 \\ 0 & 3 & 21 \\ 0 & 0 & 0 \end{bmatrix}$

   (b) $\mathbf{L} = \begin{bmatrix} 2 & 0 & 0 \\ -1 & 1 & 0 \\ 1 & -3 & 1 \end{bmatrix} \quad \mathbf{U} = \begin{bmatrix} 2 & -1 & 1 \\ 0 & 1 & -3 \\ 0 & 0 & 1 \end{bmatrix}$

3. Utilize the results of LU decomposition

   $$\mathbf{A} = \mathbf{LU} = \begin{bmatrix} 1 & 0 & 0 \\ 3/2 & 1 & 0 \\ 1/2 & 11/13 & 1 \end{bmatrix} \begin{bmatrix} 2 & -3 & -1 \\ 0 & 13/2 & -7/2 \\ 0 & 0 & 32/13 \end{bmatrix}$$

   to solve $\mathbf{Ax} = \mathbf{b}$, where $\mathbf{b}^T = \begin{bmatrix} 1 & -1 & 2 \end{bmatrix}$.

4. Use Gauss elimination to solve the equations $\mathbf{Ax} = \mathbf{b}$, where

   $$\mathbf{A} = \begin{bmatrix} 2 & -3 & -1 \\ 3 & 2 & -5 \\ 2 & 4 & -1 \end{bmatrix} \quad \mathbf{b} = \begin{bmatrix} 3 \\ -9 \\ -5 \end{bmatrix}$$

5. Solve the equations $\mathbf{AX} = \mathbf{B}$ by Gauss elimination, where

   $$\mathbf{A} = \begin{bmatrix} 2 & 0 & -1 & 0 \\ 0 & 1 & 2 & 0 \\ -1 & 2 & 0 & 1 \\ 0 & 0 & 1 & -2 \end{bmatrix} \quad \mathbf{B} = \begin{bmatrix} 1 & 0 \\ 0 & 0 \\ 0 & 1 \\ 0 & 0 \end{bmatrix}$$

Systems of Linear Algebraic Equations

6. Solve the equations $\mathbf{Ax} = \mathbf{b}$ by Gauss elimination, where

$$\mathbf{A} = \begin{bmatrix} 0 & 0 & 2 & 1 & 2 \\ 0 & 1 & 0 & 2 & -1 \\ 1 & 2 & 0 & -2 & 0 \\ 0 & 0 & 0 & -1 & 1 \\ 0 & 1 & -1 & 1 & -1 \end{bmatrix} \qquad \mathbf{b} = \begin{bmatrix} 1 \\ 1 \\ -4 \\ -2 \\ -1 \end{bmatrix}$$

Hint: reorder the equations before solving.

7. Find $\mathbf{L}$ and $\mathbf{U}$ so that

$$\mathbf{A} = \mathbf{LU} = \begin{bmatrix} 4 & -1 & 0 \\ -1 & 4 & -1 \\ 0 & -1 & 4 \end{bmatrix}$$

using (a) Doolittle's decomposition; (b) Choleski's decomposition.

8. Use Doolittle's decomposition method to solve $\mathbf{Ax} = \mathbf{b}$, where

$$\mathbf{A} = \begin{bmatrix} -3 & 6 & -4 \\ 9 & -8 & 24 \\ -12 & 24 & -26 \end{bmatrix} \qquad \mathbf{b} = \begin{bmatrix} -3 \\ 65 \\ -42 \end{bmatrix}$$

9. Solve the equations $\mathbf{Ax} = \mathbf{b}$ by Doolittle's decomposition method, where

$$\mathbf{A} = \begin{bmatrix} 2.34 & -4.10 & 1.78 \\ -1.98 & 3.47 & -2.22 \\ 2.36 & -15.17 & 6.18 \end{bmatrix} \qquad \mathbf{b} = \begin{bmatrix} 0.02 \\ -0.73 \\ -6.63 \end{bmatrix}$$

10. Solve the equations $\mathbf{AX} = \mathbf{B}$ by Doolittle's decomposition method, where

$$\mathbf{A} = \begin{bmatrix} 4 & -3 & 6 \\ 8 & -3 & 10 \\ -4 & 12 & -10 \end{bmatrix} \qquad \mathbf{B} = \begin{bmatrix} 1 & 0 \\ 0 & 1 \\ 0 & 0 \end{bmatrix}$$

11. Solve the equations $\mathbf{Ax} = \mathbf{b}$ by Choleski's decomposition method, where

$$\mathbf{A} = \begin{bmatrix} 1 & 1 & 1 \\ 1 & 2 & 2 \\ 1 & 2 & 3 \end{bmatrix} \qquad \mathbf{b} = \begin{bmatrix} 1 \\ 3/2 \\ 3 \end{bmatrix}$$

12. Solve the equations

$$\begin{bmatrix} 4 & -2 & -3 \\ 12 & 4 & -10 \\ -16 & 28 & 18 \end{bmatrix} \begin{bmatrix} x_1 \\ x_2 \\ x_3 \end{bmatrix} = \begin{bmatrix} 1.1 \\ 0 \\ -2.3 \end{bmatrix}$$

by Doolittle's decomposition method.

2.3 LU Decomposition Methods

13. Determine **L** that results from Choleski's decomposition of the diagonal matrix

$$\mathbf{A} = \begin{bmatrix} \alpha_1 & 0 & 0 & \cdots \\ 0 & \alpha_2 & 0 & \cdots \\ 0 & 0 & \alpha_3 & \cdots \\ \vdots & \vdots & \vdots & \ddots \end{bmatrix}$$

14. ■ Modify the function `gaussElimin` so that it will work with m constant vectors. Test the program by solving $\mathbf{AX} = \mathbf{B}$, where

$$\mathbf{A} = \begin{bmatrix} 2 & -1 & 0 \\ -1 & 2 & -1 \\ 0 & -1 & 1 \end{bmatrix} \quad \mathbf{B} = \begin{bmatrix} 1 & 0 & 0 \\ 0 & 1 & 0 \\ 0 & 0 & 1 \end{bmatrix}$$

15. ■ A well-known example of an ill-conditioned matrix is the *Hilbert matrix*

$$\mathbf{A} = \begin{bmatrix} 1 & 1/2 & 1/3 & \cdots \\ 1/2 & 1/3 & 1/4 & \cdots \\ 1/3 & 1/4 & 1/5 & \cdots \\ \vdots & \vdots & \vdots & \ddots \end{bmatrix}$$

Write a program that specializes in solving the equations $\mathbf{Ax} = \mathbf{b}$ by Doolittle's decomposition method, where **A** is the Hilbert matrix of arbitrary size $n \times n$, and

$$b_i = \sum_{j=1}^{n} A_{ij}$$

The program should have no input apart from n. By running the program, determine the largest n for which the solution is within 6 significant figures of the exact solution

$$\mathbf{x} = \begin{bmatrix} 1 & 1 & 1 & \cdots \end{bmatrix}^T$$

16. ■ Write a function for the solution phase of Choleski's decomposition method. Test the function by solving the equations $\mathbf{Ax} = \mathbf{b}$, where

$$\mathbf{A} = \begin{bmatrix} 4 & -2 & 2 \\ -2 & 2 & -4 \\ 2 & -4 & 11 \end{bmatrix} \quad \mathbf{b} = \begin{bmatrix} 6 \\ -10 \\ 27 \end{bmatrix}$$

Use the function `choleski` for the decomposition phase.

17. ■ Determine the coefficients of the polynomial $y = a_0 + a_1 x + a_2 x^2 + a_3 x^3$ that pass through the points (0, 10), (1, 35), (3, 31) and (4, 2).

18. ■ Determine the 4th degree polynomial $y(x)$ that passes through the points (0, −1), (1, 1), (3, 3), (5, 2) and (6, −2).

19. ■ Find the 4th degree polynomial $y(x)$ that passes through the points $(0, 1)$, $(0.75, -0.25)$ and $(1, 1)$, and has zero curvature at $(0, 1)$ and $(1, 1)$.

20. ■ Solve the equations $\mathbf{Ax} = \mathbf{b}$, where

$$\mathbf{A} = \begin{bmatrix} 3.50 & 2.77 & -0.76 & 1.80 \\ -1.80 & 2.68 & 3.44 & -0.09 \\ 0.27 & 5.07 & 6.90 & 1.61 \\ 1.71 & 5.45 & 2.68 & 1.71 \end{bmatrix} \quad \mathbf{b} = \begin{bmatrix} 7.31 \\ 4.23 \\ 13.85 \\ 11.55 \end{bmatrix}$$

By computing $|\mathbf{A}|$ and $\mathbf{Ax}$ comment on the accuracy of the solution.

2.4 Symmetric and Banded Coefficient Matrices

Introduction

Engineering problems often lead to coefficient matrices that are *sparsely populated*, meaning that most elements of the matrix are zero. If all the nonzero terms are clustered about the leading diagonal, then the matrix is said to be *banded*. An example of a banded matrix is

$$\mathbf{A} = \begin{bmatrix} X & X & 0 & 0 & 0 \\ X & X & X & 0 & 0 \\ 0 & X & X & X & 0 \\ 0 & 0 & X & X & X \\ 0 & 0 & 0 & X & X \end{bmatrix}$$

where X's denote the nonzero elements that form the populated band (some of these elements may be zero). All the elements lying outside the band are zero. The matrix shown above has a bandwidth of three, since there are at most three nonzero elements in each row (or column). Such a matrix is called *tridiagonal*.

If a banded matrix is decomposed in the form $\mathbf{A} = \mathbf{LU}$, both $\mathbf{L}$ and $\mathbf{U}$ will retain the banded structure of $\mathbf{A}$. For example, if we decomposed the matrix shown above, we would get

$$\mathbf{L} = \begin{bmatrix} X & 0 & 0 & 0 & 0 \\ X & X & 0 & 0 & 0 \\ 0 & X & X & 0 & 0 \\ 0 & 0 & X & X & 0 \\ 0 & 0 & 0 & X & X \end{bmatrix} \quad \mathbf{U} = \begin{bmatrix} X & X & 0 & 0 & 0 \\ 0 & X & X & 0 & 0 \\ 0 & 0 & X & X & 0 \\ 0 & 0 & 0 & X & X \\ 0 & 0 & 0 & 0 & X \end{bmatrix}$$

The banded structure of a coefficient matrix can be exploited to save storage and computation time. If the coefficient matrix is also symmetric, further economies are

2.4 Symmetric and Banded Coefficient Matrices

possible. In this section we show how the methods of solution discussed previously can be adapted for banded and symmetric coefficient matrices.

Tridiagonal Coefficient Matrix

Consider the solution of $\mathbf{A}\mathbf{x} = \mathbf{b}$ by Doolittle's decomposition, where $\mathbf{A}$ is the $n \times n$ tridiagonal matrix

$$\mathbf{A} = \begin{bmatrix} d_1 & e_1 & 0 & 0 & \cdots & 0 \\ c_1 & d_2 & e_2 & 0 & \cdots & 0 \\ 0 & c_2 & d_3 & e_3 & \cdots & 0 \\ 0 & 0 & c_3 & d_4 & \cdots & 0 \\ \vdots & \vdots & \vdots & \vdots & \ddots & \vdots \\ 0 & 0 & \cdots & 0 & c_{n-1} & d_n \end{bmatrix}$$

As the notation implies, we are storing the nonzero elements of $\mathbf{A}$ in the vectors

$$\mathbf{c} = \begin{bmatrix} c_1 \\ c_2 \\ \vdots \\ c_{n-1} \end{bmatrix} \quad \mathbf{d} = \begin{bmatrix} d_1 \\ d_2 \\ \vdots \\ d_{n-1} \\ d_n \end{bmatrix} \quad \mathbf{e} = \begin{bmatrix} e_1 \\ e_2 \\ \vdots \\ e_{n-1} \end{bmatrix}$$

The resulting saving of storage can be significant. For example, a 100×100 tridiagonal matrix, containing 10,000 elements, can be stored in only $99 + 100 + 99 = 298$ locations, which represents a compression ratio of about 33:1.

Let us now apply LU decomposition to the coefficient matrix. We reduce row k by getting rid of c_{k-1} with the elementary operation

$$\text{row } k \leftarrow \text{row } k - (c_{k-1}/d_{k-1}) \times \text{row } (k-1), \quad k = 2, 3, \ldots, n$$

The corresponding change in d_k is

$$d_k \leftarrow d_k - (c_{k-1}/d_{k-1})e_{k-1} \tag{2.21}$$

whereas e_k is not affected. To finish up with Doolittle's decomposition of the form $[\mathbf{L}\backslash\mathbf{U}]$, we store the multiplier $\lambda = c_{k-1}/d_{k-1}$ in the location previously occupied by c_{k-1}:

$$c_{k-1} \leftarrow c_{k-1}/d_{k-1} \tag{2.22}$$

Systems of Linear Algebraic Equations

Thus the decomposition algorithm is

```
for k in range(1,n):
    lam = c[k-1]/d[k-1]
    d[k] = d[k] - lam*e[k-1]
    c[k-1] = lam
```

Next we look at the solution phase, i.e., the solution of the $\mathbf{Ly} = \mathbf{b}$, followed by $\mathbf{Ux} = \mathbf{y}$. The equations $\mathbf{Ly} = \mathbf{b}$ can be portrayed by the augmented coefficient matrix

$$\begin{bmatrix} \mathbf{L} \mid \mathbf{b} \end{bmatrix} = \begin{bmatrix} 1 & 0 & 0 & 0 & \cdots & 0 & b_1 \\ c_1 & 1 & 0 & 0 & \cdots & 0 & b_2 \\ 0 & c_2 & 1 & 0 & \cdots & 0 & b_3 \\ 0 & 0 & c_3 & 1 & \cdots & 0 & b_4 \\ \vdots & \vdots & \vdots & \vdots & \cdots & \vdots & \vdots \\ 0 & 0 & \cdots & 0 & c_{n-1} & 1 & b_n \end{bmatrix}$$

Note that the original contents of $\mathbf{c}$ were destroyed and replaced by the multipliers during the decomposition. The solution algorithm for $\mathbf{y}$ by forward substitution is

```
y[0] = b[0]
for k in range(1,n):
    y[k] = b[k] - c[k-1]*y[k-1]
```

The augmented coefficient matrix representing $\mathbf{Ux} = \mathbf{y}$ is

$$\begin{bmatrix} \mathbf{U} \mid \mathbf{y} \end{bmatrix} = \begin{bmatrix} d_1 & e_1 & 0 & \cdots & 0 & 0 & y_1 \\ 0 & d_2 & e_2 & \cdots & 0 & 0 & y_2 \\ 0 & 0 & d_3 & \cdots & 0 & 0 & y_3 \\ \vdots & \vdots & \vdots & & \vdots & \vdots & \vdots \\ 0 & 0 & 0 & \cdots & d_{n-1} & e_{n-1} & y_{n-1} \\ 0 & 0 & 0 & \cdots & 0 & d_n & y_n \end{bmatrix}$$

Note again that the contents of $\mathbf{d}$ were altered from the original values during the decomposition phase (but $\mathbf{e}$ was unchanged). The solution for $\mathbf{x}$ is obtained by back substitution using the algorithm

```
x[n-1] = y[n-1]/d[n-1]
for k in range(n-2,-1,-1):
    x[k] = (y[k] - e[k]*x[k+1])/d[k]
end do
```

2.4 Symmetric and Banded Coefficient Matrices

■ LUdecomp3

This module contains the functions LUdecomp3 and LUsolve3 for the decomposition and solution phases of a tridiagonal matrix. In LUsolve3, the vector **y** writes over the constant vector **b** during forward substitution. Similarly, the solution vector **x** overwrites **y** in the back substitution process. In other words, **b** contains the solution upon exit from LUsolve3.

```
## module LUdecomp3
''' c,d,e = LUdecomp3(c,d,e).
    LU decomposition of tridiagonal matrix [c\d\e]. On output
    {c},{d} and {e} are the diagonals of the decomposed matrix.

    x = LUsolve3(c,d,e,b).
    Solves [c\d\e]{x} = {b}, where {c}, {d} and {e} are the
    vectors returned from LUdecomp3.
'''

def LUdecomp3(c,d,e):
    n = len(d)
    for k in range(1,n):
        lam = c[k-1]/d[k-1]
        d[k] = d[k] - lam*e[k-1]
        c[k-1] = lam
    return c,d,e

def LUsolve3(c,d,e,b):
    n = len(d)
    for k in range(1,n):
        b[k] = b[k] - c[k-1]*b[k-1]
    b[n-1] = b[n-1]/d[n-1]
    for k in range(n-2,-1,-1):
        b[k] = (b[k] - e[k]*b[k+1])/d[k]
    return b
```

Symmetric Coefficient Matrices

More often than not, coefficient matrices that arise in engineering problems are symmetric as well as banded. Therefore, it is worthwhile to discover special properties of such matrices and learn how to utilize them in the construction of efficient algorithms.

If the matrix **A** is symmetric, then the LU decomposition can be presented in the form

$$\mathbf{A} = \mathbf{LU} = \mathbf{LDL}^T \tag{2.23}$$

where **D** is a diagonal matrix. An example is Choleski's decomposition $\mathbf{A} = \mathbf{LL}^T$ that was discussed in the previous section (in this case $\mathbf{D} = \mathbf{I}$). For Doolittle's decomposition we have

$$\mathbf{U} = \mathbf{DL}^T = \begin{bmatrix} D_1 & 0 & 0 & \cdots & 0 \\ 0 & D_2 & 0 & \cdots & 0 \\ 0 & 0 & D_3 & \cdots & 0 \\ \vdots & \vdots & \vdots & \cdots & \vdots \\ 0 & 0 & 0 & \cdots & D_n \end{bmatrix} \begin{bmatrix} 1 & L_{21} & L_{31} & \cdots & L_{n1} \\ 0 & 1 & L_{32} & \cdots & L_{n2} \\ 0 & 0 & 1 & \cdots & L_{n3} \\ \vdots & \vdots & \vdots & \cdots & \vdots \\ 0 & 0 & 0 & \cdots & 1 \end{bmatrix}$$

which gives

$$\mathbf{U} = \begin{bmatrix} D_1 & D_1 L_{21} & D_1 L_{31} & \cdots & D_1 L_{n1} \\ 0 & D_2 & D_2 L_{32} & \cdots & D_2 L_{n2} \\ 0 & 0 & D_3 & \cdots & D_3 L_{3n} \\ \vdots & \vdots & \vdots & \cdots & \vdots \\ 0 & 0 & 0 & \cdots & D_n \end{bmatrix} \tag{2.24}$$

We now see that during decomposition of a symmetric matrix only **U** has to be stored, since **D** and **L** can be easily recovered from **U**. Thus Gauss elimination, which results in an upper triangular matrix of the form shown in Eq. (2.24), is sufficient to decompose a symmetric matrix.

There is an alternative storage scheme that can be employed during **LU** decomposition. The idea is to arrive at the matrix

$$\mathbf{U}^* = \begin{bmatrix} D_1 & L_{21} & L_{31} & \cdots & L_{n1} \\ 0 & D_2 & L_{32} & \cdots & L_{n2} \\ 0 & 0 & D_3 & \cdots & L_{n3} \\ \vdots & \vdots & \vdots & \ddots & \vdots \\ 0 & 0 & 0 & \cdots & D_n \end{bmatrix} \tag{2.25}$$

Here **U** can be recovered from $U_{ij} = D_i L_{ji}$. It turns out that this scheme leads to a computationally more efficient solution phase; therefore, we adopt it for symmetric, banded matrices.

2.4 Symmetric and Banded Coefficient Matrices

Symmetric, Pentadiagonal Coefficient Matrix

We encounter pentadiagonal (bandwidth = 5) coefficient matrices in the solution of fourth-order, ordinary differential equations by finite differences. Often these matrices are symmetric, in which case an $n \times n$ coefficient matrix has the form

$$\mathbf{A} = \begin{bmatrix} d_1 & e_1 & f_1 & 0 & 0 & 0 & \cdots & 0 \\ e_1 & d_2 & e_2 & f_2 & 0 & 0 & \cdots & 0 \\ f_1 & e_2 & d_3 & e_3 & f_3 & 0 & \cdots & 0 \\ 0 & f_2 & e_3 & d_4 & e_4 & f_4 & \cdots & 0 \\ \vdots & \vdots & \vdots & \vdots & \vdots & \vdots & \ddots & \vdots \\ 0 & \cdots & 0 & f_{n-4} & e_{n-3} & d_{n-2} & e_{n-2} & f_{n-2} \\ 0 & \cdots & 0 & 0 & f_{n-3} & e_{n-2} & d_{n-1} & e_{n-1} \\ 0 & \cdots & 0 & 0 & 0 & f_{n-2} & e_{n-1} & d_n \end{bmatrix} \quad (2.26)$$

As in the case of tridiagonal matrices, we store the nonzero elements in the three vectors

$$\mathbf{d} = \begin{bmatrix} d_1 \\ d_2 \\ \vdots \\ d_{n-2} \\ d_{n-1} \\ d_n \end{bmatrix} \quad \mathbf{e} = \begin{bmatrix} e_1 \\ e_2 \\ \vdots \\ e_{n-2} \\ e_{n-1} \end{bmatrix} \quad \mathbf{f} = \begin{bmatrix} f_1 \\ f_2 \\ \vdots \\ f_{n-2} \end{bmatrix}$$

Let us now look at the solution of the equations $\mathbf{Ax} = \mathbf{b}$ by Doolittle's decomposition. The first step is to transform $\mathbf{A}$ to upper triangular form by Gauss elimination. If elimination has progressed to the stage where the kth row has become the pivot row, we have the following situation:

$$\mathbf{A} = \begin{bmatrix} \ddots & \vdots & \vdots & \vdots & \vdots & \vdots & \vdots & \vdots \\ \cdots & 0 & d_k & e_k & f_k & 0 & 0 & 0 & \cdots \\ \cdots & 0 & e_k & d_{k+1} & e_{k+1} & f_{k+1} & 0 & 0 & \cdots \\ \cdots & 0 & f_k & e_{k+1} & d_{k+2} & e_{k+2} & f_{k+2} & 0 & \cdots \\ \cdots & 0 & 0 & f_{k+1} & e_{k+2} & d_{k+3} & e_{k+3} & f_{k+3} & \cdots \\ & \vdots & \vdots & \vdots & \vdots & \vdots & \vdots & \vdots & \ddots \end{bmatrix} \leftarrow$$

The elements e_k and f_k below the pivot row (the kth row) are eliminated by the operations

$$\text{row } (k+1) \leftarrow \text{row } (k+1) - (e_k/d_k) \times \text{row } k$$

$$\text{row } (k+2) \leftarrow \text{row } (k+2) - (f_k/d_k) \times \text{row } k$$

The only terms (other than those being eliminated) that are changed by the above operations are

$$d_{k+1} \leftarrow d_{k+1} - (e_k/d_k)e_k$$
$$e_{k+1} \leftarrow e_{k+1} - (e_k/d_k)f_k \qquad (2.27a)$$
$$d_{k+2} \leftarrow d_{k+2} - (f_k/d_k)f_k$$

Storage of the multipliers in the *upper* triangular portion of the matrix results in

$$e_k \leftarrow e_k/d_k \qquad f_k \leftarrow f_k/d_k \qquad (2.27b)$$

At the conclusion of the elimination phase the matrix has the form (do not confuse **d**, **e** and **f** with the original contents of **A**)

$$\mathbf{U}^* = \begin{bmatrix} d_1 & e_1 & f_1 & 0 & \cdots & 0 \\ 0 & d_2 & e_2 & f_2 & \cdots & 0 \\ 0 & 0 & d_3 & e_3 & \cdots & 0 \\ \vdots & \vdots & \vdots & \vdots & \cdots & \vdots \\ 0 & 0 & \cdots & 0 & d_{n-1} & e_{n-1} \\ 0 & 0 & \cdots & 0 & 0 & d_n \end{bmatrix}$$

Next comes the solution phase. The equations $\mathbf{Ly} = \mathbf{b}$ have the augmented coefficient matrix

$$\begin{bmatrix} \mathbf{L} \mid \mathbf{b} \end{bmatrix} = \begin{bmatrix} 1 & 0 & 0 & 0 & \cdots & 0 & b_1 \\ e_1 & 1 & 0 & 0 & \cdots & 0 & b_2 \\ f_1 & e_2 & 1 & 0 & \cdots & 0 & b_3 \\ 0 & f_2 & e_3 & 1 & \cdots & 0 & b_4 \\ \vdots & \vdots & \vdots & \vdots & \cdots & \vdots & \vdots \\ 0 & 0 & 0 & f_{n-2} & e_{n-1} & 1 & b_n \end{bmatrix}$$

Solution by forward substitution yields

$$y_1 = b_1$$
$$y_2 = b_2 - e_1 y_1 \qquad (2.28)$$
$$\vdots$$
$$y_k = b_k - f_{k-2} y_{k-2} - e_{k-1} y_{k-1}, \quad k = 3, 4, \ldots, n$$

2.4 Symmetric and Banded Coefficient Matrices

The equations to be solved by back substitution, namely $\mathbf{U}\mathbf{x} = \mathbf{y}$, have the augmented coefficient matrix

$$\begin{bmatrix} \mathbf{U} \mid \mathbf{y} \end{bmatrix} = \begin{bmatrix} d_1 & d_1 e_1 & d_1 f_1 & 0 & \cdots & 0 & \bigm| & y_1 \\ 0 & d_2 & d_2 e_2 & d_2 f_2 & \cdots & 0 & \bigm| & y_2 \\ 0 & 0 & d_3 & d_3 e_3 & \cdots & 0 & \bigm| & y_3 \\ \vdots & \vdots & \vdots & \vdots & \cdots & \vdots & \bigm| & \vdots \\ 0 & 0 & \cdots & 0 & d_{n-1} & d_{n-1} e_{n-1} & \bigm| & y_{n-1} \\ 0 & 0 & \cdots & 0 & 0 & d_n & \bigm| & y_n \end{bmatrix}$$

the solution of which is obtained by back substitution:

$$x_n = y_n/d_n$$

$$x_{n-1} = y_{n-1}/d_{n-1} - e_{n-1}x_n$$

$$x_k = y_k/d_k - e_k x_{k+1} - f_k x_{k+2}, \quad k = n-2, n-3, \ldots, 1 \tag{2.29}$$

■ LUdecomp5

The function LUdecomp5 below decomposes a symmetric, pentadiagonal matrix A of the form $\mathbf{A} = [\mathbf{f}\backslash\mathbf{e}\backslash\mathbf{d}\backslash\mathbf{e}\backslash\mathbf{f}]$. The original vectors **d**, **e** and **f** are destroyed and replaced by the vectors of the decomposed matrix. After decomposition, the solution of $\mathbf{A}\mathbf{x} = \mathbf{b}$ can be obtained by LUsolve5. During forward substitution, the original **b** is replaced by **y**. Similarly, **y** is written over by **x** in the back substitution phase, so that **b** contains the solution vector upon exit from LUsolve5.

```
## module LUdecomp5
''' d,e,f = LUdecomp5(d,e,f).
    LU decomposition of symetric pentadiagonal matrix
    [f\e\d\e\f]. On output {d},{e} and {f} are the
    diagonals of the decomposed matrix.

    x = LUsolve5(d,e,f,b).
    Solves [f\e\d\e\f]{x} = {b}, where {d}, {e} and {f}
    are the vectors returned from LUdecomp5.
'''
def LUdecomp5(d,e,f):
    n = len(d)
    for k in range(n-2):
        lam = e[k]/d[k]
        d[k+1] = d[k+1] - lam*e[k]
```

Systems of Linear Algebraic Equations

```
            e[k+1] = e[k+1] - lam*f[k]
            e[k] = lam
            lam = f[k]/d[k]
            d[k+2] = d[k+2] - lam*f[k]
            f[k] = lam
        lam = e[n-2]/d[n-2]
        d[n-1] = d[n-1] - lam*e[n-2]
        e[n-2] = lam
        return d,e,f

def LUsolve5(d,e,f,b):
    n = len(d)
    b[1] = b[1] - e[0]*b[0]
    for k in range(2,n):
        b[k] = b[k] - e[k-1]*b[k-1] - f[k-2]*b[k-2]
    b[n-1] = b[n-1]/d[n-1]
    b[n-2] = b[n-2]/d[n-2] - e[n-2]*b[n-1]
    for k in range(n-3,-1,-1):
        b[k] = b[k]/d[k] - e[k]*b[k+1] - f[k]*b[k+2]
    return b
```

EXAMPLE 2.9

As a result of Gauss elimination, a symmetric matrix **A** was transformed to the upper triangular form

$$\mathbf{U} = \begin{bmatrix} 4 & -2 & 1 & 0 \\ 0 & 3 & -3/2 & 1 \\ 0 & 0 & 3 & -3/2 \\ 0 & 0 & 0 & 35/12 \end{bmatrix}$$

Determine the original matrix **A**.

Solution First we find **L** in the decomposition **A** = **LU**. Dividing each row of **U** by its diagonal element yields

$$\mathbf{L}^T = \begin{bmatrix} 1 & -1/2 & 1/4 & 0 \\ 0 & 1 & -1/2 & 1/3 \\ 0 & 0 & 1 & -1/2 \\ 0 & 0 & 0 & 1 \end{bmatrix}$$

2.4 Symmetric and Banded Coefficient Matrices

Therefore, $\mathbf{A} = \mathbf{LU}$ becomes

$$\mathbf{A} = \begin{bmatrix} 1 & 0 & 0 & 0 \\ -1/2 & 1 & 0 & 0 \\ 1/4 & -1/2 & 1 & 0 \\ 0 & 1/3 & -1/2 & 1 \end{bmatrix} \begin{bmatrix} 4 & -2 & 1 & 0 \\ 0 & 3 & -3/2 & 1 \\ 0 & 0 & 3 & -3/2 \\ 0 & 0 & 0 & 35/12 \end{bmatrix}$$

$$= \begin{bmatrix} 4 & -2 & 1 & 0 \\ -2 & 4 & -2 & 1 \\ 1 & -2 & 4 & -2 \\ 0 & 1 & -2 & 4 \end{bmatrix}$$

EXAMPLE 2.10
Determine $\mathbf{L}$ and $\mathbf{D}$ that result from Doolittle's decomposition $\mathbf{A} = \mathbf{LDL}^T$ of the symmetric matrix

$$\mathbf{A} = \begin{bmatrix} 3 & -3 & 3 \\ -3 & 5 & 1 \\ 3 & 1 & 10 \end{bmatrix}$$

Solution We use Gauss elimination, storing the multipliers in the *upper* triangular portion of $\mathbf{A}$. At the completion of elimination, the matrix will have the form of $\mathbf{U}^*$ in Eq. (2.25).

The terms to be eliminated in the first pass are A_{21} and A_{31} using the elementary operations

$$\text{row } 2 \leftarrow \text{row } 2 - (-1) \times \text{row } 1$$
$$\text{row } 3 \leftarrow \text{row } 3 - (1) \times \text{row } 1$$

Storing the multipliers (-1 and 1) in the locations occupied by A_{12} and A_{13}, we get

$$\mathbf{A}' = \begin{bmatrix} 3 & -1 & 1 \\ 0 & 2 & 4 \\ 0 & 4 & 7 \end{bmatrix}$$

The second pass is the operation

$$\text{row } 3 \leftarrow \text{row } 3 - 2 \times \text{row } 2$$

which yields, after overwriting A_{23} with the multiplier 2

$$\mathbf{A}'' = [\mathbf{0}\backslash\mathbf{D}\backslash\mathbf{L}^T] = \begin{bmatrix} 3 & -1 & 1 \\ 0 & 2 & 2 \\ 0 & 0 & -1 \end{bmatrix}$$

Hence

$$L = \begin{bmatrix} 1 & 0 & 0 \\ -1 & 1 & 0 \\ 1 & 2 & 1 \end{bmatrix} \quad D = \begin{bmatrix} 3 & 0 & 0 \\ 0 & 2 & 0 \\ 0 & 0 & -1 \end{bmatrix}$$

EXAMPLE 2.11

Utilize the functions `LUdecmp3` and `LUsolve3` to solve $\mathbf{Ax} = \mathbf{b}$, where

$$\mathbf{A} = \begin{bmatrix} 2 & -1 & 0 & 0 & 0 \\ -1 & 2 & -1 & 0 & 0 \\ 0 & -1 & 2 & -1 & 0 \\ 0 & 0 & -1 & 2 & -1 \\ 0 & 0 & 0 & -1 & 2 \end{bmatrix} \quad \mathbf{b} = \begin{bmatrix} 5 \\ -5 \\ 4 \\ -5 \\ 5 \end{bmatrix}$$

Solution

```
#!/usr/bin/python
## example2_11
from numarray import array,ones
from LUdecomp3 import *

d = ones((5))*2.0
c = ones((4))*(-1.0)
b = array([5.0, -5.0, 4.0, -5.0, 5.0])
e = c.copy()
c,d,e = LUdecomp3(c,d,e)
x = LUsolve3(c,d,e,b)
print ''\nx =\n'',x
raw_input(''\nPress return to exit'')
```

The output is:

```
x =
[ 2. -1.  1. -1.  2.]
```

2.5 Pivoting

Introduction

Sometimes the order in which the equations are presented to the solution algorithm has a profound effect on the results. For example, consider the equations

$$\begin{aligned} 2x_1 - x_2 &= 1 \\ -x_1 + 2x_2 - x_3 &= 0 \\ -x_2 + x_3 &= 0 \end{aligned}$$

2.5 Pivoting

The corresponding augmented coefficient matrix is

$$\left[\mathbf{A} \mid \mathbf{b}\right] = \begin{bmatrix} 2 & -1 & 0 & | & 1 \\ -1 & 2 & -1 & | & 0 \\ 0 & -1 & 1 & | & 0 \end{bmatrix} \quad \text{(a)}$$

Equations (a) are in the "right order" in the sense that we would have no trouble obtaining the correct solution $x_1 = x_2 = x_3 = 1$ by Gauss elimination or LU decomposition. Now suppose that we exchange the first and third equations, so that the augmented coefficient matrix becomes

$$\left[\mathbf{A} \mid \mathbf{b}\right] = \begin{bmatrix} 0 & -1 & 1 & | & 0 \\ -1 & 2 & -1 & | & 0 \\ 2 & -1 & 0 & | & 1 \end{bmatrix} \quad \text{(b)}$$

Since we did not change the equations (only their order was altered), the solution is still $x_1 = x_2 = x_3 = 1$. However, Gauss elimination fails immediately due to the presence of the zero pivot element (the element A_{11}).

The above example demonstrates that it is sometimes essential to reorder the equations during the elimination phase. The reordering, or *row pivoting*, is also required if the pivot element is not zero, but very small in comparison to other elements in the pivot row, as demonstrated by the following set of equations:

$$\left[\mathbf{A} \mid \mathbf{b}\right] = \begin{bmatrix} \varepsilon & -1 & 1 & | & 0 \\ -1 & 2 & -1 & | & 0 \\ 2 & -1 & 0 & | & 1 \end{bmatrix} \quad \text{(c)}$$

These equations are the same as Eqs. (b), except that the small number ε replaces the zero element A_{11} in Eq. (b). Therefore, if we let $\varepsilon \to 0$, the solutions of Eqs. (b) and (c) should become identical. After the first phase of Gauss elimination, the augmented coefficient matrix becomes

$$\left[\mathbf{A'} \mid \mathbf{b'}\right] = \begin{bmatrix} \varepsilon & -1 & 1 & | & 0 \\ 0 & 2 - 1/\varepsilon & -1 + 1/\varepsilon & | & 0 \\ 0 & -1 + 2/\varepsilon & -2/\varepsilon & | & 1 \end{bmatrix} \quad \text{(d)}$$

Because the computer works with a fixed word length, all numbers are rounded off to a finite number of significant figures. If ε is very small, then $1/\varepsilon$ is huge, and an element such as $2 - 1/\varepsilon$ is rounded to $-1/\varepsilon$. Therefore, for sufficiently small ε, the Eqs. (d) are actually stored as

$$\left[\mathbf{A'} \mid \mathbf{b'}\right] = \begin{bmatrix} \varepsilon & -1 & 1 & | & 0 \\ 0 & -1/\varepsilon & 1/\varepsilon & | & 0 \\ 0 & 2/\varepsilon & -2/\varepsilon & | & 1 \end{bmatrix}$$

Because the second and third equations obviously contradict each other, the solution process fails again. This problem would not arise if the first and second, or the first and the third equations were interchanged in Eqs. (c) before the elimination.

The last example illustrates the extreme case where ε was so small that roundoff errors resulted in total failure of the solution. If we were to make ε somewhat bigger so that the solution would not "bomb" any more, the roundoff errors might still be large enough to render the solution unreliable. Again, this difficulty could be avoided by pivoting.

Diagonal Dominance

An $n \times n$ matrix $\mathbf{A}$ is said to be *diagonally dominant* if each diagonal element is larger than the sum of the other elements in the same row (we are talking here about absolute values). Thus diagonal dominance requires that

$$|A_{ii}| > \sum_{\substack{j=1 \\ j \neq i}}^{n} |A_{ij}| \quad (i = 1, 2, ..., n) \tag{2.30}$$

For example, the matrix

$$\begin{bmatrix} -2 & 4 & -1 \\ 1 & -1 & 3 \\ 4 & -2 & 1 \end{bmatrix}$$

is not diagonally dominant, but if we rearrange the rows in the following manner

$$\begin{bmatrix} 4 & -2 & 1 \\ -2 & 4 & -1 \\ 1 & -1 & 3 \end{bmatrix}$$

then we have diagonal dominance.

It can be shown that if the coefficient matrix of the equations $\mathbf{Ax} = \mathbf{b}$ is diagonally dominant, then the solution does not benefit from pivoting; that is, the equations are already arranged in the optimal order. It follows that the strategy of pivoting should be to reorder the equations so that the coefficient matrix is as close to diagonal dominance as possible. This is the principle behind scaled row pivoting, discussed next.

Gauss Elimination with Scaled Row Pivoting

Consider the solution of $\mathbf{Ax} = \mathbf{b}$ by Gauss elimination with row pivoting. Recall that pivoting aims at improving diagonal dominance of the coefficient matrix, i.e., making the pivot element as large as possible in comparison to other elements in the pivot

2.5 Pivoting

row. The comparison is made easier if we establish an array **s** with the elements

$$s_i = \max_j |A_{ij}|, \quad i = 1, 2, \ldots, n \tag{2.31}$$

Thus s_i, called the *scale factor* of row i, contains the absolute value of the largest element in the ith row of **A**. The vector **s** can be obtained with the algorithm

```
for i in range(n):
    s[i] = max(abs(a[i,:]))
```

The *relative size* of an element A_{ij} (that is, relative to the largest element in the ith row) is defined as the ratio

$$r_{ij} = \frac{|A_{ij}|}{s_i} \tag{2.32}$$

Suppose that the elimination phase has reached the stage where the kth row has become the pivot row. The augmented coefficient matrix at this point is shown below.

$$\begin{bmatrix} A_{11} & A_{12} & A_{13} & A_{14} & \cdots & A_{1n} & b_1 \\ 0 & A_{22} & A_{23} & A_{24} & \cdots & A_{2n} & b_2 \\ 0 & 0 & A_{33} & A_{34} & \cdots & A_{3n} & b_3 \\ \vdots & \vdots & \vdots & \vdots & \cdots & \vdots & \vdots \\ 0 & \cdots & 0 & A_{kk} & \cdots & A_{kn} & b_k \\ \vdots & \cdots & \vdots & \vdots & \cdots & \vdots & \vdots \\ 0 & \cdots & 0 & A_{nk} & \cdots & A_{nn} & b_n \end{bmatrix} \leftarrow$$

We don't automatically accept A_{kk} as the next pivot element, but look in the kth column below A_{kk} for a "better" pivot. The best choice is the element A_{pk} that has the largest relative size; that is, we choose p such that

$$r_{pk} = \max_j r_{jk}, \quad j \geq k$$

If we find such an element, then we interchange the rows k and p, and proceed with the elimination pass as usual. Note that the corresponding row interchange must also be carried out in the scale factor array **s**. The algorithm that does all this is

```
for k in range(0,n-1):

    # Find row containing element with largest relative size
    p = int(argmax(abs(a[k:n,k])/s[k:n])) + k

    # If this element is very small, matrix is singular
    if abs(a[p,k]) < tol: error.err('Matrix is singular')
```

```
        # Check whether rows k and p must be interchanged
        if p != k:
          # Interchange rows if needed
            swap.swapRows(b,k,p)
            swap.swapRows(s,k,p)
            swap.swapRows(a,k,p)
    # Proceed with elimination
```

The Python statement `int(argmax(v))` returns the index of the largest element in the vector v. The algorithms for exchanging rows (and columns) are included in the module `swap` shown below.

■ swap

The function `swapRows` interchanges rows *i* and *j* of a matrix or vector **v**, whereas `swapCols` interchanges columns *i* and *j* of a matrix.

```
## module swap
''' swapRows(v,i,j).
    Swaps rows i and j of vector or matrix [v].

    swapCols(v,i,j).
    Swaps columns i and j of matrix [v].
'''
def swapRows(v,i,j):
    if len(v.getshape()) == 1:
        v[i],v[j] = v[j],v[i]
    else:
        temp = v[i].copy()
        v[i] = v[j]
        v[j] = temp

def swapCols(v,i,j):
    temp = v[:,j].copy()
    v[:,j] = v[:,i]
    v[:,i] = temp
```

2.5 Pivoting

■ gaussPivot

The function gaussPivot performs Gauss elimination with row pivoting. Apart from row swapping, the elimination and solution phases are identical to gaussElimin in Art. 2.2.

```
## module gaussPivot
''' x = gaussPivot(a,b,tol=1.0e-9).
    Solves [a]{x} = {b} by Gauss elimination with
    scaled row pivoting
'''
from numarray import *
import swap
import error

def gaussPivot(a,b,tol=1.0e-9):
    n = len(b)
    
  # Set up scale factors
    s = zeros((n),type=Float64)
    for i in range(n):
        s[i] = max(abs(a[i,:]))
        
    for k in range(0,n-1):
    
      # Row interchange, if needed
        p = int(argmax(abs(a[k:n,k])/s[k:n])) + k
        if abs(a[p,k]) < tol:
            error.err('Matrix is singular')
        if p != k:
            swap.swapRows(b,k,p)
            swap.swapRows(s,k,p)
            swap.swapRows(a,k,p)
            
      # Elimination
        for i in range(k+1,n):
            if a[i,k] != 0.0:
                lam = a[i,k]/a[k,k]
                a[i,k+1:n] = a [i,k+1:n] - lam*a[k,k+1:n]
                b[i] = b[i] - lam*b[k]
```

```
            if abs(a[n-1,n-1]) < tol:
                error.err('Matrix is singular')

        # Back substitution
        for k in range(n-1,-1,-1):
            b[k] = (b[k] - dot(a[k,k+1:n],b[k+1:n]))/a[k,k]
        return b
```

■ LUpivot

The Gauss elimination algorithm can be changed to Doolittle's decomposition with minor changes. The most important of these is keeping a record of the row interchanges during the decomposition phase. In LUdecomp this record is kept in the array seq. Initially seq contains [0, 1, 2, ...]. Whenever two rows are interchanged, the corresponding interchange is also carried out in seq. Thus seq shows the order in which of the original rows have been rearranged. This information is passed on to the solution phase (LUsolve), which rearranges the elements of the constant vector in the same order before proceeding to forward and back substitutions.

```
## module LUpivot
''' a,seq = LUdecomp(a,tol=1.0e-9).
    LU decomposition of matrix [a] using scaled row pivoting.
    The returned matrix [a] = [L\U] contains [U] in the upper
    triangle and the nondiagonal terms of [L] in the lower triangle.
    Note that [L][U] is a row-wise permutation of the original [a];
    the permutations are recorded in the vector {seq}.

    x = LUsolve(a,b,seq).
    Solves [L][U]{x} = {b}, where the matrix [a] = [L\U] and the
    permutation vector {seq} are returned from LUdecomp.
'''
from numarray import argmax,abs,dot,zeros,Float64,array
import swap
import error

def LUdecomp(a,tol=1.0e-9):
    n = len(a)
    seq = array(range(n))
```

2.5 Pivoting

```
    # Set up scale factors
    s = zeros((n),type=Float64)
    for i in range(n):
        s[i] = max(abs(a[i,:]))

    for k in range(0,n-1):

      # Row interchange, if needed
        p = int(argmax(abs(a[k:n,k])/s[k:n])) + k
        if abs(a[p,k]) <  tol:
            error.err('Matrix is singular')
        if p != k:
            swap.swapRows(s,k,p)
            swap.swapRows(a,k,p)
            swap.swapRows(seq,k,p)

      # Elimination
        for i in range(k+1,n):
            if a[i,k] != 0.0:
                lam = a[i,k]/a[k,k]
                a[i,k+1:n] = a[i,k+1:n] - lam*a[k,k+1:n]
                a[i,k] = lam
    return a,seq

def LUsolve(a,b,seq):
    n = len(a)

  # Rearrange constant vector; store it in [x]
    x = b.copy()
    for i in range(n):
        x[i] = b[seq[i]]

  # Solution
    for k in range(1,n):
        x[k] = x[k] - dot(a[k,0:k],x[0:k])
    for k in range(n-1,-1,-1):
        x[k] = (x[k] - dot(a[k,k+1:n],x[k+1:n]))/a[k,k]
    return x
```

When to Pivot

Pivoting has a couple of drawbacks. One of these is the increased time of computation; the other is the destruction of the symmetry and banded structure of the coefficient matrix. The latter is of particular concern in engineering computing, where the coefficient matrices are frequently banded and symmetric, a property that is utilized in the solution, as seen in the previous chapter. Fortunately, these matrices are often diagonally dominant as well, so that they would not benefit from pivoting anyway.

There are no infallible rules for determining when pivoting should be used. Experience indicates that pivoting is likely to be counterproductive if the coefficient matrix is banded. Positive definite and, to a lesser degree, symmetric matrices also seldom gain from pivoting. And we should not forget that pivoting is not the only means of controlling roundoff errors—there is also double precision arithmetic.

It should be strongly emphasized that the above rules of thumb are only meant for equations that stem from real engineering problems. It is not difficult to concoct "textbook" examples that do not conform to these rules.

EXAMPLE 2.12
Employ Gauss elimination with scaled row pivoting to solve the equations $\mathbf{Ax} = \mathbf{b}$, where

$$\mathbf{A} = \begin{bmatrix} 2 & -2 & 6 \\ -2 & 4 & 3 \\ -1 & 8 & 4 \end{bmatrix} \qquad \mathbf{b} = \begin{bmatrix} 16 \\ 0 \\ -1 \end{bmatrix}$$

Solution The augmented coefficient matrix and the scale factor array are

$$\begin{bmatrix} \mathbf{A} \mid \mathbf{b} \end{bmatrix} = \begin{bmatrix} 2 & -2 & 6 & | & 16 \\ -2 & 4 & 3 & | & 0 \\ -1 & 8 & 4 & | & -1 \end{bmatrix} \qquad \mathbf{s} = \begin{bmatrix} 6 \\ 4 \\ 8 \end{bmatrix}$$

Note that $\mathbf{s}$ contains the absolute value of the largest element in each row of $\mathbf{A}$. At this stage, all the elements in the first column of $\mathbf{A}$ are potential pivots. To determine the best pivot element, we calculate the relative sizes of the elements in the first column:

$$\begin{bmatrix} r_{11} \\ r_{21} \\ r_{31} \end{bmatrix} = \begin{bmatrix} |A_{11}|/s_1 \\ |A_{21}|/s_2 \\ |A_{31}|/s_3 \end{bmatrix} = \begin{bmatrix} 1/3 \\ 1/2 \\ 1/8 \end{bmatrix}$$

Since r_{21} is the largest element, we conclude that A_{21} makes the best pivot element. Therefore, we exchange rows 1 and 2 of the augmented coefficient matrix and the

2.5 Pivoting

scale factor array, obtaining

$$[A \mid b] = \begin{bmatrix} -2 & 4 & 3 & | & 0 \\ 2 & -2 & 6 & | & 16 \\ -1 & 8 & 4 & | & -1 \end{bmatrix} \leftarrow \qquad s = \begin{bmatrix} 4 \\ 6 \\ 8 \end{bmatrix}$$

Now the first pass of Gauss elimination is carried out (the arrow points to the pivot row), yielding

$$[A' \mid b'] = \begin{bmatrix} -2 & 4 & 3 & | & 0 \\ 0 & 2 & 9 & | & 16 \\ 0 & 6 & 5/2 & | & -1 \end{bmatrix} \qquad s = \begin{bmatrix} 4 \\ 6 \\ 8 \end{bmatrix}$$

The potential pivot elements for the next elimination pass are A'_{22} and A'_{32}. We determine the "winner" from

$$\begin{bmatrix} * \\ r_{22} \\ r_{32} \end{bmatrix} = \begin{bmatrix} * \\ |A_{22}|/s_2 \\ |A_{32}|/s_3 \end{bmatrix} = \begin{bmatrix} * \\ 1/3 \\ 3/4 \end{bmatrix}$$

Note that r_{12} is irrelevant, since row 1 already acted as the pivot row. Therefore, it is excluded from further consideration. As r_{32} is larger than r_{22}, the third row is the better pivot row. After interchanging rows 2 and 3, we have

$$[A' \mid b'] = \begin{bmatrix} -2 & 4 & 3 & | & 0 \\ 0 & 6 & 5/2 & | & -1 \\ 0 & 2 & 9 & | & 16 \end{bmatrix} \leftarrow \qquad s = \begin{bmatrix} 4 \\ 8 \\ 6 \end{bmatrix}$$

The second elimination pass now yields

$$[A'' \mid b''] = [U \mid c] = \begin{bmatrix} -2 & 4 & 3 & | & 0 \\ 0 & 6 & 5/2 & | & -1 \\ 0 & 0 & 49/6 & | & 49/3 \end{bmatrix}$$

This completes the elimination phase. It should be noted that U is the matrix that would result from LU decomposition of the following row-wise permutation of A (the ordering of rows is the same as achieved by pivoting):

$$\begin{bmatrix} -2 & 4 & 3 \\ -1 & 8 & 4 \\ 2 & -2 & 6 \end{bmatrix}$$

Since the solution of $Ux = c$ by back substitution is not affected by pivoting, we skip the detailed of the computation. The result is $x^T = \begin{bmatrix} 1 & -1 & 2 \end{bmatrix}$.

Alternate Solution

It it not necessary to physically exchange equations during pivoting. We could accomplish Gauss elimination just as well by keeping the equations in place. The elimination would then proceed as follows (for the sake of brevity, we skip repeating the details of choosing the pivot equation):

$$\begin{bmatrix} \mathbf{A} \mid \mathbf{b} \end{bmatrix} = \begin{bmatrix} 2 & -2 & 6 & \mid & 16 \\ -2 & 4 & 3 & \mid & 0 \\ -1 & 8 & 4 & \mid & -1 \end{bmatrix} \leftarrow$$

$$\begin{bmatrix} \mathbf{A}' \mid \mathbf{b}' \end{bmatrix} = \begin{bmatrix} 0 & 2 & 9 & \mid & 16 \\ -2 & 4 & 3 & \mid & 0 \\ 0 & 6 & 5/2 & \mid & -1 \end{bmatrix} \leftarrow$$

$$\begin{bmatrix} \mathbf{A}'' \mid \mathbf{b}'' \end{bmatrix} = \begin{bmatrix} 0 & 0 & 49/6 & \mid & 49/3 \\ -2 & 4 & 3 & \mid & 0 \\ 0 & 6 & 5/2 & \mid & -1 \end{bmatrix}$$

But now the back substitution phase is a little more involved, since the order in which the equations must be solved has become scrambled. In hand computations this is not a problem, because we can determine the order by inspection. Unfortunately, "by inspection" does not work on a computer. To overcome this difficulty, we have to maintain an integer array **p** that keeps track of the row permutations during the elimination phase. The contents of **p** indicate the order in which the pivot rows were chosen. In this example, we would have at the end of Gauss elimination

$$\mathbf{p} = \begin{bmatrix} 2 \\ 3 \\ 1 \end{bmatrix}$$

showing that row 2 was the pivot row in the first elimination pass, followed by row 3 in the second pass. The equations are solved by back substitution in the reverse order: Eq. (1) is solved first for x_3, then Eq. (3) is solved for x_2, and finally Eq. (2) yields x_1.

By dispensing with swapping of equations, the scheme outlined above would probably result in a faster (and more complex) algorithm than gaussPivot, but the number of equations would have to be quite large before the difference becomes noticeable.

2.5 Pivoting

PROBLEM SET 2.2

1. Solve the equations $\mathbf{Ax} = \mathbf{b}$ by utilizing Doolittle's decomposition, where

$$\mathbf{A} = \begin{bmatrix} 3 & -3 & 3 \\ -3 & 5 & 1 \\ 3 & 1 & 5 \end{bmatrix} \quad \mathbf{b} = \begin{bmatrix} 9 \\ -7 \\ 12 \end{bmatrix}$$

2. Use Doolittle's decomposition to solve $\mathbf{Ax} = \mathbf{b}$, where

$$\mathbf{A} = \begin{bmatrix} 4 & 8 & 20 \\ 8 & 13 & 16 \\ 20 & 16 & -91 \end{bmatrix} \quad \mathbf{b} = \begin{bmatrix} 24 \\ 18 \\ -119 \end{bmatrix}$$

3. Determine $\mathbf{L}$ and $\mathbf{D}$ that result from Doolittle's decomposition of the symmetric matrix

$$\mathbf{A} = \begin{bmatrix} 2 & -2 & 0 & 0 & 0 \\ -2 & 5 & -6 & 0 & 0 \\ 0 & -6 & 16 & 12 & 0 \\ 0 & 0 & 12 & 39 & -6 \\ 0 & 0 & 0 & -6 & 14 \end{bmatrix}$$

4. Solve the tridiagonal equations $\mathbf{Ax} = \mathbf{b}$ by Doolittle's decomposition method, where

$$\mathbf{A} = \begin{bmatrix} 6 & 2 & 0 & 0 & 0 \\ -1 & 7 & 2 & 0 & 0 \\ 0 & -2 & 8 & 2 & 0 \\ 0 & 0 & 3 & 7 & -2 \\ 0 & 0 & 0 & 3 & 5 \end{bmatrix} \quad \mathbf{b} = \begin{bmatrix} 2 \\ -3 \\ 4 \\ -3 \\ 1 \end{bmatrix}$$

5. Use Gauss elimination with scaled row pivoting to solve

$$\begin{bmatrix} 4 & -2 & 1 \\ -2 & 1 & -1 \\ -2 & 3 & 6 \end{bmatrix} \begin{bmatrix} x_1 \\ x_2 \\ x_3 \end{bmatrix} = \begin{bmatrix} 2 \\ -1 \\ 0 \end{bmatrix}$$

6. Solve $\mathbf{Ax} = \mathbf{b}$ by Gauss elimination with scaled row pivoting, where

$$\mathbf{A} = \begin{bmatrix} 2.34 & -4.10 & 1.78 \\ -1.98 & 3.47 & -2.22 \\ 2.36 & -15.17 & 6.81 \end{bmatrix} \quad \mathbf{b} = \begin{bmatrix} 0.02 \\ -0.73 \\ -6.63 \end{bmatrix}$$

Systems of Linear Algebraic Equations

7. Solve the equations

$$\begin{bmatrix} 2 & -1 & 0 & 0 \\ 0 & 0 & -1 & 1 \\ 0 & -1 & 2 & -1 \\ -1 & 2 & -1 & 0 \end{bmatrix} \begin{bmatrix} x_1 \\ x_2 \\ x_3 \\ x_4 \end{bmatrix} = \begin{bmatrix} 1 \\ 0 \\ 0 \\ 0 \end{bmatrix}$$

by Gauss elimination with scaled row pivoting.

8. ■ Solve the equations

$$\begin{bmatrix} 0 & 2 & 5 & -1 \\ 2 & 1 & 3 & 0 \\ -2 & -1 & 3 & 1 \\ 3 & 3 & -1 & 2 \end{bmatrix} \begin{bmatrix} x_1 \\ x_2 \\ x_3 \\ x_4 \end{bmatrix} = \begin{bmatrix} -3 \\ 3 \\ -2 \\ 5 \end{bmatrix}$$

9. ■ Solve the symmetric, tridiagonal equations

$$4x_1 - x_2 = 9$$

$$-x_{i-1} + 4x_i - x_{i+1} = 5, \quad i = 2, \ldots, n-1$$

$$-x_{n-1} + 4x_n = 5$$

with $n = 10$.

10. ■ Solve the equations $\mathbf{Ax} = \mathbf{b}$, where

$$\mathbf{A} = \begin{bmatrix} 1.3174 & 2.7250 & 2.7250 & 1.7181 \\ 0.4002 & 0.8278 & 1.2272 & 2.5322 \\ 0.8218 & 1.5608 & 0.3629 & 2.9210 \\ 1.9664 & 2.0011 & 0.6532 & 1.9945 \end{bmatrix} \quad \mathbf{b} = \begin{bmatrix} 8.4855 \\ 4.9874 \\ 5.6665 \\ 6.6152 \end{bmatrix}$$

11. ■ Solve the equations

$$\begin{bmatrix} 10 & -2 & -1 & 2 & 3 & 1 & -4 & 7 \\ 5 & 11 & 3 & 10 & -3 & 3 & 3 & -4 \\ 7 & 12 & 1 & 5 & 3 & -12 & 2 & 3 \\ 8 & 7 & -2 & 1 & 3 & 2 & 2 & 4 \\ 2 & -15 & -1 & 1 & 4 & -1 & 8 & 3 \\ 4 & 2 & 9 & 1 & 12 & -1 & 4 & 1 \\ -1 & 4 & -7 & -1 & 1 & 1 & -1 & -3 \\ -1 & 3 & 4 & 1 & 3 & -4 & 7 & 6 \end{bmatrix} \begin{bmatrix} x_1 \\ x_2 \\ x_3 \\ x_4 \\ x_5 \\ x_6 \\ x_7 \\ x_8 \end{bmatrix} = \begin{bmatrix} 0 \\ 12 \\ -5 \\ 3 \\ -25 \\ -26 \\ 9 \\ -7 \end{bmatrix}$$

12. ■ The system shown in Fig. (a) consists of *n* linear springs that support *n* masses. The spring stiffnesses are denoted by k_i, the weights of the masses are W_i and x_i are the displacements of the masses (measured from the positions where the springs are undeformed). The so-called *displacement formulation* is obtained by

2.5 Pivoting

writing the equilibrium equation of each mass and substituting $F_i = k_i(x_{i+1} - x_i)$ for the spring forces. The result is the symmetric, tridiagonal set of equations

$$(k_1 + k_2)x_1 - k_2 x_2 = W_1$$

$$-k_i x_{i-1} + (k_i + k_{i+1})x_i - k_{i+1} x_{i+1} = W_i, \quad i = 2, 3, \ldots, n-1$$

$$-k_n x_{n-1} + k_n x_n = W_n$$

Write a program that solves these equations for given values of n, **k** and **W**. Run the program with $n = 5$ and

$$k_1 = k_2 = k_3 = 10 \text{ N/mm} \qquad k_4 = k_5 = 5 \text{ N/mm}$$
$$W_1 = W_3 = W_5 = 100 \text{ N} \qquad W_2 = W_4 = 50 \text{ N}$$

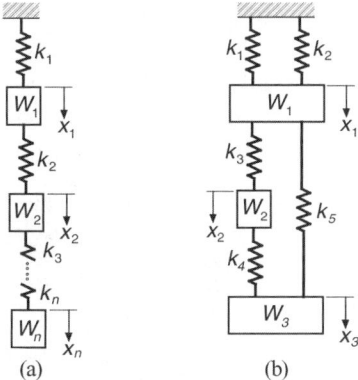

(a)　　(b)

13. ■ The displacement formulation for the mass–spring system shown in Fig. (b) results in the following equilibrium equations of the masses:

$$\begin{bmatrix} k_1 + k_2 + k_3 + k_5 & -k_3 & -k_5 \\ -k_3 & k_3 + k_4 & -k_4 \\ -k_5 & -k_4 & k_4 + k_5 \end{bmatrix} \begin{bmatrix} x_1 \\ x_2 \\ x_3 \end{bmatrix} = \begin{bmatrix} W_1 \\ W_2 \\ W_3 \end{bmatrix}$$

where k_i are the spring stiffnesses, W_i represent the weights of the masses, and x_i are the displacements of the masses from the undeformed configuration of the system. Write a program that solves these equations, given **k** and **W**. Use the program to find the displacements if

$$k_1 = k_3 = k_4 = k \qquad k_2 = k_5 = 2k$$
$$W_1 = W_3 = 2W \qquad W_2 = W$$

14. ■

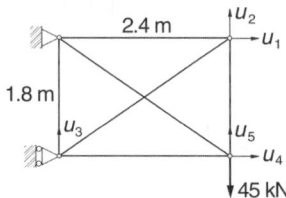

The displacement formulation for a plane truss is similar to that of a mass–spring system. The differences are: (1) the stiffnesses of the members are $k_i = (EA/L)_i$, where E is the modulus of elasticity, A represents the cross-sectional area and L is the length of the member; (2) there are two components of displacement at each joint. For the statically indeterminate truss shown the displacement formulation yields the symmetric equations $\mathbf{Ku} = \mathbf{p}$, where

$$\mathbf{K} = \begin{bmatrix} 27.58 & 7.004 & -7.004 & 0.0000 & 0.0000 \\ 7.004 & 29.57 & -5.253 & 0.0000 & -24.32 \\ -7.004 & -5.253 & 29.57 & 0.0000 & 0.0000 \\ 0.0000 & 0.0000 & 0.0000 & 27.58 & -7.004 \\ 0.0000 & -24.32 & 0.0000 & -7.004 & 29.57 \end{bmatrix} \text{MN/m}$$

$$\mathbf{p} = \begin{bmatrix} 0 & 0 & 0 & 0 & -45 \end{bmatrix}^T \text{kN}$$

Determine the displacements u_i of the joints.

15. ■

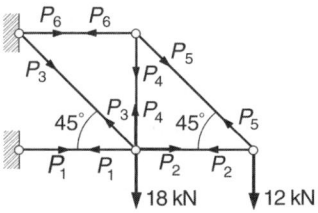

In the *force formulation* of a truss, the unknowns are the member forces P_i. For the statically determinate truss shown, the equilibrium equations of the joints are:

$$\begin{bmatrix} -1 & 1 & -1/\sqrt{2} & 0 & 0 & 0 \\ 0 & 0 & 1/\sqrt{2} & 1 & 0 & 0 \\ 0 & -1 & 0 & 0 & -1/\sqrt{2} & 0 \\ 0 & 0 & 0 & 0 & 1/\sqrt{2} & 0 \\ 0 & 0 & 0 & 0 & 1/\sqrt{2} & 1 \\ 0 & 0 & 0 & -1 & -1/\sqrt{2} & 0 \end{bmatrix} \begin{bmatrix} P_1 \\ P_2 \\ P_3 \\ P_4 \\ P_5 \\ P_6 \end{bmatrix} = \begin{bmatrix} 0 \\ 18 \\ 0 \\ 12 \\ 0 \\ 0 \end{bmatrix}$$

where the units of P_i are kN. (a) Solve the equations as they are with a computer program. (b) Rearrange the rows and columns so as to obtain a lower triangular coefficient matrix, and then solve the equations by back substitution using a calculator.

16. ■

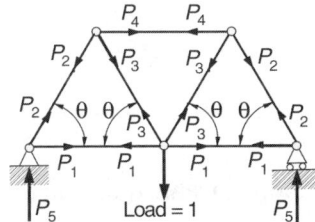

The force formulation of the symmetric truss shown results in the joint equilibrium equations

$$\begin{bmatrix} c & 1 & 0 & 0 & 0 \\ 0 & s & 0 & 0 & 1 \\ 0 & 0 & 2s & 0 & 0 \\ 0 & -c & c & 1 & 0 \\ 0 & s & s & 0 & 0 \end{bmatrix} \begin{bmatrix} P_1 \\ P_2 \\ P_3 \\ P_4 \\ P_5 \end{bmatrix} = \begin{bmatrix} 0 \\ 0 \\ 1 \\ 0 \\ 0 \end{bmatrix}$$

where $s = \sin\theta$, $c = \cos\theta$ and P_i are the unknown forces. Write a program that computes the forces, given the angle θ. Run the program with $\theta = 53°$.

17. ■

The electrical network shown can be viewed as consisting of three loops. Applying Kirchoff's law ($\sum$voltage drops = $\sum$voltage sources) to each loop yields the following equations for the loop currents i_1, i_2 and i_3:

$$5i_1 + 15(i_1 - i_3) = 220 \text{ V}$$

$$R(i_2 - i_3) + 5i_2 + 10i_2 = 0$$

$$20i_3 + R(i_3 - i_2) + 15(i_3 - i_1) = 0$$

Compute the three loop currents for $R = 5$, 10 and 20 Ω.

18.

Determine the loop currents i_1 to i_4 in the electrical network shown.

19. ■ Consider the n simultaneous equations $\mathbf{Ax} = \mathbf{b}$, where

$$A_{ij} = (i+j)^2 \qquad b_i = \sum_{j=0}^{n-1} A_{ij}, \quad i = 0, 1, \ldots, n-1, \quad j = 0, 1, \ldots, n-1$$

The solution is $\mathbf{x} = \begin{bmatrix} 1 & 1 & \cdots & 1 \end{bmatrix}^T$. Write a program that solves these equations for any given n (pivoting is recommended). Run the program with $n = 2, 3$ and 4, and comment on the results.

*2.6 Matrix Inversion

Computing the inverse of a matrix and solving simultaneous equations are related tasks. The most economical way to invert an $n \times n$ matrix $\mathbf{A}$ is to solve the equations

$$\mathbf{AX} = \mathbf{I} \tag{2.33}$$

where $\mathbf{I}$ is the $n \times n$ identity matrix. The solution $\mathbf{X}$, also of size $n \times n$, will be the inverse of $\mathbf{A}$. The proof is simple: after we premultiply both sides of Eq. (2.33) by $\mathbf{A}^{-1}$ we have $\mathbf{A}^{-1}\mathbf{AX} = \mathbf{A}^{-1}\mathbf{I}$, which reduces to $\mathbf{X} = \mathbf{A}^{-1}$.

Inversion of large matrices should be avoided whenever possible due its high cost. As seen from Eq. (2.33), inversion of $\mathbf{A}$ is equivalent to solving $\mathbf{Ax}_i = \mathbf{b}_i$ with $i = 1, 2, \ldots, n$, where $\mathbf{b}_i$ is the ith column of $\mathbf{I}$. If LU decomposition is employed in the solution, the solution phase (forward and back substitution) must be repeated n times, once for each $\mathbf{b}_i$. Since the cost of computation is proportional to n^3 for the decomposition phase and n^2 for each vector of the solution phase, the cost of inversion is considerably more expensive than the solution of $\mathbf{Ax} = \mathbf{b}$ (single constant vector $\mathbf{b}$).

Matrix inversion has another serious drawback—a banded matrix loses its structure during inversion. In other words, if $\mathbf{A}$ is banded or otherwise sparse, then $\mathbf{A}^{-1}$ is fully populated. However, the inverse of a triangular matrix remains triangular.

2.6 Matrix Inversion

EXAMPLE 2.13
Write a function that inverts a matrix using LU decomposition with pivoting. Test the function by inverting

$$A = \begin{bmatrix} 0.6 & -0.4 & 1.0 \\ -0.3 & 0.2 & 0.5 \\ 0.6 & -1.0 & 0.5 \end{bmatrix}$$

Solution The function matInv listed below uses the decomposition and solution procedures in the module LUpivot.

```
#!/usr/bin/python
## example2_13
from numarray import array,identity, matrixmultiply
from LUpivot import *

def matInv(a):
    n = len(a[0])
    aInv = identity(n)*1.0
    a,seq = LUdecomp(a)
    for i in range(n):
        aInv[:,i] = LUsolve(a,aInv[:,i],seq)
    return aInv

a = array([[ 0.6, -0.4,  1.0],\
           [-0.3,  0.2,  0.5],\
           [ 0.6, -1.0,  0.5]])
aOrig = a.copy()   # Save original [a]
aInv = matInv(a)   # Invert [a] (original [a] is destroyed)
print ''\naInv =\n'',aInv
print ''\nCheck: a*aInv =\n'', matrixmultiply(aOrig,aInv)
raw_input(''\nPress return to exit'')
```

The output is

```
aInv =
[[ 1.66666667 -2.22222222 -1.11111111]
 [ 1.25       -0.83333333 -1.66666667]
 [ 0.5         1.          0.        ]]

Check: a*aInv =
[[  1.00000000e+00  -4.44089210e-16  -1.11022302e-16]
 [  0.00000000e+00   1.00000000e+00   5.55111512e-17]
 [  0.00000000e+00  -3.33066907e-16   1.00000000e+00]]
```

EXAMPLE 2.14
Invert the matrix

$$A = \begin{bmatrix} 2 & -1 & 0 & 0 & 0 & 0 \\ -1 & 2 & -1 & 0 & 0 & 0 \\ 0 & -1 & 2 & -1 & 0 & 0 \\ 0 & 0 & -1 & 2 & -1 & 0 \\ 0 & 0 & 0 & -1 & 2 & -1 \\ 0 & 0 & 0 & 0 & -1 & 5 \end{bmatrix}$$

Solution Since the matrix is tridiagonal, we solve $AX = I$ using the functions in the module LUdecomp3 (LU decomposition of tridiagonal matrices).

```
#!/usr/bin/python
## example2_14
from numarray import array,ones,identity,Float64
from LUdecomp3 import *

n = 6
d = ones((n))*2.0
e = ones((n-1))*(-1.0)
c = e.copy()
d[n-1] = 5.0
aInv = identity(n)*1.0
c,d,e = LUdecomp3(c,d,e)
for i in range(n):
    aInv[:,i] = LUsolve3(c,d,e,aInv[:,i])
print ''\nThe inverse matrix is:\n'',aInv
raw_input(''\nPress return to exit'')
```

Running the program results in the following output:

```
The inverse matrix is:
[[ 0.84   0.68   0.52   0.36   0.2    0.04]
 [ 0.68   1.36   1.04   0.72   0.4    0.08]
 [ 0.52   1.04   1.56   1.08   0.6    0.12]
 [ 0.36   0.72   1.08   1.44   0.8    0.16]
 [ 0.2    0.4    0.6    0.8    1.     0.2 ]
 [ 0.04   0.08   0.12   0.16   0.2    0.24]]]
```

Note that A is tridiagonal, whereas A^{-1} is fully populated.

*2.7 Iterative Methods

Introduction

So far, we have discussed only direct methods of solution. The common characteristic of these methods is that they compute the solution with a finite number of operations. Moreover, if the computer were capable of infinite precision (no roundoff errors), the solution would be exact.

Iterative, or *indirect methods*, start with an initial guess of the solution **x** and then repeatedly improve the solution until the change in **x** becomes negligible. Since the required number of iterations can be large, the indirect methods are, in general, slower than their direct counterparts. However, iterative methods do have the following advantages that make them attractive for certain problems:

1. It is feasible to store only the nonzero elements of the coefficient matrix. This makes it possible to deal with very large matrices that are sparse, but not necessarily banded. In many problems, there is no need to store the coefficient matrix at all.
2. Iterative procedures are self-correcting, meaning that roundoff errors (or even arithmetic mistakes) in one iterative cycle are corrected in subsequent cycles.

A serious drawback of iterative methods is that they do not always converge to the solution. It can be shown that convergence is guaranteed only if the coefficient matrix is diagonally dominant. The initial guess for **x** plays no role in determining whether convergence takes place—if the procedure converges for one starting vector, it would do so for any starting vector. The initial guess affects only the number of iterations that are required for convergence.

Gauss–Seidel Method

The equations **Ax** = **b** are in scalar notation

$$\sum_{j=1}^{n} A_{ij} x_j = b_i, \quad i = 1, 2, \ldots, n$$

Extracting the term containing x_i from the summation sign yields

$$A_{ii} x_i + \sum_{\substack{j=1 \\ j \neq i}}^{n} A_{ij} x_j = b_i, \quad i = 1, 2, \ldots, n$$

Systems of Linear Algebraic Equations

Solving for x_i, we get

$$x_i = \frac{1}{A_{ii}} \left(b_i - \sum_{\substack{j=1 \\ j \neq i}}^{n} A_{ij} x_j \right), \quad i = 1, 2, \ldots, n$$

The last equation suggests the following iterative scheme

$$x_i \leftarrow \frac{1}{A_{ii}} \left(b_i - \sum_{\substack{j=1 \\ j \neq i}}^{n} A_{ij} x_j \right), \quad i = 1, 2, \ldots, n \quad (2.34)$$

We start by choosing the starting vector **x**. If a good guess for the solution is not available, **x** can be chosen randomly. Equation (2.34) is then used to recompute each element of **x**, always using the latest available values of x_j. This completes one iteration cycle. The procedure is repeated until the changes in **x** between successive iteration cycles become sufficiently small.

Convergence of the Gauss–Seidel method can be improved by a technique known as *relaxation*. The idea is to take the new value of x_i as a weighted average of its previous value and the value predicted by Eq. (2.34). The corresponding iterative formula is

$$x_i \leftarrow \frac{\omega}{A_{ii}} \left(b_i - \sum_{\substack{j=1 \\ j \neq i}}^{n} A_{ij} x_j \right) + (1 - \omega) x_i, \quad i = 1, 2, \ldots, n \quad (2.35)$$

where the weight ω is called the *relaxation factor*. It can be seen that if $\omega = 1$, no relaxation takes place, since Eqs. (2.34) and (2.35) produce the same result. If $\omega < 1$, Eq. (2.35) represents interpolation between the old x_i and the value given by Eq. (2.34). This is called *underrelaxation*. In cases where $\omega > 1$, we have extrapolation, or *overrelaxation*.

There is no practical method of determining the optimal value of ω beforehand; however, a good estimate can be computed during run time. Let $\Delta x^{(k)} = \left| \mathbf{x}^{(k-1)} - \mathbf{x}^{(k)} \right|$ be the magnitude of the change in **x** during the kth iteration (carried out without relaxation, i.e., with $\omega = 1$). If k is sufficiently large (say $k \geq 5$), it can be shown[4] that an approximation of the optimal value of ω is

$$\omega_{\text{opt}} \approx \frac{2}{1 + \sqrt{1 - \left(\Delta x^{(k+p)} / \Delta x^{(k)} \right)^{1/p}}} \quad (2.36)$$

where p is a positive integer.

---

[4] See, for example, Terrence J. Akai, *Applied Numerical Methods for Engineers*, John Wiley & Sons (1994), p. 100.

2.7 Iterative Methods

The essential elements of a Gauss–Seidel algorithm with relaxation are:

1. Carry out k iterations with $\omega = 1$ ($k = 10$ is reasonable). After the kth iteration record $\Delta x^{(k)}$.
2. Perform an additional p iterations and record $\Delta x^{(k+p)}$ for the last iteration.
3. Perform all subsequent iterations with $\omega = \omega_{opt}$, where ω_{opt} is computed from Eq. (2.36).

■ gaussSeidel

The function gaussSeidel is an implementation of the Gauss–Seidel method with relaxation. It automatically computes ω_{opt} from Eq. (2.36) using $k = 10$ and $p = 1$. The user must provide the function iterEqs that computes the improved x from the iterative formulas in Eq. (2.35)—see Example 2.17. The function returns the solution vector x, the number of iterations carried out and the value of ω_{opt} used.

```
## module gaussSeidel
''' x,numIter,omega = gaussSeidel(iterEqs,x,tol = 1.0e-9)
    Gauss-Seidel method for solving [A]{x} = {b}.
    The matrix [A] should be sparse. User must supply the
    function iterEqs(x,omega) that returns the improved {x},
    given the current {x} ('omega' is the relaxation factor).
'''
from numarray import dot
from math import sqrt

def gaussSeidel(iterEqs,x,tol = 1.0e-9):
    omega = 1.0
    k = 10
    p = 1
    for i in range(1,501):
        xOld = x.copy()
        x = iterEqs(x,omega)
        dx = sqrt(dot(x-xOld,x-xOld))
        if dx < tol: return x,i,omega
      # Compute of relaxation factor after k+p iterations
        if i == k: dx1 = dx
        if i == k + p:
            dx2 = dx
            omega = 2.0/(1.0 + sqrt(1.0 - (dx2/dx1)**(1.0/p)))
    print 'Gauss-Seidel failed to converge'
```

Conjugate Gradient Method

Consider the problem of finding the vector $\mathbf{x}$ that minimizes the scalar function

$$f(\mathbf{x}) = \frac{1}{2}\mathbf{x}^T\mathbf{A}\mathbf{x} - \mathbf{b}^T\mathbf{x} \tag{2.37}$$

where the matrix $\mathbf{A}$ is *symmetric* and *positive definite*. Because $f(\mathbf{x})$ is minimized when its gradient $\nabla f = \mathbf{A}\mathbf{x} - \mathbf{b}$ is zero, we see that minimization is equivalent to solving

$$\mathbf{A}\mathbf{x} = \mathbf{b} \tag{2.38}$$

Gradient methods accomplish the minimization by iteration, starting with an initial vector $\mathbf{x}_0$. Each iterative cycle k computes a refined solution

$$\mathbf{x}_{k+1} = \mathbf{x}_k + \alpha_k \mathbf{s}_k \tag{2.39}$$

The *step length* α_k is chosen so that $\mathbf{x}_{k+1}$ minimizes $f(\mathbf{x}_{k+1})$ in the *search direction* $\mathbf{s}_k$. That is, $\mathbf{x}_{k+1}$ must satisfy Eq. (2.38):

$$\mathbf{A}(\mathbf{x}_k + \alpha_k \mathbf{s}_k) = \mathbf{b} \tag{a}$$

Introducing the *residual*

$$\mathbf{r}_k = \mathbf{b} - \mathbf{A}\mathbf{x}_k \tag{2.40}$$

Eq. (a) becomes $\alpha \mathbf{A}\mathbf{s}_k = \mathbf{r}_k$. Premultiplying both sides by $\mathbf{s}_k^T$ and solving for α_k, we obtain

$$\alpha_k = \frac{\mathbf{s}_k^T \mathbf{r}_k}{\mathbf{s}_k^T \mathbf{A} \mathbf{s}_k} \tag{2.41}$$

We are still left with the problem of determining the search direction $\mathbf{s}_k$. Intuition tells us to choose $\mathbf{s}_k = -\nabla f = \mathbf{r}_k$, since this is the direction of the largest negative change in $f(\mathbf{x})$. The resulting procedure is known as the *method of steepest descent*. It is not a popular algorithm since its convergence can be slow. The more efficient conjugate gradient method uses the search direction

$$\mathbf{s}_{k+1} = \mathbf{r}_{k+1} + \beta_k \mathbf{s}_k \tag{2.42}$$

The constant β_k is chosen so that the two successive search directions are *conjugate* to each other, meaning

$$\mathbf{s}_{k+1}^T \mathbf{A} \mathbf{s}_k = 0 \tag{b}$$

2.7 Iterative Methods

The great attraction of conjugate gradients is that minimization in one conjugate direction does not undo previous minimizations (minimizations do not interfere with one another).

Substituting $\mathbf{s}_{k+1}$ from Eq. (2.42) into Eq. (b), we get

$$\left(\mathbf{r}_{k+1}^T + \beta_k \mathbf{s}_k^T\right) \mathbf{A} \mathbf{s}_k = 0$$

which yields

$$\beta_k = -\frac{\mathbf{r}_{k+1}^T \mathbf{A} \mathbf{s}_k}{\mathbf{s}_k^T \mathbf{A} \mathbf{s}_k} \tag{2.43}$$

Here is the outline of the conjugate gradient algorithm:

- Choose $\mathbf{x}_0$ (any vector will do, but one close to solution results in fewer iterations)
- $\mathbf{r}_0 \leftarrow \mathbf{b} - \mathbf{A}\mathbf{x}_0$
- $\mathbf{s}_0 \leftarrow \mathbf{r}_0$ (lacking a previous search direction, choose the direction of steepest descent)
- do with $k = 0, 1, 2, \ldots$

$$\alpha_k \leftarrow \frac{\mathbf{s}_k^T \mathbf{r}_k}{\mathbf{s}_k^T \mathbf{A} \mathbf{s}_k}$$

$$\mathbf{x}_{k+1} \leftarrow \mathbf{x}_k + \alpha_k \mathbf{s}_k$$

$$\mathbf{r}_{k+1} \leftarrow \mathbf{b} - \mathbf{A}\mathbf{x}_{k+1}$$

if $|\mathbf{r}_{k+1}| \leq \varepsilon$ exit loop (ε is the error tolerance)

$$\beta_k \leftarrow -\frac{\mathbf{r}_{k+1}^T \mathbf{A} \mathbf{s}_k}{\mathbf{s}_k^T \mathbf{A} \mathbf{s}_k}$$

$$\mathbf{s}_{k+1} \leftarrow \mathbf{r}_{k+1} + \beta_k \mathbf{s}_k$$

end do

It can be shown that the residual vectors $\mathbf{r}_1, \mathbf{r}_2, \mathbf{r}_3, \ldots$ produced by the algorithm are mutually orthogonal; that is, $\mathbf{r}_i \cdot \mathbf{r}_j = 0, i \neq j$. Now suppose that we have carried out enough iterations to have computed the whole set of n residual vectors. The residual resulting from the next iteration must be the null vector ($\mathbf{r}_{n+1} = \mathbf{0}$), indicating that the solution has been obtained. It thus appears that the conjugate gradient algorithm is not an iterative method at all, since it reaches the exact solution after n computational cycles. In practice, however, convergence is usually achieved in less than n iterations.

The conjugate gradient method is not competitive with direct methods in the solution of small sets of equations. Its strength lies in the handling of large, sparse systems (where most elements of $\mathbf{A}$ are zero). It is important to note that $\mathbf{A}$ enters the algorithm only through its multiplication by a vector; that is, in the form $\mathbf{A}\mathbf{v}$, where $\mathbf{v}$ is a vector (either $\mathbf{x}_{k+1}$ or $\mathbf{s}_k$). If $\mathbf{A}$ is sparse, it is possible to write an efficient subroutine for the multiplication and pass it, rather than $\mathbf{A}$ itself, to the conjugate gradient algorithm.

■ conjGrad

The function `conjGrad` shown below implements the conjugate gradient algorithm. The maximum allowable number of iterations is set to *n* (the number of unknowns). Note that `conjGrad` calls the function `Av` which returns the product **Av**. This function must be supplied by the user (see Example 2.18). We must also supply the starting vector x_0 and the constant (right-hand-side) vector **b**. The function returns the solution vector **x** and the number of iterations:

```
## module conjGrad
''' x, numIter = conjGrad(Av,x,b,tol=1.0e-9)
    Conjugate gradient method for solving [A]{x} = {b}.
    The matrix [A] should be sparse. User must supply
    the function Av(v) that returns the vector [A]{v}.
'''
from numarray import dot
from math import sqrt

def conjGrad(Av,x,b,tol=1.0e-9):
    n = len(b)
    r = b - Av(x)
    s = r.copy()
    for i in range(n):
        u = Av(s)
        alpha = dot(s,r)/dot(s,u)
        x = x + alpha*s
        r = b - Av(x)
        if(sqrt(dot(r,r))) < tol:
            break
        else:
            beta = -dot(r,u)/dot(s,u)
            s = r + beta*s
    return x,i
```

EXAMPLE 2.15

Solve the equations

$$\begin{bmatrix} 4 & -1 & 1 \\ -1 & 4 & -2 \\ 1 & -2 & 4 \end{bmatrix} \begin{bmatrix} x_1 \\ x_2 \\ x_3 \end{bmatrix} = \begin{bmatrix} 12 \\ -1 \\ 5 \end{bmatrix}$$

by the Gauss–Seidel method without relaxation.

2.7 Iterative Methods

Solution With the given data, the iteration formulas in Eq. (2.34) become

$$x_1 = \frac{1}{4}(12 + x_2 - x_3)$$

$$x_2 = \frac{1}{4}(-1 + x_1 + 2x_3)$$

$$x_3 = \frac{1}{4}(5 - x_1 + 2x_2)$$

Choosing the starting values $x_1 = x_2 = x_3 = 0$, we have for the first iteration

$$x_1 = \frac{1}{4}(12 + 0 - 0) = 3$$

$$x_2 = \frac{1}{4}[-1 + 3 + 2(0)] = 0.5$$

$$x_3 = \frac{1}{4}[5 - 3 + 2(0.5)] = 0.75$$

The second iteration yields

$$x_1 = \frac{1}{4}(12 + 0.5 - 0.75) = 2.9375$$

$$x_2 = \frac{1}{4}[-1 + 2.9375 + 2(0.75)] = 0.85938$$

$$x_3 = \frac{1}{4}[5 - 2.9375 + 2(0.85938)] = 0.94531$$

and the third iteration results in

$$x_1 = \frac{1}{4}(12 + 0.85938 - 0.94531) = 2.97852$$

$$x_2 = \frac{1}{4}[-1 + 2.97852 + 2(0.94531)] = 0.96729$$

$$x_3 = \frac{1}{4}[5 - 2.97852 + 2(0.96729)] = 0.98902$$

After five more iterations the results would agree with the exact solution $x_1 = 3$, $x_2 = x_3 = 1$ within five decimal places.

EXAMPLE 2.16
Solve the equations in Example 2.15 by the conjugate gradient method.

Solution The conjugate gradient method should converge after three iterations. Choosing again for the starting vector

$$\mathbf{x}_0 = \begin{bmatrix} 0 & 0 & 0 \end{bmatrix}^T$$

the computations outlined in the text proceed as follows:

First iteration

$$\mathbf{r}_0 = \mathbf{b} - \mathbf{A}\mathbf{x}_0 = \begin{bmatrix} 12 \\ -1 \\ 5 \end{bmatrix} - \begin{bmatrix} 4 & -1 & 1 \\ -1 & 4 & -2 \\ 1 & -2 & 4 \end{bmatrix} \begin{bmatrix} 0 \\ 0 \\ 0 \end{bmatrix} = \begin{bmatrix} 12 \\ -1 \\ 5 \end{bmatrix}$$

$$\mathbf{s}_0 = \mathbf{r}_0 = \begin{bmatrix} 12 \\ -1 \\ 5 \end{bmatrix}$$

$$\mathbf{A}\mathbf{s}_0 = \begin{bmatrix} 4 & -1 & 1 \\ -1 & 4 & -2 \\ 1 & -2 & 4 \end{bmatrix} \begin{bmatrix} 12 \\ -1 \\ 5 \end{bmatrix} = \begin{bmatrix} 54 \\ -26 \\ 34 \end{bmatrix}$$

$$\alpha_0 = \frac{\mathbf{s}_0^T \mathbf{r}_0}{\mathbf{s}_0^T \mathbf{A}\mathbf{s}_0} = \frac{12^2 + (-1)^2 + 5^2}{12(54) + (-1)(-26) + 5(34)} = 0.201\,42$$

$$\mathbf{x}_1 = \mathbf{x}_0 + \alpha_0 \mathbf{s}_0 = \begin{bmatrix} 0 \\ 0 \\ 0 \end{bmatrix} + 0.201\,42 \begin{bmatrix} 12 \\ -1 \\ 5 \end{bmatrix} = \begin{bmatrix} 2.417\,04 \\ -0.201\,42 \\ 1.007\,10 \end{bmatrix}$$

Second iteration

$$\mathbf{r}_1 = \mathbf{b} - \mathbf{A}\mathbf{x}_1 = \begin{bmatrix} 12 \\ -1 \\ 5 \end{bmatrix} - \begin{bmatrix} 4 & -1 & 1 \\ -1 & 4 & -2 \\ 1 & -2 & 4 \end{bmatrix} \begin{bmatrix} 2.417\,04 \\ -0.201\,42 \\ 1.007\,10 \end{bmatrix} = \begin{bmatrix} 1.123\,32 \\ 4.236\,92 \\ -1.848\,28 \end{bmatrix}$$

$$\beta_0 = -\frac{\mathbf{r}_1^T \mathbf{A}\mathbf{s}_0}{\mathbf{s}_0^T \mathbf{A}\mathbf{s}_0} = -\frac{1.123\,32(54) + 4.236\,92(-26) - 1.848\,28(34)}{12(54) + (-1)(-26) + 5(34)} = 0.133\,107$$

$$\mathbf{s}_1 = \mathbf{r}_1 + \beta_0 \mathbf{s}_0 = \begin{bmatrix} 1.123\,32 \\ 4.236\,92 \\ -1.848\,28 \end{bmatrix} + 0.133\,107 \begin{bmatrix} 12 \\ -1 \\ 5 \end{bmatrix} = \begin{bmatrix} 2.720\,76 \\ 4.103\,80 \\ -1.182\,68 \end{bmatrix}$$

$$\mathbf{A}\mathbf{s}_1 = \begin{bmatrix} 4 & -1 & 1 \\ -1 & 4 & -2 \\ 1 & -2 & 4 \end{bmatrix} \begin{bmatrix} 2.720\,76 \\ 4.103\,80 \\ -1.182\,68 \end{bmatrix} = \begin{bmatrix} 5.596\,56 \\ 16.059\,80 \\ -10.217\,60 \end{bmatrix}$$

2.7 Iterative Methods

$$\alpha_1 = \frac{\mathbf{s}_1^T \mathbf{r}_1}{\mathbf{s}_1^T \mathbf{A}\mathbf{s}_1}$$

$$= \frac{2.720\,76(1.123\,32) + 4.103\,80(4.236\,92) + (-1.182\,68)(-1.848\,28)}{2.720\,76(5.596\,56) + 4.103\,80(16.059\,80) + (-1.182\,68)(-10.217\,60)}$$

$$= 0.24276$$

$$\mathbf{x}_2 = \mathbf{x}_1 + \alpha_1 \mathbf{s}_1 = \begin{bmatrix} 2.417\,04 \\ -0.201\,42 \\ 1.007\,10 \end{bmatrix} + 0.24276 \begin{bmatrix} 2.720\,76 \\ 4.103\,80 \\ -1.182\,68 \end{bmatrix} = \begin{bmatrix} 3.077\,53 \\ 0.794\,82 \\ 0.719\,99 \end{bmatrix}$$

Third iteration

$$\mathbf{r}_2 = \mathbf{b} - \mathbf{A}\mathbf{x}_2 = \begin{bmatrix} 12 \\ -1 \\ 5 \end{bmatrix} - \begin{bmatrix} 4 & -1 & 1 \\ -1 & 4 & -2 \\ 1 & -2 & 4 \end{bmatrix} \begin{bmatrix} 3.077\,53 \\ 0.794\,82 \\ 0.719\,99 \end{bmatrix} = \begin{bmatrix} -0.235\,29 \\ 0.338\,23 \\ 0.632\,15 \end{bmatrix}$$

$$\beta_1 = -\frac{\mathbf{r}_2^T \mathbf{A}\mathbf{s}_1}{\mathbf{s}_1^T \mathbf{A}\mathbf{s}_1}$$

$$= -\frac{(-0.235\,29)(5.596\,56) + 0.338\,23(16.059\,80) + 0.632\,15(-10.217\,60)}{2.720\,76(5.596\,56) + 4.103\,80(16.059\,80) + (-1.182\,68)(-10.217\,60)}$$

$$= 0.0251\,452$$

$$\mathbf{s}_2 = \mathbf{r}_2 + \beta_1 \mathbf{s}_1 = \begin{bmatrix} -0.235\,29 \\ 0.338\,23 \\ 0.632\,15 \end{bmatrix} + 0.025\,1452 \begin{bmatrix} 2.720\,76 \\ 4.103\,80 \\ -1.182\,68 \end{bmatrix} = \begin{bmatrix} -0.166\,876 \\ 0.441\,421 \\ 0.602\,411 \end{bmatrix}$$

$$\mathbf{A}\mathbf{s}_2 = \begin{bmatrix} 4 & -1 & 1 \\ -1 & 4 & -2 \\ 1 & -2 & 4 \end{bmatrix} \begin{bmatrix} -0.166\,876 \\ 0.441\,421 \\ 0.602\,411 \end{bmatrix} = \begin{matrix} -0.506\,514 \\ 0.727\,738 \\ 1.359\,930 \end{matrix}$$

$$\alpha_2 = \frac{\mathbf{r}_2^T \mathbf{s}_2}{\mathbf{s}_2^T \mathbf{A}\mathbf{s}_2}$$

$$= \frac{(-0.235\,29)(-0.166\,876) + 0.338\,23(0.441\,421) + 0.632\,15(0.602\,411)}{(-0.166\,876)(-0.506\,514) + 0.441\,421(0.727\,738) + 0.602\,411(1.359\,930)}$$

$$= 0.464\,80$$

$$\mathbf{x}_3 = \mathbf{x}_2 + \alpha_2 \mathbf{s}_2 = \begin{bmatrix} 3.077\,53 \\ 0.794\,82 \\ 0.719\,99 \end{bmatrix} + 0.464\,80 \begin{bmatrix} -0.166\,876 \\ 0.441\,421 \\ 0.602\,411 \end{bmatrix} = \begin{bmatrix} 2.999\,97 \\ 0.999\,99 \\ 0.999\,99 \end{bmatrix}$$

The solution $\mathbf{x}_3$ is correct to almost five decimal places. The small discrepancy is caused by roundoff errors in the computations.

EXAMPLE 2.17
Write a computer program to solve the following n simultaneous equations by the Gauss–Seidel method with relaxation (the program should work with any value of n)[5].

$$\begin{bmatrix} 2 & -1 & 0 & 0 & \cdots & 0 & 0 & 0 & 1 \\ -1 & 2 & -1 & 0 & \cdots & 0 & 0 & 0 & 0 \\ 0 & -1 & 2 & -1 & \cdots & 0 & 0 & 0 & 0 \\ \vdots & \vdots & \vdots & \vdots & & \vdots & \vdots & \vdots & \vdots \\ 0 & 0 & 0 & 0 & \cdots & -1 & 2 & -1 & 0 \\ 0 & 0 & 0 & 0 & \cdots & 0 & -1 & 2 & -1 \\ 1 & 0 & 0 & 0 & \cdots & 0 & 0 & -1 & 2 \end{bmatrix} \begin{bmatrix} x_1 \\ x_2 \\ x_3 \\ \vdots \\ x_{n-2} \\ x_{n-1} \\ x_n \end{bmatrix} = \begin{bmatrix} 0 \\ 0 \\ 0 \\ \vdots \\ 0 \\ 0 \\ 1 \end{bmatrix}$$

Run the program with $n = 20$. The exact solution can be shown to be $x_i = -n/4 + i/2$, $i = 1, 2, \ldots, n$.

Solution In this case the iterative formulas in Eq. (2.35) are

$$x_1 = \omega(x_2 - x_n)/2 + (1 - \omega)x_1$$
$$x_i = \omega(x_{i-1} + x_{i+1})/2 + (1 - \omega)x_i, \quad i = 2, 3, \ldots, n-1 \quad \text{(a)}$$
$$x_n = \omega(1 - x_1 + x_{n-1})/2 + (1 - \omega)x_n$$

These formulas are evaluated in the function iterEqs.

```
#!/usr/bin/python
## example2_17
from numarray import zeros,Float64
from gaussSeidel import *

def iterEqs(x,omega):
    n = len(x)
    x[0] = omega*(x[1] - x[n-1])/2.0 + (1.0 - omega)*x[0]
    for i in range(1,n-1):
        x[i] = omega*(x[i-1] + x[i+1])/2.0 + (1.0 - omega)*x[i]
    x[n-1] = omega*(1.0 - x[0] + x[n-2])/2.0 \
           + (1.0 - omega)*x[n-1]
    return x

n = eval(raw_input(''Number of equations ==> ''))
x = zeros((n),type=Float64)
x,numIter,omega = gaussSeidel(iterEqs,x)
```

---

[5] Equations of this form are called *cyclic* tridiagonal. They occur in the finite difference formulation of second-order differential equations with periodic boundary conditions.

2.7 Iterative Methods

```
print ''\nNumber of iterations ='',numIter
print ''\nRelaxation factor ='',omega
print ''\nThe solution is:\n'',x
raw_input(''\nPress return to exit'')
```

The output from the program is:

```
Number of equations ==> 20

Number of iterations = 259

Relaxation factor = 1.70545231071

The solution is:
[-4.50000000e+00 -4.00000000e+00 -3.50000000e+00 -3.00000000e+00
 -2.50000000e+00 -2.00000000e+00 -1.50000000e+00 -9.99999997e-01
 -4.99999998e-01  2.14046747e-09  5.00000002e-01  1.00000000e+00
  1.50000000e+00  2.00000000e+00  2.50000000e+00  3.00000000e+00
  3.50000000e+00  4.00000000e+00  4.50000000e+00  5.00000000e+00]
```

The convergence is very slow, because the coefficient matrix lacks diagonal dominance—substituting the elements of **A** into Eq. (2.30) produces an equality rather than the desired inequality. If we were to change each diagonal term of the coefficient matrix from 2 to 4, **A** would be diagonally dominant and the solution would converge in about 20 iterations.

EXAMPLE 2.18
Solve Example 2.17 with the conjugate gradient method, also using $n = 20$.

Solution The program shown below utilizes the function `conjGrad`. The solution vector **x** is initialized to zero in the program, which also sets up the constant vector **b**. The function `Av(v)` returns the product **Av**, where **A** is the coefficient matrix and **v** is a vector. For the given **A**, the components of the vector **Av** are

$$(\mathbf{Av})_1 = 2v_1 - v_2 + v_n$$

$$(\mathbf{Av})_i = -v_{i-1} + 2v_i - v_{i+1}, \quad i = 2, 3, \ldots, n-1$$

$$(\mathbf{Av})_n = -v_{n-1} + 2v_n + v_1$$

which are evaluated by the function `Av(v)`.

```
#!/usr/bin/python
## example2_18
```

```
from numarray import zeros,Float64,sqrt
from conjGrad import *

def Ax(v):
    n = len(v)
    Ax = zeros((n),type=Float64)
    Ax[0] = 2.0*v[0] - v[1] + v[n-1]
    Ax[1:n-1] = -v[0:n-2] + 2.0*v[1:n-1] - v[2:n]
    Ax[n-1] = -v[n-2] + 2.0*v[n-1] + v[0]
    return Ax

n = eval(raw_input(''Number of equations ==> ''))
b = zeros((n),type=Float64)
b[n-1] = 1.0
x = zeros((n),type=Float64)
x,numIter = conjGrad(Ax,x,b)
print ''\nThe solution is:\n'',x
print ''\nNumber of iterations ='',numIter
raw_input(''\nPress return to exit'')
```

Running the program results in

```
Number of equations ==> 20

The solution is:
[-4.5 -4.  -3.5 -3.  -2.5 -2.  -1.5 -1.  -0.5  0.   0.5  1.   1.5
  2.   2.5  3.   3.5  4.   4.5  5. ]

Number of iterations = 9
```

Note that convergence was reached in only 9 iterations, whereas 259 iterations were required in the Gauss–Seidel method.

PROBLEM SET 2.3

1. Let

$$\mathbf{A} = \begin{bmatrix} 3 & -1 & 2 \\ 0 & 1 & 3 \\ -2 & 2 & -4 \end{bmatrix} \quad \mathbf{B} = \begin{bmatrix} 0 & 1 & 3 \\ 3 & -1 & 2 \\ -2 & 2 & -4 \end{bmatrix}$$

2.7 Iterative Methods

(note that **B** is obtained by interchanging the first two rows of **A**). Knowing that

$$\mathbf{A}^{-1} = \begin{bmatrix} 0.5 & 0 & 0.25 \\ 0.3 & 0.4 & 0.45 \\ -0.1 & 0.2 & -0.15 \end{bmatrix}$$

determine $\mathbf{B}^{-1}$.

2. Invert the triangular matrices

$$\mathbf{A} = \begin{bmatrix} 2 & 4 & 3 \\ 0 & 6 & 5 \\ 0 & 0 & 2 \end{bmatrix} \qquad \mathbf{B} = \begin{bmatrix} 2 & 0 & 0 \\ 3 & 4 & 0 \\ 4 & 5 & 6 \end{bmatrix}$$

3. Invert the triangular matrix

$$\mathbf{A} = \begin{bmatrix} 1 & 1/2 & 1/4 & 1/8 \\ 0 & 1 & 1/3 & 1/9 \\ 0 & 0 & 1 & 1/4 \\ 0 & 0 & 0 & 1 \end{bmatrix}$$

4. Invert the following matrices:

(a) $\mathbf{A} = \begin{bmatrix} 1 & 2 & 4 \\ 1 & 3 & 9 \\ 1 & 4 & 16 \end{bmatrix}$ (b) $\mathbf{B} = \begin{bmatrix} 4 & -1 & 0 \\ -1 & 4 & -1 \\ 0 & -1 & 4 \end{bmatrix}$

5. Invert the matrix

$$\mathbf{A} = \begin{bmatrix} 4 & -2 & 1 \\ -2 & 1 & -1 \\ 1 & -2 & 4 \end{bmatrix}$$

6. ■ Invert the following matrices with any method:

$$\mathbf{A} = \begin{bmatrix} 5 & -3 & -1 & 0 \\ -2 & 1 & 1 & 1 \\ 3 & -5 & 1 & 2 \\ 0 & 8 & -4 & -3 \end{bmatrix} \qquad \mathbf{B} = \begin{bmatrix} 4 & -1 & 0 & 0 \\ -1 & 4 & -1 & 0 \\ 0 & -1 & 4 & -1 \\ 0 & 0 & -1 & 4 \end{bmatrix}$$

7. ■ Invert the matrix with any method:

$$\mathbf{A} = \begin{bmatrix} 1 & 3 & -9 & 6 & 4 \\ 2 & -1 & 6 & 7 & 1 \\ 3 & 2 & -3 & 15 & 5 \\ 8 & -1 & 1 & 4 & 2 \\ 11 & 1 & -2 & 18 & 7 \end{bmatrix}$$

and comment on the reliability of the result.

8. ■ The joint displacements **u** of the plane truss in Prob. 14, Problem Set 2.2 are related to the applied joint forces **p** by

$$\mathbf{Ku} = \mathbf{p} \tag{a}$$

where

$$\mathbf{K} = \begin{bmatrix} 27.580 & 7.004 & -7.004 & 0.000 & 0.000 \\ 7.004 & 29.570 & -5.253 & 0.000 & -24.320 \\ -7.004 & -5.253 & 29.570 & 0.000 & 0.000 \\ 0.000 & 0.000 & 0.000 & 27.580 & -7.004 \\ 0.000 & -24.320 & 0.000 & -7.004 & 29.570 \end{bmatrix} \text{MN/m}$$

is called the *stiffness matrix* of the truss. If Eq. (a) is inverted by multiplying each side by $\mathbf{K}^{-1}$, we obtain $\mathbf{u} = \mathbf{K}^{-1}\mathbf{p}$, where $\mathbf{K}^{-1}$ is known as the *flexibility matrix*. The physical meaning of the elements of the flexibility matrix is: K_{ij}^{-1} = displacements u_i ($i = 1, 2, \ldots, 5$) produced by the unit load $p_j = 1$. Compute (a) the flexibility matrix of the truss; (b) the displacements of the joints due to the load $p_5 = -45$ kN (the load shown in Prob. 14, Problem Set 2.2).

9. ■ Invert the matrices

$$\mathbf{A} = \begin{bmatrix} 3 & -7 & 45 & 21 \\ 12 & 11 & 10 & 17 \\ 6 & 25 & -80 & -24 \\ 17 & 55 & -9 & 7 \end{bmatrix} \quad \mathbf{B} = \begin{bmatrix} 1 & 1 & 1 & 1 \\ 1 & 2 & 2 & 2 \\ 2 & 3 & 4 & 4 \\ 4 & 5 & 6 & 7 \end{bmatrix}$$

10. ■ Write a program for inverting an $n \times n$ lower triangular matrix. The inversion procedure should contain only forward substitution. Test the program by inverting the matrix

$$\mathbf{A} = \begin{bmatrix} 36 & 0 & 0 & 0 \\ 18 & 36 & 0 & 0 \\ 9 & 12 & 36 & 0 \\ 5 & 4 & 9 & 36 \end{bmatrix}$$

11. Use the Gauss–Seidel method to solve

$$\begin{bmatrix} -2 & 5 & 9 \\ 7 & 1 & 1 \\ -3 & 7 & -1 \end{bmatrix} \begin{bmatrix} x_1 \\ x_2 \\ x_3 \end{bmatrix} = \begin{bmatrix} 1 \\ 6 \\ -26 \end{bmatrix}$$

2.7 Iterative Methods

12. Solve the following equations with the Gauss–Seidel method:

$$\begin{bmatrix} 12 & -2 & 3 & 1 \\ -2 & 15 & 6 & -3 \\ 1 & 6 & 20 & -4 \\ 0 & -3 & 2 & 9 \end{bmatrix} \begin{bmatrix} x_1 \\ x_2 \\ x_3 \\ x_4 \end{bmatrix} = \begin{bmatrix} 0 \\ 0 \\ 20 \\ 0 \end{bmatrix}$$

13. Use the Gauss–Seidel method with relaxation to solve $\mathbf{Ax} = \mathbf{b}$, where

$$\mathbf{A} = \begin{bmatrix} 4 & -1 & 0 & 0 \\ -1 & 4 & -1 & 0 \\ 0 & -1 & 4 & -1 \\ 0 & 0 & -1 & 3 \end{bmatrix} \qquad \mathbf{b} = \begin{bmatrix} 15 \\ 10 \\ 10 \\ 10 \end{bmatrix}$$

Take $x_i = b_i/A_{ii}$ as the starting vector and use $\omega = 1.1$ for the relaxation factor.

14. Solve the equations

$$\begin{bmatrix} 2 & -1 & 0 \\ -1 & 2 & -1 \\ 0 & -1 & 1 \end{bmatrix} \begin{bmatrix} x_1 \\ x_2 \\ x_3 \end{bmatrix} = \begin{bmatrix} 1 \\ 1 \\ 1 \end{bmatrix}$$

by the conjugate gradient method. Start with $\mathbf{x} = \mathbf{0}$.

15. Use the conjugate gradient method to solve

$$\begin{bmatrix} 3 & 0 & -1 \\ 0 & 4 & -2 \\ -1 & -2 & 5 \end{bmatrix} \begin{bmatrix} x_1 \\ x_2 \\ x_3 \end{bmatrix} = \begin{bmatrix} 4 \\ 10 \\ -10 \end{bmatrix}$$

starting with $\mathbf{x} = \mathbf{0}$.

16. ■ Solve the simultaneous equations $\mathbf{Ax} = \mathbf{b}$ and $\mathbf{Bx} = \mathbf{b}$ by the Gauss–Seidel method with relaxation, where

$$\mathbf{b} = \begin{bmatrix} 10 & -8 & 10 & 10 & -8 & 10 \end{bmatrix}^T$$

$$\mathbf{A} = \begin{bmatrix} 3 & -2 & 1 & 0 & 0 & 0 \\ -2 & 4 & -2 & 1 & 0 & 0 \\ 1 & -2 & 4 & -2 & 1 & 0 \\ 0 & 1 & -2 & 4 & -2 & 1 \\ 0 & 0 & 1 & -2 & 4 & -2 \\ 0 & 0 & 0 & 1 & -2 & 3 \end{bmatrix}$$

$$\mathbf{B} = \begin{bmatrix} 3 & -2 & 1 & 0 & 0 & 1 \\ -2 & 4 & -2 & 1 & 0 & 0 \\ 1 & -2 & 4 & -2 & 1 & 0 \\ 0 & 1 & -2 & 4 & -2 & 1 \\ 0 & 0 & 1 & -2 & 4 & -2 \\ 1 & 0 & 0 & 1 & -2 & 3 \end{bmatrix}$$

Note that **A** is not diagonally dominant, but that does not necessarily preclude convergence.

17. ■ Modify the program in Example 2.17 (Gauss–Seidel method) so that it will solve the following equations:

$$\begin{bmatrix} 4 & -1 & 0 & 0 & \cdots & 0 & 0 & 0 & 1 \\ -1 & 4 & -1 & 0 & \cdots & 0 & 0 & 0 & 0 \\ 0 & -1 & 4 & -1 & \cdots & 0 & 0 & 0 & 0 \\ \vdots & \vdots & \vdots & \vdots & \cdots & \vdots & \vdots & \vdots & \vdots \\ 0 & 0 & 0 & 0 & \cdots & -1 & 4 & -1 & 0 \\ 0 & 0 & 0 & 0 & \cdots & 0 & -1 & 4 & -1 \\ 1 & 0 & 0 & 0 & \cdots & 0 & 0 & -1 & 4 \end{bmatrix} \begin{bmatrix} x_1 \\ x_2 \\ x_3 \\ \vdots \\ x_{n-2} \\ x_{n-1} \\ x_n \end{bmatrix} = \begin{bmatrix} 0 \\ 0 \\ 0 \\ \vdots \\ 0 \\ 0 \\ 100 \end{bmatrix}$$

Run the program with $n = 20$ and compare the number of iterations with Example 2.17.

18. ■ Modify the program in Example 2.18 to solve the equations in Prob. 17 by the conjugate gradient method. Run the program with $n = 20$.

19. ■

The edges of the square plate are kept at the temperatures shown. Assuming steady-state heat conduction, the differential equation governing the temperature T in the interior is

$$\frac{\partial^2 T}{\partial x^2} + \frac{\partial^2 T}{\partial y^2} = 0$$

If this equation is approximated by finite differences using the mesh shown, we obtain the following algebraic equations for temperatures at the mesh points:

$$\begin{bmatrix} -4 & 1 & 0 & 1 & 0 & 0 & 0 & 0 & 0 \\ 1 & -4 & 1 & 0 & 1 & 0 & 0 & 0 & 0 \\ 0 & 1 & -4 & 0 & 0 & 1 & 0 & 0 & 0 \\ 1 & 0 & 0 & -4 & 1 & 0 & 1 & 0 & 0 \\ 0 & 1 & 0 & 1 & -4 & 1 & 0 & 1 & 0 \\ 0 & 0 & 1 & 0 & 1 & -4 & 0 & 0 & 1 \\ 0 & 0 & 0 & 1 & 0 & 0 & -4 & 1 & 0 \\ 0 & 0 & 0 & 0 & 1 & 0 & 1 & -4 & 1 \\ 0 & 0 & 0 & 0 & 0 & 1 & 0 & 1 & -4 \end{bmatrix} \begin{bmatrix} T_1 \\ T_2 \\ T_3 \\ T_4 \\ T_5 \\ T_6 \\ T_7 \\ T_8 \\ T_9 \end{bmatrix} = - \begin{bmatrix} 0 \\ 0 \\ 100 \\ 0 \\ 0 \\ 100 \\ 200 \\ 200 \\ 300 \end{bmatrix}$$

Solve these equations with the conjugate gradient method.

*2.8 Other Methods

A matrix can be decomposed in numerous ways, some of which are generally useful, whereas others find use in special applications. The most important of the latter are the QR factorization and the singular value decomposition.

The *QR decomposition* of a matrix **A** is

$$\mathbf{A} = \mathbf{QR}$$

where **Q** is an orthogonal matrix (recall that the matrix **Q** is orthogonal if $\mathbf{Q}^{-1} = \mathbf{Q}^T$) and **R** is an upper triangular matrix. Unlike LU factorization, QR decomposition does not require pivoting to sustain stability, but it does involve about twice as many operations. Due to its relative inefficiency, the QR factorization is not used as a general-purpose tool, but finds its niche in applications that put a premium on stabiliy (e.g., solution of eigenvalue problems).

The *singular value decomposition* is useful in dealing with singular or ill-conditioned matrices. Here the factorization is

$$\mathbf{A} = \mathbf{U\Lambda V}^T$$

where **U** and **V** are orthogonal matrices and

$$\mathbf{\Lambda} = \begin{bmatrix} \lambda_1 & 0 & 0 & \cdots \\ 0 & \lambda_2 & 0 & \cdots \\ 0 & 0 & \lambda_3 & \cdots \\ \vdots & \vdots & \vdots & \ddots \end{bmatrix}$$

is a diagonal matrix. The elements λ_i of $\mathbf{\Lambda}$ can be shown to be positive or zero. If $\mathbf{A}$ is symmetric and positive definite, then the λ's are the eigenvalues of $\mathbf{A}$. A nice characteristic of the singular value decomposition is that it works even if $\mathbf{A}$ is singular or ill-conditioned. The conditioning of $\mathbf{A}$ can be diagnosed from magnitudes of the λ's: the matrix is singular if one or more of the λ's are zero, and it is ill-conditioned if the condition number

$$\text{cond}(\mathbf{A}) = \lambda_{\max}/\lambda_{\min}$$

is very large.

3 Interpolation and Curve Fitting

Given the $n+1$ data points (x_i, y_i), $i = 0, 1, \ldots, n$, estimate $y(x)$.

3.1 Introduction

Discrete data sets, or tables of the form

x_0	x_1	x_2	$\cdots$	x_n
y_0	y_1	y_2	$\cdots$	y_n

are commonly involved in technical calculations. The source of the data may be experimental observations or numerical computations. There is a distinction between interpolation and curve fitting. In interpolation we construct a curve *through* the data points. In doing so, we make the implicit assumption that the data points are accurate and distinct. Curve fitting is applied to data that contain scatter (noise), usually due to measurement errors. Here we want to find a smooth curve that *approximates* the data in some sense. Thus the curve does not necessarily hit the data points. The difference between interpolation and curve fitting is illustrated in Fig. 3.1.

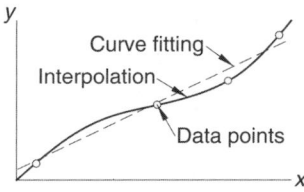

Figure 3.1. Interpolation and curve fitting of data.

3.2 Polynomial Interpolation

Lagrange's Method

The simplest form of an interpolant is a polynomial. It is always possible to construct a *unique* polynomial of degree n that passes through $n+1$ distinct data points. One means of obtaining this polynomial is the formula of Lagrange

$$P_n(x) = \sum_{i=0}^{n} y_i \ell_i(x) \tag{3.1a}$$

where the subscript n denotes the degree of the polynomial and

$$\ell_i(x) = \frac{x - x_0}{x_i - x_0} \cdot \frac{x - x_1}{x_i - x_1} \cdots \frac{x - x_{i-1}}{x_i - x_{i-1}} \cdot \frac{x - x_{i+1}}{x_i - x_{i+1}} \cdots \frac{x - x_n}{x_i - x_n}$$

$$= \prod_{\substack{j=0 \\ j \neq i}}^{n} \frac{x - x_j}{x_i - x_j}, \quad i = 0, 1, \ldots, n \tag{3.1b}$$

are called the *cardinal functions*.

For example, if $n = 1$, the interpolant is the straight line $P_1(x) = y_0 \ell_0(x) + y_1 \ell_1(x)$, where

$$\ell_0(x) = \frac{x - x_1}{x_0 - x_1} \qquad \ell_1(x) = \frac{x - x_0}{x_1 - x_0}$$

With $n = 2$, interpolation is parabolic: $P_2(x) = y_0 \ell_0(x) + y_1 \ell_1(x) + y_2 \ell_2(x)$, where now

$$\ell_0(x) = \frac{(x - x_1)(x - x_2)}{(x_0 - x_1)(x_0 - x_2)}$$

$$\ell_1(x) = \frac{(x - x_0)(x - x_2)}{(x_1 - x_0)(x_1 - x_2)}$$

$$\ell_2(x) = \frac{(x - x_0)(x - x_1)}{(x_2 - x_0)(x_2 - x_1)}$$

The cardinal functions are polynomials of degree n and have the property

$$\ell_i(x_j) = \begin{cases} 0 \text{ if } i \neq j \\ 1 \text{ if } i = j \end{cases} = \delta_{ij} \tag{3.2}$$

where δ_{ij} is the Kronecker delta. This property is illustrated in Fig. 3.2 for three-point interpolation ($n = 2$) with $x_0 = 0$, $x_1 = 2$ and $x_2 = 3$.

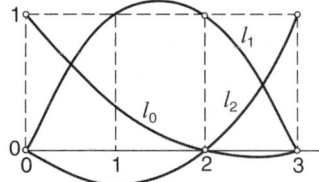

Figure 3.2. Example of quadratic cardinal functions.

3.2 Polynomial Interpolation

To prove that the interpolating polynomial passes through the data points, we substitute $x = x_j$ into Eq. (3.1a) and then utilize Eq. (3.2). The result is

$$P_n(x_j) = \sum_{i=0}^{n} y_i \ell_i(x_j) = \sum_{i=0}^{n} y_i \delta_{ij} = y_j$$

It can be shown that the error in polynomial interpolation is

$$f(x) - P_n(x) = \frac{(x - x_0)(x - x_1) \cdots (x - x_n)}{(n+1)!} f^{(n+1)}(\xi) \qquad (3.3)$$

where ξ lies somewhere in the interval (x_0, x_n); its value is otherwise unknown. It is instructive to note that the farther a data point is from x, the more it contributes to the error at x.

Newton's Method

Although Lagrange's method is conceptually simple, it does not lend itself to an efficient algorithm. A better computational procedure is obtained with Newton's method, where the interpolating polynomial is written in the form

$$P_n(x) = a_0 + (x - x_0)a_1 + (x - x_0)(x - x_1)a_2 + \cdots + (x - x_0)(x - x_1) \cdots (x - x_{n-1})a_n$$

This polynomial lends itself to an efficient evaluation procedure. Consider, for example, four data points ($n = 3$). Here the interpolating polynomial is

$$P_3(x) = a_0 + (x - x_0)a_1 + (x - x_0)(x - x_1)a_2 + (x - x_0)(x - x_1)(x - x_2)a_3$$
$$= a_0 + (x - x_0)\{a_1 + (x - x_1)[a_2 + (x - x_2)a_3]\}$$

which can be evaluated backwards with the following recurrence relations:

$$P_0(x) = a_3$$
$$P_1(x) = a_2 + (x - x_2)P_0(x)$$
$$P_2(x) = a_1 + (x - x_1)P_1(x)$$
$$P_3(x) = a_0 + (x - x_0)P_2(x)$$

For arbitrary n we have

$$P_0(x) = a_n \qquad P_k(x) = a_{n-k} + (x - x_{n-k})P_{k-1}(x), \quad k = 1, 2, \ldots, n \qquad (3.4)$$

Denoting the x-coordinate array of the data points by `xData` and the degree of the polynomial by n, we have the following algorithm for computing $P_n(x)$:

```
p = a[n]
for k in range(1,n+1):
    p = a[n-k] + (x - xData[n-k])*p
```

The coefficients of P_n are determined by forcing the polynomial to pass through each data point: $y_i = P_n(x_i), \quad i = 0, 1, \ldots, n$. This yields the simultaneous equations

$$y_0 = a_0$$
$$y_1 = a_0 + (x_1 - x_0)a_1$$
$$y_2 = a_0 + (x_2 - x_0)a_1 + (x_2 - x_0)(x_2 - x_1)a_2 \quad \text{(a)}$$
$$\vdots$$
$$y_n = a_0 + (x_n - x_0)a_1 + \cdots + (x_n - x_0)(x_n - x_1)\cdots(x_n - x_{n-1})a_n$$

Introducing the *divided differences*

$$\nabla y_i = \frac{y_i - y_0}{x_i - x_0}, \quad i = 1, 2, \ldots, n$$
$$\nabla^2 y_i = \frac{\nabla y_i - \nabla y_1}{x_i - x_1}, \quad i = 2, 3, \ldots, n$$
$$\nabla^3 y_i = \frac{\nabla^2 y_i - \nabla^2 y_2}{x_i - x_2}, \quad i = 3, 4, \ldots n \quad (3.5)$$
$$\vdots$$
$$\nabla^n y_n = \frac{\nabla^{n-1} y_n - \nabla^{n-1} y_{n-1}}{x_n - x_{n-1}}$$

the solution of Eqs. (a) is

$$a_0 = y_0 \quad a_1 = \nabla y_1 \quad a_2 = \nabla^2 y_2 \quad \cdots \quad a_n = \nabla^n y_n \quad (3.6)$$

If the coefficients are computed by hand, it is convenient to work with the format in Table 3.1 (shown for $n = 4$).

x_0	y_0				
x_1	y_1	∇y_1			
x_2	y_2	∇y_2	$\nabla^2 y_2$		
x_3	y_3	∇y_3	$\nabla^2 y_3$	$\nabla^3 y_3$	
x_4	y_4	∇y_4	$\nabla^2 y_4$	$\nabla^3 y_4$	$\nabla^4 y_4$

Table 3.1

3.2 Polynomial Interpolation

The diagonal terms (y_0, ∇y_1, $\nabla^2 y_2$, $\nabla^3 y_3$ and $\nabla^4 y_4$) in the table are the coefficients of the polynomial. If the data points are listed in a different order, the entries in the table will change, but the resultant polynomial will be the same—recall that a polynomial of degree n interpolating $n+1$ distinct data points is unique.

Machine computations can be carried out within a one-dimensional array **a** employing the following algorithm (we use the notation $m = n+1 =$ number of data points):

```
a = yData.copy()
for k in range(1,m):
    for i in range(k,m):
        a[i] = (a[i] - a[k-1])/(xData[i] - xData[k-1])
```

Initially, **a** contains the y-coordinates of the data, so that it is identical to the second column in Table 3.1. Each pass through the outer loop generates the entries in the next column, which overwrite the corresponding elements of **a**. Therefore, **a** ends up containing the diagonal terms of Table 3.1, i.e., the coefficients of the polynomial.

■ newtonPoly

This module contains the two functions required for interpolation by Newton's method. Given the data point arrays xData and yData, the function coeffts returns the coefficient array **a**. After the coefficients are found, the interpolant $P_n(x)$ can be evaluated at any value of x with the function evalPoly.

```
## module newtonPoly
''' p = evalPoly(a,xData,x).
    Evaluates Newton's polynomial p at x. The coefficient
    vector {a} can be computed by the function 'coeffts'.

    a = coeffts(xData,yData).
    Computes the coefficients of Newton's polynomial.
'''
def evalPoly(a,xData,x):
    n = len(xData) - 1  # Degree of polynomial
    p = a[n]
    for k in range(1,n+1):
        p = a[n-k] + (x -xData[n-k])*p
    return p

def coeffts(xData,yData):
    m = len(xData)   # Number of data points
```

```
    a = yData.copy()
    for k in range(1,m):
        a[k:m] = (a[k:m] - a[k-1])/(xData[k:m] - xData[k-1])
    return a
```

Neville's Method

Newton's method of interpolation involves two steps: computation of the coefficients, followed by evaluation of the polynomial. This works well if the interpolation is carried out repeatedly at different values of x using the same polynomial. If only one point is to interpolated, a method that computes the interpolant in a single step, such as Neville's algorithm, is a better choice.

Let $P_k[x_i, x_{i+1}, \ldots, x_{i+k}]$ denote the polynomial of degree k that passes through the $k+1$ data points $(x_i, y_i), (x_{i+1}, y_{i+1}), \ldots, (x_{i+k}, y_{i+k})$. For a single data point, we have

$$P_0[x_i] = y_i \tag{3.7}$$

The interpolant based on two data points is

$$P_1[x_i, x_{i+1}] = \frac{(x - x_{i+1})P_0[x_i] + (x_i - x)P_0[x_{i+1}]}{x_i - x_{i+1}}$$

It is easily verified that $P_1[x_i, x_{i+1}]$ passes through the two data points; that is, $P_1[x_i, x_{i+1}] = y_i$ when $x = x_i$, and $P_1[x_i, x_{i+1}] = y_{i+1}$ when $x = x_{i+1}$.

The three-point interpolant is

$$P_2[x_i, x_{i+1}, x_{i+2}] = \frac{(x - x_{i+2})P_1[x_i, x_{i+1}] + (x_i - x)P_1[x_{i+1}, x_{i+2}]}{x_i - x_{i+2}}$$

To show that this interpolant does intersect the data points, we first substitute $x = x_i$, obtaining

$$P_2[x_i, x_{i+1}, x_{i+2}] = P_1[x_i, x_{i+1}] = y_i$$

Similarly, $x = x_{i+2}$ yields

$$P_2[x_i, x_{i+1}, x_{i+2}] = P_1[x_{i+1}, x_{i+2}] = y_{i+2}$$

Finally, when $x = x_{i+1}$ we have

$$P_1[x_i, x_{i+1}] = P_1[x_{i+1}, x_{i+2}] = y_{i+1}$$

so that

$$P_2[x_i, x_{i+1}, x_{i+2}] = \frac{(x_{i+1} - x_{i+2})y_{i+1} + (x_i - x_{i+1})y_{i+1}}{x_i - x_{i+2}} = y_{i+1}$$

3.2 Polynomial Interpolation

Having established the pattern, we can now deduce the general recursive formula:

$$P_k[x_i, x_{i+1}, \ldots, x_{i+k}]$$
$$= \frac{(x - x_{i+k})P_{k-1}[x_i, x_{i+1}, \ldots, x_{i+k-1}] + (x_i - x)P_{k-1}[x_{i+1}, x_{i+2}, \ldots, x_{i+k}]}{x_i - x_{i+k}} \quad (3.8)$$

Given the value of x, the computations can be carried out in the following tabular format (shown for four data points):

	$k = 0$	$k = 1$	$k = 2$	$k = 3$
x_0	$P_0[x_0] = y_0$	$P_1[x_0, x_1]$	$P_2[x_0, x_1, x_2]$	$P_3[x_0, x_1, x_2, x_3]$
x_1	$P_0[x_1] = y_1$	$P_1[x_1, x_2]$	$P_2[x_1, x_2, x_3]$	
x_2	$P_0[x_2] = y_2$	$P_1[x_2, x_3]$		
x_3	$P_0[x_3] = y_3$			

Table 3.2

If the number of data points is m, the algorithm that computes the elements of the table is

```
y = yData.copy()
for k in range (1,m):
    for i in range(m-k):
        y[i] = ((x - xData[i+k])*y[i] + (xData[i] - x)*y[i+1])/ \
               (xData[i] - xData[i+k])
```

This algorithm works with the one-dimensional array **y**, which initially contains the y-values of the data (the second column in Table 3.2). Each pass through the outer loop computes the elements of **y** in the next column, which overwrite the previous entries. At the end of the procedure, **y** contains the diagonal terms of the table. The value of the interpolant (evaluated at x) that passes through all the data points is the first element of **y**.

■ neville

The following function implements Neville's method; it returns $P_n(x)$

```
## module neville
''' p = neville(xData,yData,x).
    Evaluates the polynomial interpolant p(x) that passes
    trough the specified data points by Neville's method.
'''
```

```
def neville(xData,yData,x):
    m = len(xData)     # number of data points
    y = yData.copy()
    for k in range(1,m):
        y[0:m-k] = ((x - xData[k:m])*y[0:m-k] +      \
                    (xData[0:m-k] - x)*y[1:m-k+1])/  \
                    (xData[0:m-k] - xData[k:m])
    return y[0]
```

Limitations of Polynomial Interpolation

Polynomial interpolation should be carried out with the fewest feasible number of data points. Linear interpolation, using the nearest two points, is often sufficient if the data points are closely spaced. Three to six nearest-neighbor points produce good results in most cases. An interpolant intersecting more than six points must be viewed with suspicion. The reason is that the data points that are far from the point of interest do not contribute to the accuracy of the interpolant. In fact, they can be detrimental.

The danger of using too many points is illustrated in Fig. 3.3. There are 11 equally spaced data points represented by the circles. The solid line is the interpolant, a polynomial of degree ten, that intersects all the points. As seen in the figure, a polynomial of such a high degree has a tendency to oscillate excessively between the data points. A much smoother result would be obtained by using a cubic interpolant spanning four nearest-neighbor points.

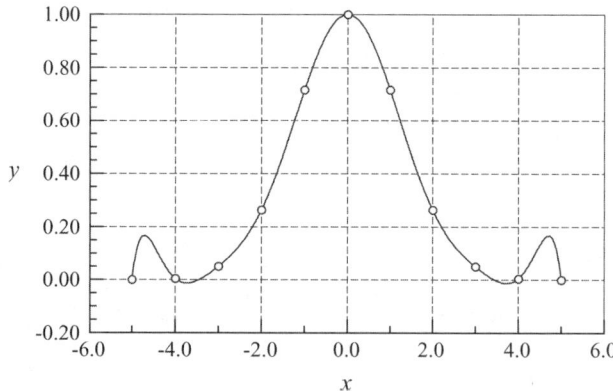

Figure 3.3. Polynomial interpolant displaying oscillations.

Polynomial extrapolation (interpolating outside the range of data points) is dangerous. As an example, consider Fig. 3.4. There are six data points, shown as circles.

3.2 Polynomial Interpolation

The fifth-degree interpolating polynomial is represented by the solid line. The interpolant looks fine within the range of data points, but drastically departs from the obvious trend when $x > 12$. Extrapolating y at $x = 14$, for example, would be absurd in this case.

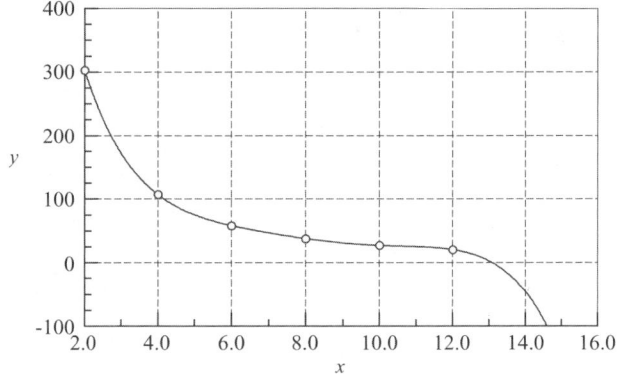

Figure 3.4. Extrapolation may not follow the trend of data.

If extrapolation cannot be avoided, the following two measures can be useful:

- Plot the data and visually verify that the extrapolated value makes sense.
- Use a low-order polynomial based on nearest-neighbor data points. A linear or quadratic interpolant, for example, would yield a reasonable estimate of $y(14)$ for the data in Fig. 3.4.
- Work with a plot of $\log x$ vs. $\log y$, which is usually much smoother than the x–y curve, and thus safer to extrapolate. Frequently this plot is almost a straight line. This is illustrated in Fig. 3.5, which represents the logarithmic plot of the data in Fig. 3.4.

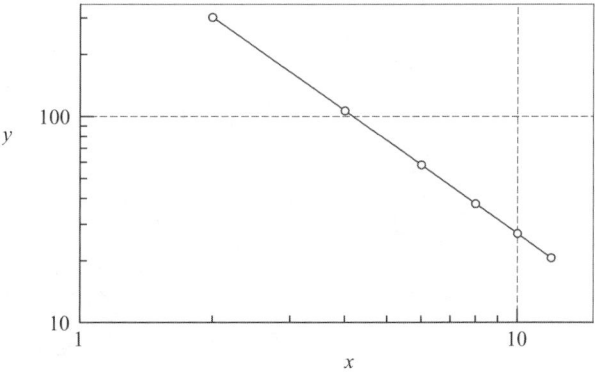

Figure 3.5. Logarithmic plot of the data in Fig. 3.4.

EXAMPLE 3.1
Given the data points

x	0	2	3
y	7	11	28

use Lagrange's method to determine y at $x = 1$.

Solution

$$\ell_0 = \frac{(x - x_1)(x - x_2)}{(x_0 - x_1)(x_0 - x_2)} = \frac{(1 - 2)(1 - 3)}{(0 - 2)(0 - 3)} = \frac{1}{3}$$

$$\ell_1 = \frac{(x - x_0)(x - x_2)}{(x_1 - x_0)(x_1 - x_2)} = \frac{(1 - 0)(1 - 3)}{(2 - 0)(2 - 3)} = 1$$

$$\ell_2 = \frac{(x - x_0)(x - x_1)}{(x_2 - x_0)(x_2 - x_1)} = \frac{(1 - 0)(1 - 2)}{(3 - 0)(3 - 2)} = -\frac{1}{3}$$

$$y = y_0 \ell_0 + y_1 \ell_1 + y_2 \ell_2 = \frac{7}{3} + 11 - \frac{28}{3} = 4$$

EXAMPLE 3.2
The data points

x	-2	1	4	-1	3	-4
y	-1	2	59	4	24	-53

lie on a polynomial. Determine the degree of this polynomial by constructing the divided difference table, similar to Table 3.1.

Solution

i	x_i	y_i	∇y_i	$\nabla^2 y_i$	$\nabla^3 y_i$	$\nabla^4 y_i$	$\nabla^5 y_i$
0	-2	-1					
1	1	2	1				
2	4	59	10	3			
3	-1	4	5	-2	1		
4	3	24	5	2	1	0	
5	-4	-53	26	-5	1	0	0

Here are a few sample calculations used in arriving at the figures in the table:

$$\nabla y_2 = \frac{y_2 - y_0}{x_2 - x_0} = \frac{59 - (-1)}{4 - (-2)} = 10$$

$$\nabla^2 y_2 = \frac{\nabla y_2 - \nabla y_1}{x_2 - x_1} = \frac{10 - 1}{4 - 1} = 3$$

$$\nabla^3 y_5 = \frac{\nabla^2 y_5 - \nabla^2 y_2}{x_5 - x_2} = \frac{-5 - 3}{-4 - 4} = 1$$

3.2 Polynomial Interpolation

From the table we see that the last nonzero coefficient (last nonzero diagonal term) of Newton's polynomial is $\nabla^3 y_3$, which is the coefficient of the cubic term. Hence the polynomial is a cubic.

EXAMPLE 3.3
Given the data points

x	4.0	3.9	3.8	3.7
y	-0.06604	-0.02724	0.01282	0.05383

determine the root of $y(x) = 0$ by Neville's method.

Solution This is an example of *inverse interpolation*, where the roles of x and y are interchanged. Instead of computing y at a given x, we are finding x that corresponds to a given y (in this case, $y = 0$). Employing the format of Table 3.2 (with x and y interchanged, of course), we obtain

i	y_i	$P_0[\ ] = x_i$	$P_1[\,,\,]$	$P_2[\,,\,,\,]$	$P_3[\,,\,,\,,\,]$
0	-0.06604	4.0	3.8298	3.8316	3.8317
1	-0.02724	3.9	3.8320	3.8318	
2	0.01282	3.8	3.8313		
3	0.05383	3.7			

The following are sample computations used in the table:

$$P_1[y_0, y_1] = \frac{(y - y_1) P_0[y_0] + (y_0 - y) P_0[y_1]}{y_0 - y_1}$$

$$= \frac{(0 + 0.02724)(4.0) + (-0.06604 - 0)(3.9)}{-0.06604 + 0.02724} = 3.8298$$

$$P_2[y_1, y_2, y_3] = \frac{(y - y_3) P_1[y_1, y_2] + (y_1 - y) P_1[y_2, y_3]}{y_1 - y_3}$$

$$= \frac{(0 - 0.05383)(3.8320) + (-0.02724 - 0)(3.8313)}{-0.02724 - 0.05383} = 3.8318$$

All the P's in the table are estimates of the root resulting from different orders of interpolation involving different data points. For example, $P_1[y_0, y_1]$ is the root obtained from linear interpolation based on the first two points, and $P_2[y_1, y_2, y_3]$ is the result from quadratic interpolation using the last three points. The root obtained from cubic interpolation over all four data points is $x = P_3[y_0, y_1, y_2, y_3] = 3.8317$.

EXAMPLE 3.4

The data points in the table lie on the plot of $f(x) = 4.8 \cos \dfrac{\pi x}{20}$. Interpolate this data by Newton's method at $x = 0, 0.5, 1.0, \ldots, 8.0$ and compare the results with the "exact" values $y_i = f(x_i)$.

x	0.15	2.30	3.15	4.85	6.25	7.95
y	4.79867	4.49013	4.2243	3.47313	2.66674	1.51909

Solution

```
#!/usr/bin/python
## example3_4
from numarray import array,arange
from math import pi,cos
from newtonPoly import *

xData = array([0.15,2.3,3.15,4.85,6.25,7.95])
yData = array([4.79867,4.49013,4.2243,3.47313,2.66674,1.51909])
a = coeffts(xData,yData)
print ''   x     yInterp    yExact''
print ''-----------------------''
for x in arange(0.0,8.1,0.5):
    y = evalPoly(a,xData,x)
    yExact = 4.8*cos(pi*x/20.0)
    print ''%3.1f %9.5f %9.5f''% (x,y,yExact)
raw_input(''\nPress return to exit'')
```

The results are:

```
  x      yInterp    yExact
-----------------------
0.0     4.80003    4.80000
0.5     4.78518    4.78520
1.0     4.74088    4.74090
1.5     4.66736    4.66738
2.0     4.56507    4.56507
2.5     4.43462    4.43462
3.0     4.27683    4.27683
3.5     4.09267    4.09267
4.0     3.88327    3.88328
```

4.5	3.64994	3.64995
5.0	3.39411	3.39411
5.5	3.11735	3.11735
6.0	2.82137	2.82137
6.5	2.50799	2.50799
7.0	2.17915	2.17915
7.5	1.83687	1.83688
8.0	1.48329	1.48328

3.3 Interpolation with Cubic Spline

If there are more than a few data points, a cubic spline is hard to beat as a global interpolant. It is considerably "stiffer" than a polynomial in the sense that it has less tendency to oscillate between data points.

Figure 3.6. Mechanical model of natural cubic spline.

The mechanical model of a cubic spline is shown in Fig. 3.6. It is a thin, elastic beam that is attached with pins to the data points. Because the beam is unloaded between the pins, each segment of the spline curve is a cubic polynomial—recall from beam theory that $d^4 y/dx^4 = q/(EI)$, so that $y(x)$ is a cubic since $q = 0$. At the pins, the slope and bending moment (and hence the second derivative) are continuous. There is no bending moment at the two end pins; consequently, the second derivative of the spline is zero at the end points. Since these end conditions occur naturally in the beam model, the resulting curve is known as the *natural cubic spline*. The pins, i.e., the data points, are called the *knots* of the spline.

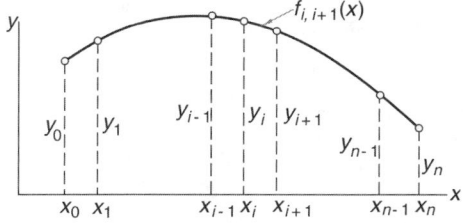

Figure 3.7. Cubic spline.

Figure 3.7 shows a cubic spline that spans $n+1$ knots. We use the notation $f_{i,i+1}(x)$ for the cubic polynomial that spans the segment between knots i and $i + 1$.

Interpolation and Curve Fitting

Note that the spline is a *piecewise cubic* curve, put together from the n cubics $f_{0,1}(x)$, $f_{1,2}(x)$, ..., $f_{n-1,n}(x)$, all of which have different coefficients.

If we denote the second derivative of the spline at knot i by k_i, continuity of second derivatives requires that

$$f''_{i-1,i}(x_i) = f''_{i,i+1}(x_i) = k_i \tag{a}$$

At this stage, each k is unknown, except for

$$k_0 = k_n = 0 \tag{3.9}$$

The starting point for computing the coefficients of $f_{i,i+1}(x)$ is the expression for $f''_{i,i+1}(x)$, which we know to be linear. Using Lagrange's two-point interpolation, we can write

$$f''_{i,i+1}(x) = k_i \ell_i(x) + k_{i+1} \ell_{i+1}(x)$$

where

$$\ell_i(x) = \frac{x - x_{i+1}}{x_i - x_{i+1}} \qquad \ell_{1+1}(x) = \frac{x - x_i}{x_{i+1} - x_i}$$

Therefore,

$$f''_{i,i+1}(x) = \frac{k_i(x - x_{i+1}) - k_{i+1}(x - x_i)}{x_i - x_{i+1}} \tag{b}$$

Integrating twice with respect to x, we obtain

$$f_{i,i+1}(x) = \frac{k_i(x - x_{i+1})^3 - k_{i+1}(x - x_i)^3}{6(x_i - x_{i+1})} + A(x - x_{i+1}) - B(x - x_i) \tag{c}$$

where A and B are constants of integration. The terms arising from the integration would usually be written as $Cx + D$. By letting $C = A - B$ and $D = -Ax_{i+1} + Bx_i$, we end up with the last two terms of Eq. (c), which are more convenient to use in the computations that follow.

Imposing the condition $f_{i,i+1}(x_i) = y_i$, we get from Eq. (c)

$$\frac{k_i(x_i - x_{i+1})^3}{6(x_i - x_{i+1})} + A(x_i - x_{i+1}) = y_i$$

Therefore,

$$A = \frac{y_i}{x_i - x_{i+1}} - \frac{k_i}{6}(x_i - x_{i+1}) \tag{d}$$

Similarly, $f_{i,i+1}(x_{i+1}) = y_{i+1}$ yields

$$B = \frac{y_{i+1}}{x_i - x_{i+1}} - \frac{k_{i+1}}{6}(x_i - x_{i+1}) \tag{e}$$

3.3 Interpolation with Cubic Spline

Substituting Eqs. (d) and (e) into Eq. (c) results in

$$f_{i,i+1}(x) = \frac{k_i}{6}\left[\frac{(x-x_{i+1})^3}{x_i-x_{i+1}} - (x-x_{i+1})(x_i-x_{i+1})\right]$$
$$-\frac{k_{i+1}}{6}\left[\frac{(x-x_i)^3}{x_i-x_{i+1}} - (x-x_i)(x_i-x_{i+1})\right] \quad (3.10)$$
$$+\frac{y_i(x-x_{i+1}) - y_{i+1}(x-x_i)}{x_i-x_{i+1}}$$

The second derivatives k_i of the spline at the interior knots are obtained from the slope continuity conditions $f'_{i-1,i}(x_i) = f'_{i,i+1}(x_i)$, where $i = 1, 2, \ldots, n-1$. After a little algebra, this results in the simultaneous equations

$$k_{i-1}(x_{i-1}-x_i) + 2k_i(x_{i-1}-x_{i+1}) + k_{i+1}(x_i-x_{i+1})$$
$$= 6\left(\frac{y_{i-1}-y_i}{x_{i-1}-x_i} - \frac{y_i-y_{i+1}}{x_i-x_{i+1}}\right), \quad i = 1, 2, \cdots, n-1 \quad (3.11)$$

Because Eqs. (3.11) have a tridiagonal coefficient matrix, they can be solved economically with the functions in module `LUdecomp3` described in Section 2.4.

If the data points are evenly spaced at intervals h, then $x_{i-1} - x_i = x_i - x_{i+1} = -h$, and the Eqs. (3.11) simplify to

$$k_{i-1} + 4k_i + k_{i+1} = \frac{6}{h^2}(y_{i-1} - 2y_i + y_{i+1}), \quad i = 1, 2, \ldots, n-1 \quad (3.12)$$

∎ cubicSpline

The first stage of cubic spline interpolation is to set up Eqs. (3.11) and solve them for the unknown k's (recall that $k_0 = k_n = 0$). This task is carried out by the function `curvatures`. The second stage is the computation of the interpolant at x from Eq. (3.10). This step can be repeated any number of times with different values of x using the function `evalSpline`. The function `findSegment` embedded in `evalSpline` finds the segment of the spline that contains x using the method of bisection. It returns the segment number; that is, the value of the subscript i in Eq. (3.10).

```
## module cubicSpline
''' k = curvatures(xData,yData).
    Returns the curvatures {k}of cubic spline at the knots.

    y = evalSpline(xData,yData,k,x).
    Evaluates cubic spline at x. The curvatures {k} can be
```

Interpolation and Curve Fitting

```
    computed with the function 'curvatures'.
''' 
from numarray import zeros,ones,Float64,array
from LUdecomp3 import *

def curvatures(xData,yData):
    n = len(xData) - 1
    c = zeros((n),type=Float64)
    d = ones((n+1),type=Float64)
    e = zeros((n),type=Float64)
    k = zeros((n+1),type=Float64)
    c[0:n-1] = xData[0:n-1] - xData[1:n]
    d[1:n] = 2.0*(xData[0:n-1] - xData[2:n+1])
    e[1:n] = xData[1:n] - xData[2:n+1]
    k[1:n] =6.0*(yData[0:n-1] - yData[1:n]) \
             /(xData[0:n-1] - xData[1:n]) \
          -6.0*(yData[1:n] - yData[2:n+1])   \
             /(xData[1:n] - xData[2:n+1])
    LUdecomp3(c,d,e)
    LUsolve3(c,d,e,k)
    return k

def evalSpline(xData,yData,k,x):

    def findSegment(xData,x):
        iLeft = 0
        iRight = len(xData)- 1
        while 1:
            if (iRight-iLeft) <= 1: return iLeft
            i =(iLeft + iRight)/2
            if x < xData[i]: iRight = i
            else: iLeft = i

    i = findSegment(xData,x)   # Find the segment spanning x
    h = xData[i] - xData[i+1]
    y = ((x - xData[i+1])**3/h - (x - xData[i+1])*h)*k[i]/6.0  \
      - ((x - xData[i])**3/h - (x - xData[i])*h)*k[i+1]/6.0    \
      + (yData[i]*(x - xData[i+1])                             \
      - yData[i+1]*(x - xData[i]))/h
    return y
```

3.3 Interpolation with Cubic Spline

EXAMPLE 3.5
Use natural cubic spline to determine y at $x = 1.5$. The data points are

x	1	2	3	4	5
y	0	1	0	1	0

Solution The five knots are equally spaced at $h = 1$. Recalling that the second derivative of a natural spline is zero at the first and last knot, we have $k_0 = k_4 = 0$. The second derivatives at the other knots are obtained from Eq. (3.12). Using $i = 1, 2, 3$ results in the simultaneous equations

$$0 + 4k_1 + k_2 = 6\,[0 - 2(1) + 0] = -12$$
$$k_1 + 4k_2 + k_3 = 6\,[1 - 2(0) + 1] = 12$$
$$k_2 + 4k_3 + 0 = 6\,[0 - 2(1) + 0] = -12$$

The solution is $k_1 = k_3 = -30/7$, $k_2 = 36/7$.

The point $x = 1.5$ lies in the segment between knots 0 and 1. The corresponding interpolant is obtained from Eq. (3.10) by setting $i = 0$. With $x_i - x_{i+1} = -h = -1$, we obtain from Eq. (3.10)

$$f_{0,1}(x) = -\frac{k_0}{6}\left[(x - x_1)^3 - (x - x_1)\right] + \frac{k_1}{6}\left[(x - x_0)^3 - (x - x_0)\right]$$
$$- [y_0(x - x_1) - y_1(x - x_0)]$$

Therefore,

$$y(1.5) = f_{0,1}(1.5)$$
$$= 0 + \frac{1}{6}\left(-\frac{30}{7}\right)\left[(1.5 - 1)^3 - (1.5 - 1)\right] - [0 - 1(1.5 - 1)]$$
$$= 0.7679$$

The plot of the interpolant, which in this case is made up of four cubic segments, is shown in the figure.

EXAMPLE 3.6
Sometimes it is preferable to replace one or both of the end conditions of the cubic spline with something other than the natural conditions. Use the end condition $f'_{0,1}(0) = 0$ (zero slope), rather than $f''_{0,1}(0) = 0$ (zero curvature), to determine the cubic spline interpolant at $x = 2.6$, given the data points

x	0	1	2	3
y	1	1	0.5	0

Solution We must first modify Eqs. (3.12) to account for the new end condition. Setting $i = 0$ in Eq. (3.10) and differentiating, we get

$$f'_{0,1}(x) = \frac{k_0}{6}\left[3\frac{(x-x_1)^2}{x_0-x_1} - (x_0 - x_1)\right] - \frac{k_1}{6}\left[3\frac{(x-x_0)^2}{x_0-x_1} - (x_0 - x_1)\right] + \frac{y_0 - y_1}{x_0 - x_1}$$

Thus the end condition $f'_{0,1}(x_0) = 0$ yields

$$\frac{k_0}{3}(x_0 - x_1) + \frac{k_1}{6}(x_0 - x_1) + \frac{y_0 - y_1}{x_0 - x_1} = 0$$

or

$$2k_0 + k_1 = -6\frac{y_0 - y_1}{(x_0 - x_1)^2}$$

From the given data we see that $y_0 = y_1 = 1$, so that the last equation becomes

$$2k_0 + k_1 = 0 \tag{a}$$

The other equations in Eq. (3.12) are unchanged. Knowing that $k_3 = 0$, we have

$$k_0 + 4k_1 + k_2 = 6\left[1 - 2(1) + 0.5\right] = -3 \tag{b}$$

$$k_1 + 4k_2 = 6\left[1 - 2(0.5) + 0\right] = 0 \tag{c}$$

The solution of Eqs. (a)–(c) is $k_0 = 0.4615$, $k_1 = -0.9231$, $k_2 = 0.2308$.

The interpolant can now be evaluated from Eq. (3.10). Substituting $i = 2$ and $x_i - x_{i+1} = -1$, we obtain

$$f_{2,3}(x) = \frac{k_2}{6}\left[-(x-x_3)^3 + (x-x_3)\right] - \frac{k_3}{6}\left[-(x-x_2)^3 + (x-x_2)\right] - y_2(x - x_3) + y_3(x - x_2)$$

Therefore,

$$y(2.6) = f_{2,3}(2.6) = \frac{0.2308}{6}\left[-(-0.4)^3 + (-0.4)\right] - 0 - 0.5(-0.4) + 0$$

$$= 0.1871$$

3.3 Interpolation with Cubic Spline

EXAMPLE 3.7

Utilize the module `cubicSpline` to write a program that interpolates between given data points with natural cubic spline. The program must be able to evaluate the interpolant for more than one value of x. As a test, use data points specified in Example 3.4 and compute the interpolant at $x = 1.5$ and $x = 4.5$ (due to symmetry, these values should be equal).

Solution

```
#!/usr/bin/python
## example3_7
from numarray import array,Float64
from cubicSpline import *

xData = array([1,2,3,4,5],type=Float64)
yData = array([0,1,0,1,0],type=Float64)
k = curvatures(xData,yData)
while 1:
    try: x = eval(raw_input(''\nx ==> ''))
    except SyntaxError: break
    print ''y ='',evalSpline(xData,yData,k,x)
raw_input(''Done. Press return to exit'')
```

Running the program produces the following result:

```
x ==> 1.5
y = 0.767857142857

x ==> 4.5
y = 0.767857142857

x ==>
Done. Press return to exit
```

PROBLEM SET 3.1

1. Given the data points

x	-1.2	0.3	1.1
y	-5.76	-5.61	-3.69

determine y at $x = 0$ using (a) Neville's method; and (b) Lagrange's method.

2. Find the zero of $y(x)$ from the following data:

x	0	0.5	1	1.5	2	2.5	3
y	1.8421	2.4694	2.4921	1.9047	0.8509	−0.4112	−1.5727

Use Lagrange's interpolation over (a) three; and (b) four nearest-neighbor data points. *Hint*: after finishing part (a), part (b) can be computed with a relatively small effort.

3. The function $y(x)$ represented by the data in Prob. 2 has a maximum at $x = 0.7679$. Compute this maximum by Neville's interpolation over four nearest-neighbor data points.

4. Use Neville's method to compute y at $x = \pi/4$ from the data points

x	0	0.5	1	1.5	2
y	−1.00	1.75	4.00	5.75	7.00

5. Given the data

x	0	0.5	1	1.5	2
y	−0.7854	0.6529	1.7390	2.2071	1.9425

find y at $x = \pi/4$ and at $\pi/2$. Use the method that you consider to be most convenient.

6. The points

x	−2	1	4	−1	3	−4
y	−1	2	59	4	24	−53

lie on a polynomial. Use the divided difference table of Newton's method to determine the degree of the polynomial.

7. Use Newton's method to find the polynomial that fits the following points:

x	−3	2	−1	3	1
y	0	5	−4	12	0

8. Use Neville's method to determine the equation of the quadratic that passes through the points

x	−1	1	3
y	17	−7	−15

3.3 Interpolation with Cubic Spline

9. The density of air ρ varies with elevation h in the following manner:

h (km)	0	3	6
ρ (kg/m^3)	1.225	0.905	0.652

   Express $\rho(h)$ as a quadratic function using Lagrange's method.

10. Determine the natural cubic spline that passes through the data points

x	0	1	2
y	0	2	1

    Note that the interpolant consists of two cubics, one valid in $0 \le x \le 1$, the other in $1 \le x \le 2$. Verify that these cubics have the same first and second derivatives at $x = 1$.

11. Given the data points

x	1	2	3	4	5
y	13	15	12	9	13

    determine the natural cubic spline interpolant at $x = 3.4$.

12. Compute the zero of the function $y(x)$ from the following data:

x	0.2	0.4	0.6	0.8	1.0
y	1.150	0.855	0.377	-0.266	-1.049

    Use inverse interpolation with the natural cubic spline. *Hint*: reorder the data so that the values of y are in ascending order.

13. Solve Example 3.6 with a cubic spline that has constant second derivatives within its first and last segments (the end segments are parabolic). The end conditions for this spline are $k_0 = k_1$ and $k_{n-1} = k_n$.

14. ■ Write a computer program for interpolation by Neville's method. The program must be able to compute the interpolant at several user-specified values of x. Test the program by determining y at $x = 1.1, 1.2$ and 1.3 from the following data:

x	-2.0	-0.1	-1.5	0.5
y	2.2796	1.0025	1.6467	1.0635
x	-0.6	2.2	1.0	1.8
y	1.0920	2.6291	1.2661	1.9896

    (Answer: $y = 1.3262, 1.3938, 1.4693$)

15. ■ The specific heat c_p of aluminum depends on temperature T as follows:[6]

T (°C)	−250	−200	−100	0	100	300
c_p (kJ/kg·K)	0.0163	0.318	0.699	0.870	0.941	1.04

Determine c_p at $T = 200°$C and $400°$C.

16. ■ Find y at $x = 0.46$ from the data

x	0	0.0204	0.1055	0.241	0.582	0.712	0.981
y	0.385	1.04	1.79	2.63	4.39	4.99	5.27

17. ■ The table shows the drag coefficient c_D of a sphere as a function of Reynolds number Re.[7] Use natural cubic spline to find c_D at Re = 5, 50, 500 and 5000. *Hint*: use log–log scale.

Re	0.2	2	20	200	2000	20 000
c_D	103	13.9	2.72	0.800	0.401	0.433

18. ■ Solve Prob. 17 using a polynomial interpolant intersecting four nearest-neighbor data points.

19. ■ The kinematic viscosity μ_k of water varies with temperature T in the following manner:

T (°C)	0	21.1	37.8	54.4	71.1	87.8	100
μ_k (10^{-3} m^2/s)	1.79	1.13	0.696	0.519	0.338	0.321	0.296

Interpolate μ_k at $T = 10°, 30°, 60°$ and $90°$C.

20. ■ The table shows how the relative density ρ of air varies with altitude h. Determine the relative density of air at 10.5 km.

h (km)	0	1.525	3.050	4.575	6.10	7.625	9.150
ρ	1	0.8617	0.7385	0.6292	0.5328	0.4481	0.3741

3.4 Least-Squares Fit

Overview

If the data are obtained from experiments, they typically contain a significant amount of random noise due to measurement errors. The task of curve fitting is to find a

---

[6] Source: Black, Z. B., and Hartley, J. G., *Thermodynamics*, Harper & Row, 1985.
[7] Source: Kreith, F., *Principles of Heat Transfer*, Harper & Row, 1973.

3.4 Least-Squares Fit

smooth curve that fits the data points "on the average." This curve should have a simple form (e.g., a low-order polynomial), so as to not reproduce the noise.

Let

$$f(x) = f(x; a_0, a_1, \ldots, a_m)$$

be the function that is to be fitted to the $n+1$ data points (x_i, y_i), $i = 0, 1, \ldots, n$. The notation implies that we have a function of x that contains $m+1$ variable parameters $a_0, a_1, \ldots, a_m$, where $m < n$. The form of $f(x)$ is determined beforehand, usually from the theory associated with the experiment from which the data is obtained. The only means of adjusting the fit is the parameters. For example, if the data represent the displacements y_i of an overdamped mass–spring system at time t_i, the theory suggests the choice $f(t) = a_0 t e^{-a_1 t}$. Thus curve fitting consists of two steps: choosing the form of $f(x)$, followed by computation of the parameters that produce the best fit to the data.

This brings us to the question: what is meant by "best" fit? If the noise is confined to the y-coordinate, the most commonly used measure is the *least-squares fit*, which minimizes the function

$$S(a_0, a_1, \ldots, a_m) = \sum_{i=0}^{n} [y_i - f(x_i)]^2 \quad (3.13)$$

with respect to each a_j. Therefore, the optimal values of the parameters are given by the solution of the equations

$$\frac{\partial S}{\partial a_k} = 0, \quad k = 0, 1, \ldots, m \quad (3.14)$$

The terms $r_i = y_i - f(x_i)$ in Eq. (3.13) are called *residuals*; they represent the discrepancy between the data points and the fitting function at x_i. The function S to be minimized is thus the sum of the squares of the residuals. Equations (3.14) are generally nonlinear in a_j and may thus be difficult to solve. Often the fitting function is chosen as a linear combination of specified functions $f_j(x)$:

$$f(x) = a_0 f_0(x) + a_1 f_1(x) + \cdots + a_m f_m(x)$$

in which case Eqs. (3.14) are linear. If the fitting function is a polynomial, we have $f_0(x) = 1$, $f_1(x) = x$, $f_2(x) = x^2$, etc.

The spread of the data about the fitting curve is quantified by the *standard deviation*, defined as

$$\sigma = \sqrt{\frac{S}{n-m}} \quad (3.15)$$

Note that if $n = m$, we have *interpolation*, not curve fitting. In that case both the numerator and the denominator in Eq. (3.15) are zero, so that σ is indeterminate.

Fitting a Straight Line

Fitting a straight line

$$f(x) = a + bx \tag{3.16}$$

to data is also known as *linear regression*. In this case the function to be minimized is

$$S(a,b) = \sum_{i=0}^{n} [y_i - f(x_i)]^2 = \sum_{i=0}^{n} (y_i - a - bx_i)^2$$

Equations (3.14) now become

$$\frac{\partial S}{\partial a} = \sum_{i=0}^{n} -2(y_i - a - bx_i) = 2\left[a(n+1) + b\sum_{i=0}^{n} x_i - \sum_{i=0}^{n} y_i\right] = 0$$

$$\frac{\partial S}{\partial b} = \sum_{i=0}^{n} -2(y_i - a - bx_i)x_i = 2\left(a\sum_{i=0}^{n} x_i + b\sum_{i=0}^{n} x_i^2 - \sum_{i=0}^{n} x_i y_i\right) = 0$$

Dividing both equations by $2(n+1)$ and rearranging terms, we get

$$a + \bar{x}b = \bar{y} \qquad \bar{x}a + \left(\frac{1}{n+1}\sum_{i=0}^{n} x_i^2\right) b = \frac{1}{n+1}\sum_{i=0}^{n} x_i y_i$$

where

$$\bar{x} = \frac{1}{n+1}\sum_{i=0}^{n} x_i \qquad \bar{y} = \frac{1}{n+1}\sum_{i=0}^{n} y_i \tag{3.17}$$

are the mean values of the x and y data. The solution for the parameters is

$$a = \frac{\bar{y}\sum x_i^2 - \bar{x}\sum x_i y_i}{\sum x_i^2 - (n+1)\bar{x}^2} \qquad b = \frac{\sum x_i y_i - \bar{x}\sum y_i}{\sum x_i^2 - (n+1)\bar{x}^2} \tag{3.18}$$

These expressions are susceptible to roundoff errors (the two terms in each numerator as well as in each denominator can be roughly equal). It is better to compute the parameters from

$$b = \frac{\sum y_i(x_i - \bar{x})}{\sum x_i(x_i - \bar{x})} \qquad a = \bar{y} - \bar{x}b \tag{3.19}$$

which are equivalent to Eqs. (3.18), but much less affected by rounding off.

Fitting Linear Forms

Consider the least-squares fit of the *linear form*

$$f(x) = a_0 f_0(x) + a_1 f_1(x) + \cdots + a_m f_m(x) = \sum_{j=0}^{m} a_j f_j(x) \tag{3.20}$$

3.4 Least-Squares Fit

where each $f_j(x)$ is a predetermined function of x, called a *basis function*. Substitution in Eq. (3.13) yields

$$S = \sum_{i=0}^{n}\left[y_i - \sum_{j=0}^{m} a_j f_j(x_i)\right]^2$$

Thus Eqs. (3.14) are

$$\frac{\partial S}{\partial a_k} = -2\left\{\sum_{i=0}^{n}\left[y_i - \sum_{j=0}^{m} a_j f_j(x_i)\right] f_k(x_i)\right\} = 0, \quad k = 0, 1, \ldots, m$$

Dropping the constant (-2) and interchanging the order of summation, we get

$$\sum_{j=0}^{m}\left[\sum_{i=0}^{n} f_j(x_i) f_k(x_i)\right] a_j = \sum_{i=0}^{n} f_k(x_i) y_i, \quad k = 0, 1, \ldots, m$$

In matrix notation these equations are

$$\mathbf{Aa} = \mathbf{b} \tag{3.21a}$$

where

$$A_{kj} = \sum_{i=0}^{n} f_j(x_i) f_k(x_i) \qquad b_k = \sum_{i=0}^{n} f_k(x_i) y_i \tag{3.21b}$$

Equations (3.21a), known as the *normal equations* of the least-squares fit, can be solved with the methods discussed in Chapter 2. Note that the coefficient matrix is symmetric, i.e., $A_{kj} = A_{jk}$.

Polynomial Fit

A commonly used linear form is a polynomial. If the degree of the polynomial is m, we have $f(x) = \sum_{j=0}^{m} a_j x^j$. Here the basis functions are

$$f_j(x) = x^j \qquad (j = 0, 1, \ldots, m) \tag{3.22}$$

so that Eqs. (3.21b) become

$$A_{kj} = \sum_{i=0}^{n} x_i^{j+k} \qquad b_k = \sum_{i=0}^{n} x_i^k y_i$$

or

$$\mathbf{A} = \begin{bmatrix} n & \sum x_i & \sum x_i^2 & \cdots & \sum x_i^m \\ \sum x_i & \sum x_i^2 & \sum x_i^3 & \cdots & \sum x_i^{m+1} \\ \vdots & \vdots & \vdots & \ddots & \vdots \\ \sum x_i^{m-1} & \sum x_i^m & \sum x_i^{m+1} & \cdots & \sum x_i^{2m} \end{bmatrix} \qquad \mathbf{b} = \begin{bmatrix} \sum y_i \\ \sum x_i y_i \\ \vdots \\ \sum x_i^m y_i \end{bmatrix} \tag{3.23}$$

where $\sum$ stands for $\sum_{i=0}^{n}$. The normal equations become progressively ill-conditioned with increasing m. Fortunately, this is of little practical consequence, because only low-order polynomials are useful in curve fitting. Polynomials of high order are not recommended, because they tend to reproduce the noise inherent in the data.

■ polyFit

The function polyFit in this module sets up and solves the normal equations for the coefficients of a polynomial of degree m. It returns the coefficients of the polynomial. To facilitate computations, the terms n, $\sum x_i$, $\sum x_i^2$, ..., $\sum x_i^{2m}$ that make up the coefficient matrix in Eq. (3.23) are first stored in the vector **s** and then inserted into **A**. The normal equations are then solved by Gauss elimination with pivoting. Following the solution, the standard deviation σ can be computed with the function stdDev. The polynomial evaluation in stdDev is carried out by the embedded function evalPoly—see Section 4.7 for an explanation of the algorithm.

```
## module polyFit
''' c = polyFit(xData,yData,m).
    Returns coefficients of the polynomial
    p(x) = c[0] + c[1]x + c[2]x^2 +...+ c[m]x^m
    that fits the specified data in the least
    squares sense.

    sigma = stdDev(c,xData,yData).
    Computes the std. deviation between p(x)
    and the data.
'''
from numarray import zeros,Float64
from math import sqrt
from gaussPivot import *

def polyFit(xData,yData,m):
    a = zeros((m+1,m+1),type=Float64)
    b = zeros((m+1),type=Float64)
    s = zeros((2*m+1),type=Float64)
    for i in range(len(xData)):
        temp = yData[i]
        for j in range(m+1):
            b[j] = b[j] + temp
            temp = temp*xData[i]
```

3.4 Least-Squares Fit

```
            temp = 1.0
            for j in range(2*m+1):
                s[j] = s[j] + temp
                temp = temp*xData[i]
        for i in range(m+1):
            for j in range(m+1):
                a[i,j] = s[i+j]
    return gaussPivot(a,b)

def stdDev(c,xData,yData):

    def evalPoly(c,x):
        m = len(c) - 1
        p = c[m]
        for j in range(m):
            p = p*x + c[m-j-1]
        return p

    n = len(xData) - 1
    m = len(c) - 1
    sigma = 0.0
    for i in range(n+1):
        p = evalPoly(c,xData[i])
        sigma = sigma + (yData[i] - p)**2
    sigma = sqrt(sigma/(n - m))
    return sigma
```

Weighting of Data

There are occasions when our confidence in the accuracy of data varies from point to point. For example, the instrument taking the measurements may be more sensitive in a certain range of data. Sometimes the data represent the results of several experiments, each carried out under different conditions. Under these circumstances we may want to assign a confidence factor, or *weight*, to each data point and minimize the sum of the squares of the *weighted residuals* $r_i = W_i [y_i - f(x_i)]$, where W_i are the weights. Hence the function to be minimized is

$$S(a_0, a_1, \ldots, a_m) = \sum_{i=0}^{n} W_i^2 [y_i - f(x_i)]^2 \qquad (3.24)$$

This procedure forces the fitting function $f(x)$ closer to the data points that have higher weights.

Weighted linear regression

If the fitting function is the straight line $f(x) = a + bx$, Eq. (3.24) becomes

$$S(a, b) = \sum_{i=0}^{n} W_i^2 (y_i - a - bx_i)^2 \qquad (3.25)$$

The conditions for minimizing S are

$$\frac{\partial S}{\partial a} = -2 \sum_{i=0}^{n} W_i^2 (y_i - a - bx_i) = 0$$

$$\frac{\partial S}{\partial b} = -2 \sum_{i=0}^{n} W_i^2 (y_i - a - bx_i) x_i = 0$$

or

$$a \sum_{i=0}^{n} W_i^2 + b \sum_{i=0}^{n} W_i^2 x_i = \sum_{i=0}^{n} W_i^2 y_i \qquad (3.26a)$$

$$a \sum_{i=0}^{n} W_i^2 x_i + b \sum_{i=0}^{n} W_i^2 x_i^2 = \sum_{i=0}^{n} W_i^2 x_i y_i \qquad (3.26b)$$

Dividing Eq. (3.26a) by $\sum W_i^2$ and introducing the *weighted averages*

$$\hat{x} = \frac{\sum W_i^2 x_i}{\sum W_i^2} \qquad \hat{y} = \frac{\sum W_i^2 y_i}{\sum W_i^2} \qquad (3.27)$$

we obtain

$$a = \hat{y} - b\hat{x} \qquad (3.28a)$$

Substituting into Eq. (3.26b) and solving for b yields after some algebra

$$b = \frac{\sum W_i^2 y_i (x_i - \hat{x})}{\sum W_i^2 x_i (x_i - \hat{x})} \qquad (3.28b)$$

Note that Eqs. (3.28) are quite similar to Eqs. (3.19) for unweighted data.

Fitting exponential functions

A special application of weighted linear regression arises in fitting various exponential functions to data. Consider as an example the fitting function

$$f(x) = ae^{bx}$$

Normally, the least-squares fit would lead to equations that are nonlinear in a and b. But if we fit $\ln y$ rather than y, the problem is transformed to linear regression: fit the function

$$F(x) = \ln f(x) = \ln a + bx$$

3.4 Least-Squares Fit

to the data points $(x_i, \ln y_i)$, $i = 0, 1, \ldots, n$. This simplification comes at a price: least-squares fit to the logarithm of the data is not quite the same as the least-squares fit to the original data. The residuals of the logarithmic fit are

$$R_i = \ln y_i - F(x_i) = \ln y_i - (\ln a + bx_i) \qquad (3.29\text{a})$$

whereas the residuals used in fitting the original data are

$$r_i = y_i - f(x_i) = y_i - ae^{bx_i} \qquad (3.29\text{b})$$

This discrepancy can be largely eliminated by weighting the logarithmic fit. From Eq. (3.29b) we obtain $\ln(r_i - y_i) = \ln(ae^{bx_i}) = \ln a + bx_i$, so that Eq. (3.29a) can be written as

$$R_i = \ln y_i - \ln(r_i - y_i) = \ln\left(1 - \frac{r_i}{y_i}\right)$$

If the residuals r_i are sufficiently small ($r_i << y_i$), we can use the approximation $\ln(1 - r_i/y_i) \approx r_i/y_i$, so that

$$R_i \approx r_i/y_i$$

We can now see that by minimizing $\sum R_i^2$, we have inadvertently introduced the weights $1/y_i$. This effect can be negated if we apply the weights $W_i = y_i$ when fitting $F(x)$ to $(\ln y_i, x_i)$. That is, minimizing

$$S = \sum_{i=0}^{n} y_i^2 R_i^2 \qquad (3.30)$$

is a good approximation to minimizing $\sum r_i^2$.

Other examples that also benefit from the weights $W_i = y_i$ are given in Table 3.3.

$f(x)$	$F(x)$	Data to be fitted by $F(x)$
axe^{bx}	$\ln[f(x)/x] = \ln a + bx$	$[x_i, \ln(y_i/x_i)]$
ax^b	$\ln f(x) = \ln a + b\ln(x)$	$(\ln x_i, \ln y_i)$

Table 3.3

EXAMPLE 3.8
Fit a straight line to the data shown and compute the standard deviation.

x	0.0	1.0	2.0	2.5	3.0
y	2.9	3.7	4.1	4.4	5.0

Solution The averages of the data are

$$\bar{x} = \frac{1}{5}\sum x_i = \frac{0.0 + 1.0 + 2.0 + 2.5 + 3.0}{5} = 1.7$$

$$\bar{y} = \frac{1}{5}\sum y_i = \frac{2.9 + 3.7 + 4.1 + 4.4 + 5.0}{5} = 4.02$$

The intercept a and slope b of the interpolant can now be determined from Eq. (3.19):

$$b = \frac{\sum y_i(x_i - \bar{x})}{\sum x_i(x_i - \bar{x})}$$

$$= \frac{2.9(-1.7) + 3.7(-0.7) + 4.1(0.3) + 4.4(0.8) + 5.0(1.3)}{0.0(-1.7) + 1.0(-0.7) + 2.0(0.3) + 2.5(0.8) + 3.0(1.3)}$$

$$= \frac{3.73}{5.8} = 0.6431$$

$$a = \bar{y} - \bar{x}b = 4.02 - 1.7(0.6431) = 2.927$$

Therefore, the regression line is $f(x) = 2.927 + 0.6431x$, which is shown in the figure together with the data points.

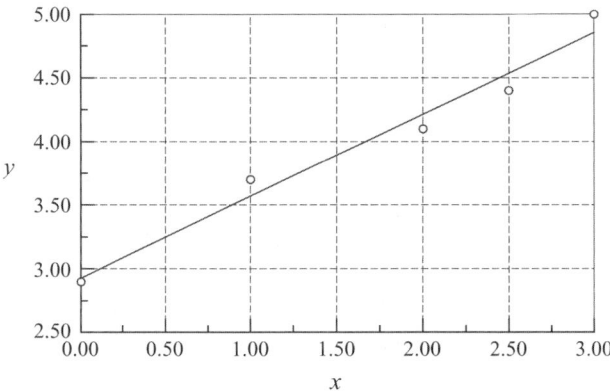

We start the evaluation of the standard deviation by computing the residuals:

x	0.000	1.000	2.000	2.500	3.000
y	2.900	3.700	4.100	4.400	5.000
$f(x)$	2.927	3.570	4.213	4.535	4.856
$y - f(x)$	−0.027	0.130	−0.113	−0.135	0.144

3.4 Least-Squares Fit

The sum of the squares of the residuals is

$$S = \sum [y_i - f(x_i)]^2$$
$$= (-0.027)^2 + (0.130)^2 + (-0.113)^2 + (-0.135)^2 + (0.144)^2 = 0.06936$$

so that the standard deviation in Eq. (3.15) becomes

$$\sigma = \sqrt{\frac{S}{n-m}} = \sqrt{\frac{0.06936}{5-2}} = 0.1520$$

EXAMPLE 3.9
Determine the parameters a and b so that $f(x) = ae^{bx}$ fits the following data in the least-squares sense.

x	1.2	2.8	4.3	5.4	6.8	7.9
y	7.5	16.1	38.9	67.0	146.6	266.2

Use two different methods: (1) fit $\ln y_i$; and (2) fit $\ln y_i$ with weights $W_i = y_i$. Compute the standard deviation in each case.

Solution of Part (1) The problem is to fit the function $\ln(ae^{bx}) = \ln a + bx$ to the data

x	1.2	2.8	4.3	5.4	6.8	7.9
$z = \ln y$	2.015	2.779	3.661	4.205	4.988	5.584

We are now dealing with linear regression, where the parameters to be found are $A = \ln a$ and b. Following the steps in Example 3.8, we get (skipping some of the arithmetic details)

$$\bar{x} = \frac{1}{6} \sum x_i = 4.733 \qquad \bar{z} = \frac{1}{6} \sum z_i = 3.872$$

$$b = \frac{\sum z_i(x_i - \bar{x})}{\sum x_i(x_i - \bar{x})} = \frac{16.716}{31.153} = 0.5366 \qquad A = \bar{z} - \bar{x}b = 1.3323$$

Therefore, $a = e^A = 3.790$ and the fitting function becomes $f(x) = 3.790e^{0.5366}$. The plots of $f(x)$ and the data points are shown in the figure.

Interpolation and Curve Fitting

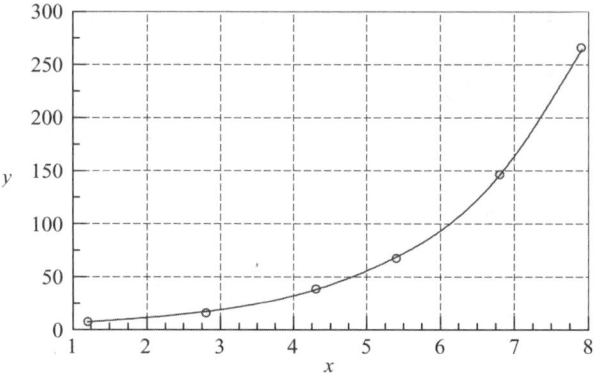

Here is the computation of standard deviation:

x	1.20	2.80	4.30	5.40	6.80	7.90
y	7.50	16.10	38.90	67.00	146.60	266.20
$f(x)$	7.21	17.02	38.07	68.69	145.60	262.72
$y - f(x)$	0.29	−0.92	0.83	−1.69	1.00	3.48

$$S = \sum [y_i - f(x_i)]^2 = 17.59$$

$$\sigma = \sqrt{\frac{S}{6-2}} = 2.10$$

As pointed out before, this is an approximate solution of the stated problem, since we did not fit y_i, but $\ln y_i$. Judging by the plot, the fit seems to be quite good.

Solution of Part (2) We again fit $\ln(ae^{bx}) = \ln a + bx$ to $z = \ln y$, but this time the weights $W_i = y_i$ are used. From Eqs. (3.27) the weighted averages of the data are (recall that we fit $z = \ln y$)

$$\hat{x} = \frac{\sum y_i^2 x_i}{\sum y_i^2} = \frac{737.5 \times 10^3}{98.67 \times 10^3} = 7.474$$

$$\hat{z} = \frac{\sum y_i^2 z_i}{\sum y_i^2} = \frac{528.2 \times 10^3}{98.67 \times 10^3} = 5.353$$

and Eqs. (3.28) yield for the parameters

$$b = \frac{\sum y_i^2 z_i (x_i - \hat{x})}{\sum y_i^2 x_i (x_i - \hat{x})} = \frac{35.39 \times 10^3}{65.05 \times 10^3} = 0.5440$$

$$\ln a = \hat{z} - b\hat{x} = 5.353 - 0.5440(7.474) = 1.287$$

3.4 Least-Squares Fit

Therefore,
$$a = e^{\ln a} = e^{1.287} = 3.622$$

so that the fitting function is $f(x) = 3.622e^{0.5440x}$. As expected, this result is somewhat different from that obtained in Part (1).

The computations of the residuals and the standard deviation are as follows:

x	1.20	2.80	4.30	5.40	6.80	7.90
y	7.50	16.10	38.90	67.00	146.60	266.20
$f(x)$	6.96	16.61	37.56	68.33	146.33	266.20
$y - f(x)$	0.54	−0.51	1.34	−1.33	0.267	0.00

$$S = \sum [y_i - f(x_i)]^2 = 4.186$$

$$\sigma = \sqrt{\frac{S}{6-2}} = 1.023$$

Observe that the residuals and standard deviation are smaller than in Part (1), indicating a better fit, as expected.

It can be shown that fitting y_i directly (which involves the solution of a transcendental equation) results in $f(x) = 3.614e^{0.5442x}$. The corresponding standard deviation is $\sigma = 1.022$, which is very close to the result in Part (2).

EXAMPLE 3.10
Write a program that fits a polynomial of arbitrary degree m to the data points shown below. Use the program to determine m that best fits this data in the least-squares sense.

x	−0.04	0.93	1.95	2.90	3.83	5.00
y	−8.66	−6.44	−4.36	−3.27	−0.88	0.87
x	5.98	7.05	8.21	9.08	10.09	
y	3.31	4.63	6.19	7.40	8.85	

Solution The program shown below prompts for m. Execution is terminated by entering an invalid character (e.g., the "return" character).

```
#!/usr/bin/python
## example3_10
```

```
from numarray import array
from polyFit import *

xData = array([-0.04,0.93,1.95,2.90,3.83,5.0,      \
               5.98,7.05,8.21,9.08,10.09])
yData = array([-8.66,-6.44,-4.36,-3.27,-0.88,0.87, \
               3.31,4.63,6.19,7.4,8.85])
while 1:
    try:
        m = eval(raw_input(''\nDegree of polynomial ==> ''))
        coeff = polyFit(xData,yData,m)
        print ''Coefficients are:\n'',coeff
        print ''Std. deviation ='',stdDev(coeff,xData,yData)
    except SyntaxError: break
raw_input(''Finished. Press return to exit'')
```

The results are:

```
Degree of polynomial ==> 1
Coefficients are:
[-7.94533287  1.72860425]
Std. deviation = 0.511278836737

Degree of polynomial ==> 2
Coefficients are:
[-8.57005662  2.15121691 -0.04197119]
Std. deviation = 0.310992072855

Degree of polynomial ==> 3
Coefficients are:
[-8.46603423e+00  1.98104441e+00  2.88447008e-03 -2.98524686e-03]
Std. deviation = 0.319481791568

Degree of polynomial ==> 4
Coefficients are:
[ -8.45673473e+00   1.94596071e+00   2.06138060e-02
     -5.82026909e-03   1.41151619e-04]
Std. deviation = 0.344858410479

Degree of polynomial ==>
Finished. Press return to exit
```

3.4 Least-Squares Fit

Because the quadratic $f(x) = -8.5700 + 2.1512x - 0.041971x^2$ produces the smallest standard deviation, it can be considered as the "best" fit to the data. But be warned—the standard deviation is not a reliable measure of the goodness-of-fit. It is always a good idea to plot the data points and $f(x)$ before final determination is made. The plot of our data indicates that the quadratic (solid line) is indeed a reasonable choice for the fitting function.

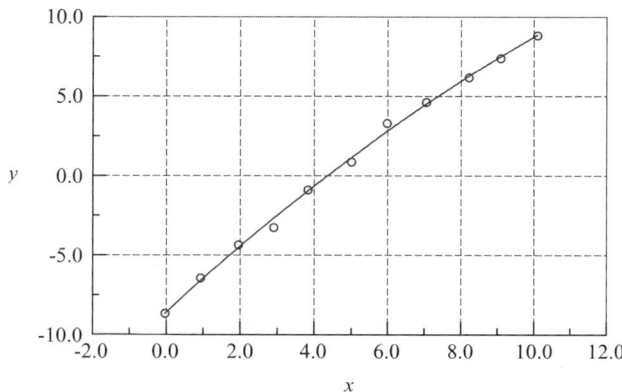

PROBLEM SET 3.2

Instructions Plot the data points and the fitting function whenever appropriate.

1. Show that the straight line obtained by least-squares fit of unweighted data always passes through the point $(\bar{x}, \bar{y})$.

2. Use linear regression to find the line that fits the data

x	-1.0	-0.5	0	0.5	1.0
y	-1.00	-0.55	0.00	0.45	1.00

and determine the standard deviation.

3. Three tensile tests were carried out on an aluminum bar. In each test the strain was measured at the same values of stress. The results were

Stress (MPa)	34.5	69.0	103.5	138.0
Strain (Test 1)	0.46	0.95	1.48	1.93
Strain (Test 2)	0.34	1.02	1.51	2.09
Strain (Test 3)	0.73	1.10	1.62	2.12

where the units of strain are mm/m. Use linear regression to estimate the modulus of elasticity of the bar (modulus of elasticity = stress/strain).

4. Solve Prob. 3 assuming that the third test was performed on an inferior machine, so that its results carry only half the weight of the other two tests.

5. ■ Fit a straight line to the following data and compute the standard deviation.

x	0	0.5	1	1.5	2	2.5
y	3.076	2.810	2.588	2.297	1.981	1.912
x	3	3.5	4	4.5	5	
y	1.653	1.478	1.399	1.018	0.794	

6. ■ The table displays the mass M and average fuel consumption ϕ of motor vehicles manufactured by Ford and Honda in 1999. Fit a straight line $\phi = a + bM$ to the data and compute the standard deviation.

Model	M (kg)	ϕ (km/liter)
Contour	1310	10.2
Crown Victoria	1810	8.1
Escort	1175	11.9
Expedition	2360	5.5
Explorer	1960	6.8
F-150	2020	6.8
Ranger	1755	7.7
Taurus	1595	8.9
Accord	1470	9.8
CR-V	1430	10.2
Civic	1110	13.2
Passport	1785	7.7

7. ■ The relative density ρ of air was measured at various altitudes h. The results were:

h (km)	0	1.525	3.050	4.575	6.10	7.625	9.150
ρ	1	0.8617	0.7385	0.6292	0.5328	0.4481	0.3741

Use a quadratic least-squares fit to determine the relative air density at $h = 10.5$ km. (This problem was solved by interpolation in Prob. 20, Problem Set 3.1.)

8. ■ The kinematic viscosity μ_k of water varies with temperature T as shown in the table. Determine the cubic that best fits the data, and use it to compute μ_k at

3.4 Least-Squares Fit

$T = 10°, 30°, 60°$ and $90°$C. (This problem was solved in Prob. 19, Problem Set 3.1 by interpolation.)

T (°C)	0	21.1	37.8	54.4	71.1	87.8	100
μ_k (10^{-3} m²/s)	1.79	1.13	0.696	0.519	0.338	0.321	0.296

9. ■ Fit a straight line and a quadratic to the data

x	1.0	2.5	3.5	4.0	1.1	1.8	2.2	3.7
y	6.008	15.722	27.130	33.772	5.257	9.549	11.098	28.828

Which is a better fit?

10. ■ The table displays thermal efficiencies of some early steam engines.[8] Determine the polynomial that provides the best fit to the data and use it to predict the thermal efficiency in the year 2000.

Year	Efficiency (%)	Type
1718	0.5	Newcomen
1767	0.8	Smeaton
1774	1.4	Smeaton
1775	2.7	Watt
1792	4.5	Watt
1816	7.5	Woolf compound
1828	12.0	Improved Cornish
1834	17.0	Improved Cornish
1878	17.2	Corliss compound
1906	23.0	Triple expansion

11. The table shows the variation of the relative thermal conductivity k of sodium with temperature T. Find the quadratic that fits the data in the least-squares sense.

T (°C)	79	190	357	524	690
k	1.00	0.932	0.839	0.759	0.693

12. Let $f(x) = ax^b$ be the least-squares fit of the data (x_i, y_i), $i = 0, 1, \ldots, n$, and let $F(x) = \ln a + b \ln x$ be the least-squares fit of $(\ln x_i, \ln y_i)$—see Table 3.3. Prove that

---

[8] Source: Singer, C., Holmyard, E. J., Hall, A. R., and Williams, T. H., *A History of Technology*, Oxford University Press, 1958.

$R_i \approx r_i/y_i$, where the residuals are $r_i = y_i - f(x_i)$ and $R_i = \ln y_i - F(x_i)$. Assume that $r_i \ll y_i$.

13. Determine a and b for which $f(x) = a\sin(\pi x/2) + b\cos(\pi x/2)$ fits the following data in the least-squares sense.

x	−0.5	−0.19	0.02	0.20	0.35	0.50
y	−3.558	−2.874	−1.995	−1.040	−0.068	0.677

14. Determine a and b so that $f(x) = ax^b$ fits the following data in the least-squares sense.

x	0.5	1.0	1.5	2.0	2.5
y	0.49	1.60	3.36	6.44	10.16

15. Fit the function $f(x) = axe^{bx}$ to the data and compute the standard deviation.

x	0.5	1.0	1.5	2.0	2.5
y	0.541	0.398	0.232	0.106	0.052

16. ■ The intensity of radiation of a radioactive substance was measured at half-year intervals. The results were:

t (years)	0	0.5	1	1.5	2	2.5
γ	1.000	0.994	0.990	0.985	0.979	0.977
t (years)	3	3.5	4	4.5	5	5.5
γ	0.972	0.969	0.967	0.960	0.956	0.952

where γ is the relative intensity of radiation. Knowing that radioactivity decays exponentially with time: $\gamma(t) = ae^{-bt}$, estimate the radioactive half-life of the substance.

3.5 Other Methods

Some data are better interpolated by *rational functions* than by polynomials. A rational function $R(x)$ is the quotient of two polynomials:

$$R(x) = \frac{P_m(x)}{Q_n(x)} = \frac{a_0 + a_1 x + a_2 x^2 + \cdots a_m x^m}{b_0 + b_1 x + b_2 x^2 + \cdots b_n x^n}$$

3.5 Other Methods

Since $R(x)$ is a ratio, its coefficients can be scaled so that one of the coefficients (usually b_0) is unity. That leaves $m + n + 1$ undetermined coefficients which can be computed by passing $R(x)$ through $m + n + 1$ data points.

The following data set is an ideal candidate for rational function interpolation:

x	0	0.6	0.8	0.95
y	0	1.3764	3.0777	12.7062

From the plot of the data points (open circles in Fig. 3.8) it appears that y tends to infinity near $x = 1$. That makes the data ill suited for polynomial interpolation. However, the rational function interpolant

$$f(x) = \frac{1.3668x - 0.7515x^2}{1 - 1.0013x}$$

(the solid line in Fig. 3.8) has no trouble handling the singularity.

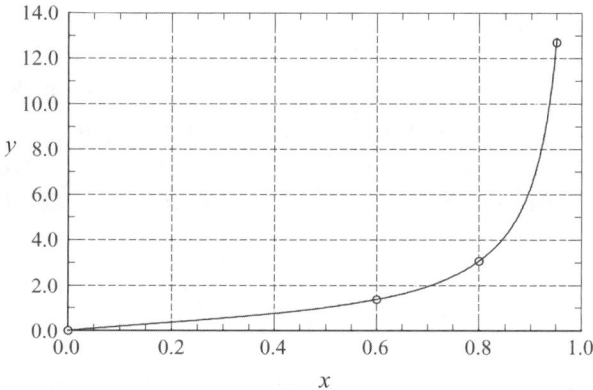

Figure 3.8. Rational function interpolation.

4 Roots of Equations

> Find the solutions of $f(x) = 0$, where the function f is given

4.1 Introduction

A common problem encountered in engineering analysis is this: given a function $f(x)$, determine the values of x for which $f(x) = 0$. The solutions (values of x) are known as the *roots* of the equation $f(x) = 0$, or the *zeroes* of the function $f(x)$.

Before proceeding further, it might be helpful to review the concept of a *function*. The equation

$$y = f(x)$$

contains three elements: an input value x, an output value y, and the rule f for computing y. The function is said to be given if the rule f is specified. In numerical computing the rule is invariably a computer algorithm. It may be a function statement, such as

$$f(x) = \cosh(x)\cos(x) - 1$$

or a complex procedure containing hundreds or thousands of lines of code. As long as the algorithm produces an output y for each input x, it qualifies as a function.

The roots of equations may be real or complex. The complex roots are seldom computed, since they rarely have physical significance. An exception is the polynomial equation

$$a_0 + a_1 x + a_1 x^2 + \cdots + a_n x^n = 0$$

where the complex roots may be meaningful (as in the analysis of damped vibrations, for example). For the time being, we will concentrate on finding the real roots of equations. Complex zeroes of polynomials are treated near the end of this chapter.

In general, an equation may have any number of (real) roots, or no roots at all. For example,

$$\sin x - x = 0$$

has a single root, namely $x = 0$, whereas

$$\tan x - x = 0$$

has an infinite number of roots ($x = 0, \pm 4.493, \pm 7.725, \ldots$).

All methods of finding roots are iterative procedures that require a starting point, i.e., an estimate of the root. This estimate can be crucial; a bad starting value may fail to converge, or it may converge to the "wrong" root (a root different from the one sought). There is no universal recipe for estimating the value of a root. If the equation is associated with a physical problem, then the context of the problem (physical insight) might suggest the approximate location of the root. Otherwise, a systematic numerical search for the roots can be carried out. One such search method is described in the next article. The roots can also be located visually by plotting the function.

It is highly advisable to go a step further and *bracket* the root (determine its lower and upper bounds) before passing the problem to a root finding algorithm. Prior bracketing is, in fact, mandatory in the methods described in this chapter.

4.2 Incremental Search Method

The approximate locations of the roots are best determined by plotting the function. Often a very rough plot, based on a few points, is sufficient to give us reasonable starting values. Another useful tool for detecting and bracketing roots is the incremental search method. It can also be adapted for computing roots, but the effort would not be worthwhile, since other methods described in this chapter are more efficient for that.

The basic idea behind the incremental search method is simple: if $f(x_1)$ and $f(x_2)$ have opposite signs, then there is at least one root in the interval (x_1, x_2). If the interval is small enough, it is likely to contain a single root. Thus the zeroes of $f(x)$ can be detected by evaluating the function at intervals Δx and looking for change in sign.

There are several potential problems with the incremental search method:

- It is possible to miss two closely spaced roots if the search increment Δx is larger than the spacing of the roots.
- A double root (two roots that coincide) will not be detected.
- Certain singularities of $f(x)$ can be mistaken for roots. For example, $f(x) = \tan x$ changes sign at $x = \pm \frac{1}{2} n\pi$, $n = 1, 3, 5, \ldots$, as shown in Fig. 4.1. However, these locations are not true zeroes, since the function does not cross the x-axis.

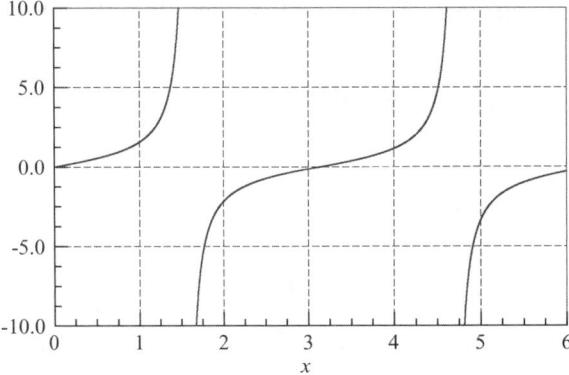

Figure 4.1. Plot of tan x.

■ rootsearch

This function searches for a zero of the user-supplied function $f(x)$ in the interval (a, b) in increments of dx. It returns the bounds (x1,x2) of the root if the search was successful; x1 = x2 = None indicates that no roots were detected. After the first root (the root closest to a) has been detected, rootsearch can be called again with a replaced by x2 in order to find the next root. This can be repeated as long as rootsearch detects a root.

```
## module rootsearch
''' x1,x2 = rootsearch(f,a,b,dx).
    Searches the interval (a,b) in increments dx for
    the bounds (x1,x2) of the smallest root of f(x).
    Returns x1 = x2 = None if no roots were detected.
'''
def rootsearch(f,a,b,dx):
    x1 = a; f1 = f(a)
    x2 = a + dx; f2 = f(x2)
    while f1*f2 > 0.0:
        if x1 >= b: return None,None
        x1 = x2; f1 = f2
        x2 = x1 + dx; f2 = f(x2)
    else:
        return x1,x2
```

EXAMPLE 4.1

Use incremental search with $\Delta x = 0.2$ to bracket the smallest positive zero of $f(x) = x^3 - 10x^2 + 5$.

Solution We evaluate $f(x)$ at intervals $\Delta x = 0.2$, staring at $x = 0$, until the function changes its sign (value of the function is of no interest to us; only its sign is relevant). This procedure yields the following results:

x	$f(x)$
0.0	5.000
0.2	4.608
0.4	3.464
0.6	1.616
0.8	-0.888

From the sign change of the function we conclude that the smallest positive zero lies between $x = 0.6$ and $x = 0.8$.

4.3 Method of Bisection

After a root of $f(x) = 0$ has been bracketed in the interval (x_1, x_2), several methods can be used to close in on it. The method of bisection accomplishes this by successively halving the interval until it becomes sufficiently small. This technique is also known as the *interval halving method*. Bisection is not the fastest method available for computing roots, but it is the most reliable. Once a root has been bracketed, bisection will always close in on it.

The method of bisection uses the same principle as incremental search: if there is a root in the interval (x_1, x_2), then $f(x_1) \cdot f(x_2) < 0$. In order to halve the interval, we compute $f(x_3)$, where $x_3 = \frac{1}{2}(x_1 + x_2)$ is the midpoint of the interval. If $f(x_2) \cdot f(x_3) < 0$, then the root must be in (x_2, x_3) and we record this by replacing the original bound x_1 by x_3. Otherwise, the root lies in (x_1, x_3), in which case x_2 is replaced by x_3. In either case, the new interval (x_1, x_2) is half the size of the original interval. The bisection is repeated until the interval has been reduced to a small value ε, so that

$$|x_2 - x_1| \leq \varepsilon$$

It is easy to compute the number of bisections required to reach a prescribed ε. The original interval Δx is reduced to $\Delta x/2$ after one bisection, $\Delta x/2^2$ after two bisections, and after n bisections it is $\Delta x/2^n$. Setting $\Delta x/2^n = \varepsilon$ and solving for n, we get

$$n = \frac{\ln(|\Delta x|/\varepsilon)}{\ln 2} \qquad (4.1)$$

■ bisect

This function uses the method of bisection to compute the root of f(x) = 0 that is known to lie in the interval (x1,x2). The number of bisections n required to reduce the interval to `tol` is computed from Eq. (4.1). By setting `switch = 1`, we force the routine to check whether the magnitude of `f(x)` decreases with each interval halving. If it does not, something may be wrong (probably the "root" is not a root at all, but a singularity) and `root = None` is returned. Since this feature is not always desirable, the default value is `switch = 0`. The function `error.err`, which we use to terminate a program, is listed in Art. 1.6.

```
## module bisect
''' root = bisect(f,x1,x2,switch=0,tol=1.0e-9).
    Finds a root of f(x) = 0 by bisection.
    The root must be bracketed in (x1,x2).
    Setting switch = 1 returns root = None if
    f(x) increases as a result of a bisection.
'''
from math import log,ceil
import error

def bisect(f,x1,x2,switch=0,epsilon=1.0e-9):
    f1 = f(x1)
    if f1 == 0.0: return x1
    f2 = f(x2)
    if f2 == 0.0: return x2
    if f1*f2 > 0.0: error.err('Root is not bracketed')
    n = ceil(log(abs(x2 - x1)/epsilon)/log(2.0))
    for i in range(n):
        x3 = 0.5*(x1 + x2); f3 = f(x3)
        if (switch == 1) and (abs(f3) >abs(f1)) \
                         and (abs(f3) > abs(f2)):
            return None
        if f3 == 0.0: return x3
        if f2*f3 < 0.0:
            x1 = x3; f1 = f3
        else:
            x2 =x3; f2 = f3
    return (x1 + x2)/2.0
```

EXAMPLE 4.2
Use bisection to find the root of $f(x) = x^3 - 10x^2 + 5 = 0$ that lies in the interval $(0.6, 0.8)$.

Solution The best way to implement the method is to use the following table. Note that the interval to be bisected is determined by the sign of $f(x)$, not its magnitude.

4.3 Method of Bisection

x	$f(x)$	Interval
0.6	1.616	—
0.8	−0.888	(0.6, 0.8)
(0.6 + 0.8)/2 = 0.7	0.443	(0.7, 0.8)
(0.8 + 0.7)/2 = 0.75	−0.203	(0.7, 0.75)
(0.7 + 0.75)/2 = 0.725	0.125	(0.725, 0.75)
(0.75 + 0.725)/2 = 0.7375	−0.038	(0.725, 0.7375)
(0.725 + 0.7375)/2 = 0.73125	0.044	(0.7375, 0.73125)
(0.7375 + 0.73125)/2 = 0.73438	0.003	(0.7375, 0.73438)
(0.7375 + 0.73438)/2 = 0.73594	−0.017	(0.73438, 0.73594)
(0.73438 + 0.73594)/2 = 0.73516	−0.007	(0.73438, 0.73516)
(0.73438 + 0.73516)/2 = 0.73477	−0.002	(0.73438, 0.73477)
(0.73438 + 0.73477)/2 = 0.73458	0.000	—

The final result $x = 0.7346$ is correct within four decimal places.

EXAMPLE 4.3
Find *all* the zeros of $f(x) = x - \tan x$ in the interval (0, 20) by the method of bisection. Utilize the functions rootsearch and bisect.

Solution Note that $\tan x$ is singular and changes sign at $x = \pi/2, 3\pi/2, \ldots$. To prevent bisect from mistaking these point for roots, we set switch = 1. The closeness of roots to the singularities is another potential problem that can be alleviated by using small Δx in rootsearch. Choosing $\Delta x = 0.01$, we arrive at the following program:

```
#!/usr/bin/python
## example4_3
from math import tan
from rootsearch import *
from bisect import *

def f(x): return x - tan(x)

a,b,dx = (0.0, 20.0, 0.01)
print ''The roots are:''
while 1:
    x1,x2 = rootsearch(f,a,b,dx)
```

```
    if x1 != None:
        a = x2
        root = bisect(f,x1,x2,1)
        if root != None: print root
    else:
        print ''\nDone''
        break
raw_input(''Press return to exit'')
```

The output from the program is:

```
The roots are:
0.0
4.4934094581
7.72525183707
10.9041216597
14.0661939129
17.2207552722

Done
```

4.4 Brent's Method

Brent's method[9] combines bisection and quadratic interpolation into an efficient root-finding algorithm. In most problems the method is much faster than bisection alone, but it can become sluggish if the function is not smooth. It is the recommended method of root solving if the derivative of the function is difficult or impossible to compute.

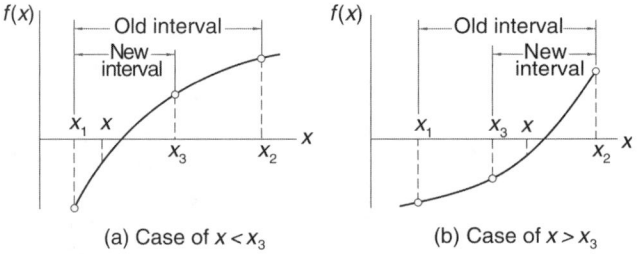

Figure 4.2. Inverse quadratic iteration.

---

[9] Brent, R. P., *Algorithms for Minimization without Derivatives*, Prentice-Hall, 1973.

4.4 Brent's Method

Brent's method assumes that a root of $f(x) = 0$ has been initially bracketed in the interval (x_1, x_2). The root-finding process starts with a bisection step that halves the interval to either (x_1, x_3) or (x_3, x_2), where $x_3 = (x_1 + x_2)/2$, as shown in Figs. 4.2(a) and (b). In the course of bisection we had to compute $f_1 = f(x_1)$, $f_2 = f(x_2)$ and $f_3 = f(x_3)$, so that we now know three points on the $f(x)$ curve (the open circles in the figure). These points allow us to carry out the next iteration of the root by *inverse quadratic interpolation* (viewing x as a quadratic function of f). If the result x of the interpolation falls inside the latest bracket (as is the case in Figs. 4.2) we accept the result. Otherwise, another round of bisection is applied.

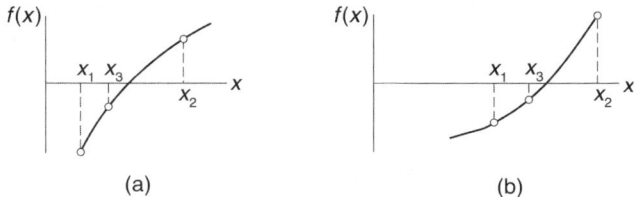

(a)   (b)

Figure 4.3. Relabeling roots after an iteration.

The next step is to relabel x as x_3 and rename the limits of the new interval x_1 and x_2 ($x_1 < x_3 < x_2$), as indicated in Figs. 4.3. We have now recovered the original sequencing of points in Figs. 4.2, but the interval (x_1, x_2) containing the root has been reduced. This completes the first iteration cycle. In the next cycle another inverse quadratic interpolation is attempted and the process is repeated until the convergence criterion $|x - x_3| < \varepsilon$ is satisfied, where ε is a prescribed error tolerance.

The inverse quadratic interpolation is carried out with Lagrange's three-point interpolant described in Section 3.2. Interchanging the roles of x and f, we have

$$x(f) = \frac{(f - f_2)(f - f_3)}{(f_1 - f_2)(f_1 - f_3)} x_1 + \frac{(f - f_1)(f - f_3)}{(f_2 - f_1)(f_2 - f_3)} x_2 + \frac{(f - f_1)(f - f_2)}{(f_3 - f_1)(f_3 - f_2)} x_3$$

Setting $f = 0$ and simplifying, we obtain for the estimate of the root

$$x = x(0) = -\frac{f_2 f_3 x_1 (f_2 - f_3) + f_3 f_1 x_2 (f_3 - f_1) + f_1 f_2 x_3 (f_1 - f_2)}{(f_1 - f_2)(f_2 - f_3)(f_3 - f_1)}$$

The change in the root is

$$\Delta x = x - x_3 = f_3 \frac{x_3(f_1 - f_2)(f_2 - f_3 + f_1) + f_2 x_1 (f_2 - f_3) + f_1 x_2 (f_3 - f_1)}{(f_2 - f_1)(f_3 - f_1)(f_2 - f_3)} \quad (4.2)$$

brent

The function brent listed below is a simplified version of the algorithm proposed by Brent. It omits some of Brent's safeguards against slow convergence; it also uses a less sophisticated convergence criterion.

```
## module brent
''' root = brent(f,a,b,tol=1.0e-9).
    Finds root of f(x) = 0 by combining quadratic interpolation
    with bisection (simplified Brent's method).
    The root must be bracketed in (a,b).
    Calls user-supplied function f(x).
'''
import error

def brent(f,a,b,tol=1.0e-9):
    x1 = a; x2 = b;
    f1 = f(x1)
    if f1 == 0.0: return x1
    f2 = f(x2)
    if f2 == 0.0: return x2
    if f1*f2 > 0.0: error.err('Root is not bracketed')
    x3 = 0.5*(a + b)
    for i in range(30):
        f3 = f(x3)
        if abs(f3) < tol: return x3
      # Tighten the brackets on the root
        if f1*f3 < 0.0: b = x3
        else: a = x3
        if (b - a) < tol*max(abs(b),1.0): return 0.5*(a + b)
      # Try quadratic interpolation
        denom = (f2 - f1)*(f3 - f1)*(f2 - f3)
        numer = x3*(f1 - f2)*(f2 - f3 + f1)    \
              + f2*x1*(f2 - f3) + f1*x2*(f3 - f1)
      # If division by zero, push x out of bounds
        try: dx = f3*numer/denom
        except ZeroDivisionError: dx = b - a
        x = x3 + dx
      # If iterpolation goes out of bounds, use bisection
        if (b - x)*(x - a) < 0.0:
```

4.4 Brent's Method

```
            dx = 0.5*(b - a)
            x = a + dx
      # Let x3 <-- x & choose new x1 and x2 so that x1 < x3 < x2
        if x < x3:
            x2 = x3; f2 = f3
        else:
            x1 = x3; f1 = f3
        x3 = x
    print 'Too many iterations in brent'
```

EXAMPLE 4.4
Determine the root of $f(x) = x^3 - 10x^2 + 5 = 0$ that lies in $(0.6, 0.8)$ with Brent's method.

Solution
Bisection The starting points are

$$x_1 = 0.6 \quad f_1 = 0.6^3 - 10(0.6)^2 + 5 = 1.616$$
$$x_2 = 0.8 \quad f_2 = 0.8^3 - 10(0.8)^2 + 5 = -0.888$$

Bisection yields the point

$$x_3 = 0.7 \quad f_3 = 0.7^3 - 10(0.7)^2 + 5 = 0.443$$

By inspecting the signs of f we conclude that the new brackets on the root are $(x_3, x_2) = (0.7, 0.8)$.

First interpolation cycle The numerator of the quotient in Eq. (4.2) is

$$\text{num} = x_3(f_1 - f_2)(f_2 - f_3 + f_1) + f_2 x_1(f_2 - f_3) + f_1 x_2(f_3 - f_1)$$
$$= 0.7(1.616 + 0.888)(-0.888 - 0.443 + 1.616)$$
$$\quad -0.888(0.6)(-0.888 - 0.443) + 1.616(0.8)(0.443 - 1.616)$$
$$= -0.307\,75$$

and the denominator is

$$\text{den} = (f_2 - f_1)(f_3 - f_1)(f_2 - f_3)$$
$$= (-0.888 - 1.616)(0.443 - 1.616)(-0.888 - 0.443) = -3.9094$$

Therefore,

$$\Delta x = f_3 \frac{\text{num}}{\text{den}} = 0.443 \frac{(-0.307\,75)}{(-3.9094)} = 0.034\,87$$

and

$$x = x_3 + \Delta x = 0.7 + 0.034\,87 = 0.734\,87$$

Since the result is within the established brackets, we accept it.

Relabel points As $x > x_3$, the points are relabeled as illustrated in Figs. 4.2(b) and 4.3(b):

$$x_1 \leftarrow x_3 = 0.7$$
$$f_1 \leftarrow f_3 = 0.443$$
$$x_3 \leftarrow x = 0.734\,87$$
$$f_3 = 0.734\,87^3 - 10(0.734\,87)^2 + 5 = -0.003\,48$$

The new brackets on the root are $(x_1, x_3) = (0.7, 0.734\,87)$

Second interpolation cycle Applying the interpolation in Eq. (4.2) again, we obtain (skipping the arithmetical details)

$$\Delta x = -0.000\,27$$
$$x = x_3 + \Delta x = 0.734\,87 - 0.000\,27 = 0.734\,60$$

Again x falls within the latest brackets, so the result is acceptable. At this stage, x is correct to five decimal places.

EXAMPLE 4.5
Compute the zero of

$$f(x) = x|\cos x| - 1$$

that lies in the interval $(0, 4)$ with Brent's method.

4.5 Newton–Raphson Method

Solution

The plot of $f(x)$ shows that this is a rather nasty function within the specified interval, containing a slope discontinuity and two local maxima. The sensible approach is to avoid the potentially troublesome regions of the function by bracketing the root as tightly as possible from a visual inspection of the plot. In this case, the interval $(a, b) = (2.0, 2.2)$ would be a good starting point for Brent's algorithm.

Is Brent's method robust enough to handle the problem with the original brackets $(0, 4)$? Well, here is the program and its output:

```
#!/usr/bin/python
## example4_5
from math import cos
from brent import *

def f(x): return x*abs(cos(x)) - 1.0

print ''root ='',brent(f,0.0,4.0)
raw_input(''Press return to exit'')

root = 2.0739328091
```

The result was obtained in only five iterations.

4.5 Newton–Raphson Method

The Newton–Raphson algorithm is the best-known method of finding roots for a good reason: it is simple and fast. The only drawback of the method is that it uses the derivative $f'(x)$ of the function as well as the function $f(x)$ itself. Therefore, the Newton–Raphson method is usable only in problems where $f'(x)$ can be readily computed.

The Newton–Raphson formula can be derived from the Taylor series expansion of $f(x)$ about x:

$$f(x_{i+1}) = f(x_i) + f'(x_i)(x_{i+1} - x_i) + O(x_{i+1} - x_i)^2 \tag{a}$$

where $\mathcal{O}(z)$ is to be read as "of the order of z"—see Appendix A1. If x_{i+1} is a root of $f(x) = 0$, Eq. (a) becomes

$$0 = f(x_i) + f'(x_i)(x_{i+1} - x_i) + O(x_{i+1} - x_i)^2 \tag{b}$$

Assuming that x_i is a close to x_{i+1}, we can drop the last term in Eq. (b) and solve for x_{i+1}. The result is the Newton–Raphson formula

$$x_{i+1} = x_i - \frac{f(x_i)}{f'(x_i)} \tag{4.3}$$

If x denotes the true value of the root, the error in x_i is $E_i = x - x_i$. It can be shown that if x_{i+1} is computed from Eq. (4.3), the corresponding error is

$$E_{i+1} = -\frac{f''(x_i)}{2 f'(x_i)} E_i^2$$

indicating that Newton–Raphson method converges *quadratically* (the error is the square of the error in the previous step). As a consequence, the number of significant figures is roughly doubled in every iteration, provided that x_i is close to the root.

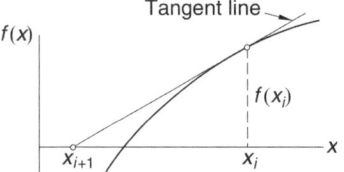

Figure 4.4. Graphical interpretation of the Newton–Raphson formula.

A graphical depiction of the Newton–Raphson formula is shown in Fig. 4.4. The formula approximates $f(x)$ by the straight line that is tangent to the curve at x_i. Thus x_{i+1} is at the intersection of the x-axis and the tangent line.

The algorithm for the Newton–Raphson method is simple: it repeatedly applies Eq. (4.3), starting with an initial value x_0, until the convergence criterion

$$|x_{i+1} - x_1| < \varepsilon$$

is reached, ε being the error tolerance. Only the latest value of x has to be stored. Here is the algorithm:

4.5 Newton–Raphson Method

1. Let x be a guess for the root of $f(x) = 0$.
2. Compute $\Delta x = -f(x)/f'(x)$.
3. Let $x \leftarrow x + \Delta x$ and repeat steps 2–3 until $|\Delta x| < \varepsilon$.

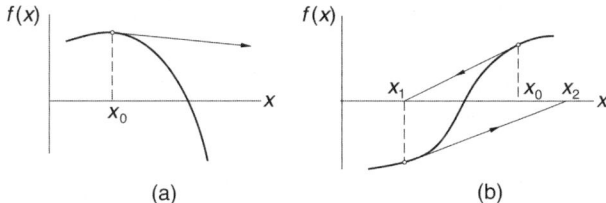

Figure 4.5. Examples where the Newton–Raphson method diverges.

Although the Newton–Raphson method converges fast near the root, its global convergence characteristics are poor. The reason is that the tangent line is not always an acceptable approximation of the function, as illustrated in the two examples in Fig. 4.5. But the method can be made nearly fail-safe by combining it with bisection, as in Brent's method.

∎ newtonRaphson

The following *safe version* of the Newton–Raphson method assumes that the root to be computed is initially bracketed in (a,b). The midpoint of the bracket is used as the initial guess of the root. The brackets are updated after each iteration. If a Newton–Raphson iteration does not stay within the brackets, it is disregarded and replaced with bisection. Since newtonRaphson uses the function f(x) as well as its derivative, function routines for both (denoted by f and df in the listing) must be provided by the user.

```
## module newtonRaphson
''' root = newtonRaphson(f,df,a,b,tol=1.0e-9).
    Finds a root of f(x) = 0 by combining the Newton-Raphson
    method with bisection. The root must be bracketed in (a,b).
    Calls user-supplied functions f(x) and its derivative df(x).
'''
def newtonRaphson(f,df,a,b,tol=1.0e-9):
    import error
    fa = f(a)
    if fa == 0.0: return a
    fb = f(b)
```

```
        if fb == 0.0: return b
        if fa*fb > 0.0: error.err('Root is not bracketed')
        x = 0.5*(a + b)
        for i in range(30):
            fx = f(x)
            if abs(fx) < tol: return x
          # Tighten the brackets on the root
            if fa*fx < 0.0:
                b = x
            else:
                a = x; fa = fx
          # Try a Newton-Raphson step
            dfx = df(x)
          # If division by zero, push x out of bounds
            try: dx = -fx/dfx
            except ZeroDivisionError: dx = b - a
            x = x + dx
          # If the result is outside the brackets, use bisection
            if (b - x)*(x - a) < 0.0:
                dx = 0.5*(b-a)
                x = a + dx
          # Check for convergence
            if abs(dx) < tol*max(abs(b),1.0): return x
        print 'Too many iterations in Newton-Raphson'
```

EXAMPLE 4.6

A root of $f(x) = x^3 - 10x^2 + 5 = 0$ lies close to $x = 0.7$. Compute this root with the Newton–Raphson method.

Solution The derivative of the function is $f'(x) = 3x^2 - 20x$, so that the Newton–Raphson formula in Eq. (4.3) is

$$x \leftarrow x - \frac{f(x)}{f'(x)} = x - \frac{x^3 - 10x^2 + 5}{3x^2 - 20x} = \frac{2x^3 - 10x^2 - 5}{x(3x - 20)}$$

It takes only two iterations to reach five decimal place accuracy:

$$x \leftarrow \frac{2(0.7)^3 - 10(0.7)^2 - 5}{0.7\,[3(0.7) - 20]} = 0.735\,36$$

$$x \leftarrow \frac{2(0.735\,36)^3 - 10(0.735\,36)^2 - 5}{0.735\,36\,[3(0.735\,36) - 20]} = 0.734\,60$$

4.5 Newton–Raphson Method

EXAMPLE 4.7
Find the smallest positive zero of
$$f(x) = x^4 - 6.4x^3 + 6.45x^2 + 20.538x - 31.752$$

Solution

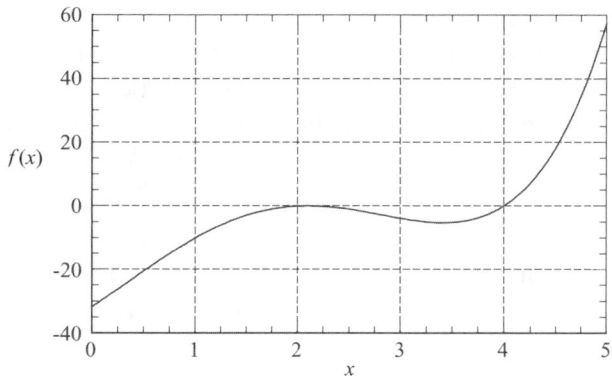

Inspecting the plot of the function, we suspect that the smallest positive zero is a double root near $x = 2$. Bisection and Brent's method would not work here, since they depend on the function changing its sign at the root. The same argument applies to the function newtonRaphson. But there no reason why the unrefined version of the Newton–Raphson method should not succeed. We used the following program, which prints the number of iterations in addition to the root:

```
#!/usr/bin/python
## example4_7

def f(x): return x**4 - 6.4*x**3 + 6.45*x**2 + 20.538*x - 31.752
def df(x): return 4.0*x**3 - 19.2*x**2 + 12.9*x + 20.538

def newtonRaphson(x,tol=1.0e-9):
    for i in range(30):
        dx = -f(x)/df(x)
        x = x + dx
        if abs(dx) < tol: return x,i
    print 'Too many iterations\n'

root,numIter = newtonRaphson(2.0)
print 'Root =',root
print 'Number of iterations =',numIter
raw_input(''Press return to exit'')
```

The output is

```
Root = 2.09999998403
Number of iterations = 23
```

The true value of the root is $x = 2.1$. It can be shown that near a multiple root the convergence of the Newton–Raphson method is linear, rather than quadratic, which explains the large number of iterations. Convergence to a multiple root can be speeded up by replacing the Newton–Raphson formula in Eq. (4.3) with

$$x_{i+1} = x_i - m\frac{f(x_i)}{f'(x_i)}$$

where m is the multiplicity of the root ($m = 2$ in this problem). After making the change in the above program, we obtained the result in only 5 iterations.

4.6 Systems of Equations

Introduction

Up to this point, we confined our attention to solving the single equation $f(x) = 0$. Let us now consider the n-dimensional version of the same problem, namely

$$\mathbf{f(x)} = \mathbf{0}$$

or, using scalar notation

$$f_1(x_1, x_2, \ldots, x_n) = 0$$
$$f_2(x_1, x_2, \ldots, x_n) = 0$$
$$\vdots$$
$$f_n(x_1, x_2, \ldots, x_n) = 0$$
(4.4)

The solution of n simultaneous, nonlinear equations is a much more formidable task than finding the root of a single equation. The trouble is the lack of a reliable method for bracketing the solution vector **x**. Therefore, we cannot provide the solution algorithm with a guaranteed good starting value of **x**, unless such a value is suggested by the physics of the problem.

The simplest and the most effective means of computing **x** is the Newton–Raphson method. It works well with simultaneous equations, provided that it is supplied with a good starting point. There are other methods that have better global convergence characteristics, but all of then are variants of the Newton–Raphson method.

Newton–Raphson Method

In order to derive the Newton–Raphson method for a system of equations, we start with the Taylor series expansion of $f_i(\mathbf{x})$ about the point $\mathbf{x}$:

$$f_i(\mathbf{x} + \Delta\mathbf{x}) = f_i(\mathbf{x}) + \sum_{j=1}^{n} \frac{\partial f_i}{\partial x_j} \Delta x_j + O(\Delta x^2) \tag{4.5a}$$

Dropping terms of order Δx^2, we can write Eq. (4.5a) as

$$\mathbf{f}(\mathbf{x} + \Delta\mathbf{x}) = \mathbf{f}(\mathbf{x}) + \mathbf{J}(\mathbf{x})\Delta\mathbf{x} \tag{4.5b}$$

where $\mathbf{J}(\mathbf{x})$ is the *Jacobian matrix* (of size $n \times n$) made up of the partial derivatives

$$J_{ij} = \frac{\partial f_i}{\partial x_j} \tag{4.6}$$

Note that Eq. (4.5b) is a linear approximation (vector $\Delta\mathbf{x}$ being the variable) of the vector-valued function $\mathbf{f}$ in the vicinity of point $\mathbf{x}$.

Let us now assume that $\mathbf{x}$ is the current approximation of the solution of $\mathbf{f}(\mathbf{x}) = \mathbf{0}$, and let $\mathbf{x} + \Delta\mathbf{x}$ be the improved solution. To find the correction $\Delta\mathbf{x}$, we set $\mathbf{f}(\mathbf{x} + \Delta\mathbf{x}) = \mathbf{0}$ in Eq. (4.5b). The result is a set of linear equations for $\Delta\mathbf{x}$:

$$\mathbf{J}(\mathbf{x})\Delta\mathbf{x} = -\mathbf{f}(\mathbf{x}) \tag{4.7}$$

The following steps constitute the Newton–Raphson method for simultaneous, nonlinear equations:

1. Estimate the solution vector $\mathbf{x}$.
2. Evaluate $\mathbf{f}(\mathbf{x})$.
3. Compute the Jacobian matrix $\mathbf{J}(\mathbf{x})$ from Eq. (4.6).
4. Set up the simultaneous equations in Eq. (4.7) and solve for $\Delta\mathbf{x}$.
5. Let $\mathbf{x} \leftarrow \mathbf{x} + \Delta\mathbf{x}$ and repeat steps 2–5.

The above process is continued until $|\Delta\mathbf{x}| < \varepsilon$, where ε is the error tolerance. As in the one-dimensional case, success of the Newton–Raphson procedure depends entirely on the initial estimate of $\mathbf{x}$. If a good starting point is used, convergence to the solution is very rapid. Otherwise, the results are unpredictable.

Because analytical derivation of each $\partial f_i / \partial x_j$ can be difficult or impractical, it is preferable to let the computer calculate the partial derivatives from the finite difference approximation

$$\frac{\partial f_i}{\partial x_j} \approx \frac{f_i(\mathbf{x} + \mathbf{e}_j h) - f_i(\mathbf{x})}{h} \qquad (4.8)$$

where h is a small increment and $\mathbf{e}_j$ represents a unit vector in the direction of x_j. This formula can be obtained from Eq. (4.5a) after dropping the terms of order Δx^2 and setting $\Delta \mathbf{x} = \mathbf{e}_j h$. We get away with the approximation in Eq. (4.8) because the Newton–Raphson method is rather insensitive to errors in $\mathbf{J}(\mathbf{x})$. By using this approximation, we also avoid the tedium of typing the expressions for $\partial f_i/\partial x_j$ into the computer code.

■ newtonRaphson2

This function is an implementation of the Newton–Raphson method. The nested function jacobian computes the Jacobian matrix from the finite difference approximation in Eq. (4.8). The simultaneous equations in Eq. (4.7) are solved by Gauss elimination with row pivoting using the function gaussPivot. listed in Section 2.5. The function subroutine f that returns the array $\mathbf{f}(\mathbf{x})$ must be supplied by the user.

```
## module newtonRaphson2
''' soln = newtonRaphson2(f,x,tol=1.0e-9).
    Solves the simultaneous equations f(x) = 0 by
    the Newton-Raphson method using {x} as the initial
    guess. Note that {f} and {x} are vectors.
'''
from numarray import zeros,Float64,dot,sqrt
from gaussPivot import *

def newtonRaphson2(f,x,tol=1.0e-9):

    def jacobian(f,x):
        h = 1.0e-4
        n = len(x)
        jac = zeros((n,n),type=Float64)
        f0 = f(x)
        for i in range(n):
            temp = x[i]
            x[i] = temp + h
            f1 = f(x)
            x[i] = temp
```

4.6 Systems of Equations

```
        jac[:,i] = (f1 - f0)/h
    return jac,f0

for i in range(30):
    jac,f0 = jacobian(f,x)
    if sqrt(dot(f0,f0)/len(x)) < tol: return x
    dx = gaussPivot(jac,-f0)
    x = x + dx
    if sqrt(dot(dx,dx)) < tol*max(abs(x),1.0): return x
print 'Too many iterations'
```

Note that the Jacobian matrix $J(x)$ is recomputed in each iterative loop. Since each calculation of $J(x)$ involves $n+1$ evaluations of $f(x)$ (n is the number of equations), the expense of computation can be high depending on n and the complexity of $f(x)$. It is often possible to save computer time by neglecting the changes in the Jacobian matrix between iterations, thus computing $J(x)$ only once. This will work provided that the initial x is sufficiently close to the solution.

EXAMPLE 4.8
Determine the points of intersection between the circle $x^2 + y^2 = 3$ and the hyperbola $xy = 1$.

Solution The equations to be solved are

$$f_1(x, y) = x^2 + y^2 - 3 = 0 \tag{a}$$

$$f_2(x, y) = xy - 1 = 0 \tag{b}$$

The Jacobian matrix is

$$J(x, y) = \begin{bmatrix} \partial f_1/\partial x & \partial f_1/\partial y \\ \partial f_2/\partial x & \partial f_2/\partial y \end{bmatrix} = \begin{bmatrix} 2x & 2y \\ y & x \end{bmatrix}$$

Thus the linear equations $J(x)\Delta x = -f(x)$ associated with the Newton–Raphson method are

$$\begin{bmatrix} 2x & 2y \\ y & x \end{bmatrix} \begin{bmatrix} \Delta x \\ \Delta y \end{bmatrix} = \begin{bmatrix} -x^2 - y^2 + 3 \\ -xy + 1 \end{bmatrix} \tag{c}$$

By plotting the circle and the hyperbola, we see that there are four points of intersection. It is sufficient, however, to find only one of these points, as the others can be deduced from symmetry. From the plot we also get a rough estimate of the coordinates of an intersection point: $x = 0.5$, $y = 1.5$, which we use as the starting values.

Roots of Equations

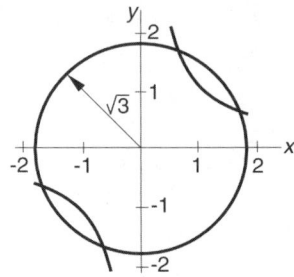

The computations then proceed as follows.

First iteration Substituting $x = 0.5$, $y = 1.5$ in Eq. (c), we get

$$\begin{bmatrix} 1.0 & 3.0 \\ 1.5 & 0.5 \end{bmatrix} \begin{bmatrix} \Delta x \\ \Delta y \end{bmatrix} = \begin{bmatrix} 0.50 \\ 0.25 \end{bmatrix}$$

the solution of which is $\Delta x = \Delta y = 0.125$. Therefore, the improved coordinates of the intersection point are

$$x = 0.5 + 0.125 = 0.625 \qquad y = 1.5 + 0.125 = 1.625$$

Second iteration Repeating the procedure using the latest values of x and y, we obtain

$$\begin{bmatrix} 1.250 & 3.250 \\ 1.625 & 0.625 \end{bmatrix} \begin{bmatrix} \Delta x \\ \Delta y \end{bmatrix} = \begin{bmatrix} -0.031250 \\ -0.015625 \end{bmatrix}$$

which yields $\Delta x = \Delta y = -0.00694$. Thus

$$x = 0.625 - 0.006\,94 = 0.618\,06 \qquad y = 1.625 - 0.006\,94 = 1.618\,06$$

Third iteration Substitution of the latest x and y into Eq. (c) yields

$$\begin{bmatrix} 1.236\,12 & 3.23612 \\ 1.618\,06 & 0.61806 \end{bmatrix} \begin{bmatrix} \Delta x \\ \Delta y \end{bmatrix} = \begin{bmatrix} -0.000\,116 \\ -0.000\,058 \end{bmatrix}$$

The solution is $\Delta x = \Delta y = -0.00003$, so that

$$x = 0.618\,06 - 0.000\,03 = 0.618\,03$$

$$y = 1.618\,06 - 0.000\,03 = 1.618\,03$$

Subsequent iterations would not change the results within five significant figures. Therefore, the coordinates of the four intersection points are

$$\pm(0.618\,03, 1.618\,03) \text{ and } \pm (1.618\,03, 0.618\,03)$$

4.6 Systems of Equations

Alternate solution If there are only a few equations, it may be possible to eliminate all but one of the unknowns. Then we would be left with a single equation which can be solved by the methods described in Sections 4.2–4.5. In this problem, we obtain from Eq. (b)

$$y = \frac{1}{x}$$

which upon substitution into Eq. (a) yields $x^2 + 1/x^2 - 3 = 0$, or

$$x^4 - 3x^2 + 1 = 0$$

The solutions of this biquadratic equation: $x = \pm 0.618\,03$ and $\pm 1.618\,03$ agree with the results obtained by the Newton–Raphson method.

EXAMPLE 4.9
Find a solution of

$$\sin x + y^2 + \ln z - 7 = 0$$
$$3x + 2^y - z^3 + 1 = 0$$
$$x + y + z - 5 = 0$$

using `newtonRaphson2`. Start with the point (1, 1, 1).

Solution Letting $x_0 = x$, $x_1 = y$ and $x_2 = z$, we obtain the following program:

```
#!/usr/bin/python
## example4_9
from numarray import zeros,array
from math import sin,log
from newtonRaphson2 import *

def f(x):
    f = zeros((len(x)),type=Float64)
    f[0] = sin(x[0]) + x[1]**2 + log(x[2]) - 7.0
    f[1] = 3.0*x[0] + 2.0**x[1] - x[2]**3 + 1.0
    f[2] = x[0] + x[1] + x[2] - 5.0
    return f

x = array([1.0, 1.0, 1.0])
print newtonRaphson2(f,x)
raw_input (''\nPress return to exit'')
```

The output from this program is

```
[0.59905376    2.3959314    2.00501484]
```

PROBLEM SET 4.1

1. Use the Newton–Raphson method and a four-function calculator ($+ - \times \div$ operations only) to compute $\sqrt[3]{75}$ with four significant figure accuracy.

2. Find the smallest positive (real) root of $x^3 - 3.23x^2 - 5.54x + 9.84 = 0$ by the method of bisection.

3. The smallest positive, nonzero root of $\cosh x \cos x - 1 = 0$ lies in the interval $(4, 5)$. Compute this root by Brent's method.

4. Solve Prob. 3 by the Newton–Raphson method.

5. A root of the equation $\tan x - \tanh x = 0$ lies in $(7.0, 7.4)$. Find this root with three decimal place accuracy by the method of bisection.

6. Determine the two roots of $\sin x + 3 \cos x - 2 = 0$ that lie in the interval $(-2, 2)$. Use the Newton–Raphson method.

7. A popular method in hand computation is the *secant formula* where the improved estimate of the root (x_{i+1}) is obtained by linear interpolation based two previous estimates (x_i and x_{i-1}):

$$x_{i+1} = x_i - \frac{x_i - x_{i-1}}{f(x_i) - f(x_{i-1})} f(x_i)$$

Solve Prob. 6 using the secant formula.

8. Draw a plot of $f(x) = \cosh x \cos x - 1$ in the range $4 \leq x \leq 8$. (a) Verify from the plot that the smallest positive, nonzero root of $f(x) = 0$ lies in the interval $(4, 5)$. (b) Show graphically that the Newton–Raphson formula would not converge to this root if it is started with $x = 4$.

9. The equation $x^3 - 1.2x^2 - 8.19x + 13.23 = 0$ has a double root close to $x = 2$. Determine this root with the Newton–Raphson method within four decimal places.

10. ■ Write a program that computes all the roots of $f(x) = 0$ in a given interval with Brent's method. Utilize the functions `rootsearch` and `brent`. You may use the program in Example 4.3 as a model. Test the program by finding the roots of $x \sin x + 3 \cos x - x = 0$ in $(-6, 6)$.

11. ■ Solve Prob. 10 with the Newton–Raphson method.

12. ■ Determine all real roots of $x^4 + 0.9x^3 - 2.3x^2 + 3.6x - 25.2 = 0$.

13. ■ Compute all positive real roots of $x^4 + 2x^3 - 7x^2 + 3 = 0$.

14. ■ Find all positive, nonzero roots of $\sin x - 0.1x = 0$.

15. ■ The natural frequencies of a uniform cantilever beam are related to the roots β_i of the frequency equation $f(\beta) = \cosh\beta \cos\beta + 1 = 0$, where

$$\beta_i^4 = (2\pi f_i)^2 \frac{mL^3}{EI}$$

$f_i = i$th natural frequency (cps)

$m =$ mass of the beam

$L =$ length of the beam

$E =$ modulus of elasticity

$I =$ moment of inertia of the cross section

Determine the lowest two frequencies of a steel beam 0.9 m. long, with a rectangular cross section 25 mm wide and 2.5 mm in. high. The mass density of steel is 7850 kg/m³ and $E = 200$ GPa.

16. ■

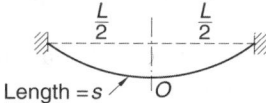

Length $=s$

A steel cable of length s is suspended as shown in the figure. The maximum tensile stress in the cable, which occurs at the supports, is

$$\sigma_{max} = \sigma_0 \cosh\beta$$

where

$$\beta = \frac{\gamma L}{2\sigma_0}$$

$\sigma_0 =$ tensile stress in the cable at O

$\gamma =$ weight of the cable per unit volume

$L =$ horizontal span of the cable

The length to span ratio of the cable is related to β by

$$\frac{s}{L} = \frac{1}{\beta} \sinh\beta$$

Find σ_{max} if $\gamma = 77 \times 10^3$ N/m³ (steel), $L = 1000$ m and $s = 1100$ m.

17. ■

The aluminum W310 × 202 (wide flange) column is subjected to an eccentric axial load P as shown. The maximum compressive stress in the column is given by the so-called *secant formula*:

$$\sigma_{max} = \bar{\sigma}\left[1 + \frac{ec}{r^2}\sec\left(\frac{L}{2r}\sqrt{\frac{\bar{\sigma}}{E}}\right)\right]$$

where

$\bar{\sigma} = P/A =$ average stress
$A = 25\,800 \text{ mm}^2 =$ cross-sectional area of the column
$e = 85 \text{ mm} =$ eccentricity of the load
$c = 170 \text{ mm} =$ half-depth of the column
$r = 142 \text{ mm} =$ radius of gyration of the cross section
$L = 7100 \text{ mm} =$ length of the column
$E = 71 \times 10^9 \text{ Pa} =$ modulus of elasticity

Determine the maximum load P that the column can carry if the maximum stress is not to exceed 120×10^6 Pa.

18. ■

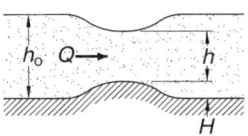

Bernoulli's equation for fluid flow in an open channel with a small bump is

$$\frac{Q^2}{2gb^2h_0^2} + h_0 = \frac{Q^2}{2gb^2h^2} + h + H$$

where

$Q = 1.2 \text{ m}^3/\text{s} =$ volume rate of flow
$g = 9.81 \text{ m/s}^2 =$ gravitational acceleration
$b = 1.8 \text{ m} =$ width of channel
$h_0 = 0.6 \text{ m} =$ upstream water level
$H = 0.075 \text{ m} =$ height of bump
$h =$ water level above the bump

Determine h.

4.6 Systems of Equations

19. ■ The speed v of a Saturn V rocket in vertical flight near the surface of earth can be approximated by

$$v = u \ln \frac{M_0}{M_0 - \dot{m}t} - gt$$

where

$u = 2510 \text{ m/s} = $ velocity of exhaust relative to the rocket
$M_0 = 2.8 \times 10^6 \text{ kg} = $ mass of rocket at liftoff
$\dot{m} = 13.3 \times 10^3 \text{ kg/s} = $ rate of fuel consumption
$g = 9.81 \text{ m/s}^2 = $ gravitational acceleration
$t = $ time measured from liftoff

Determine the time when the rocket reaches the speed of sound (335 m/s).

20. ■

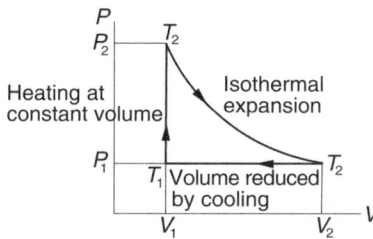

The figure shows the thermodynamic cycle of an engine. The efficiency of this engine for monoatomic gas is

$$\eta = \frac{\ln(T_2/T_1) - (1 - T_1/T_2)}{\ln(T_2/T_1) + (1 - T_1/T_2)/(\gamma - 1)}$$

where T is the absolute temperature and $\gamma = 5/3$. Find T_2/T_1 that results in 30% efficiency ($\eta = 0.3$).

21. ■ Gibb's free energy of one mole of hydrogen at temperature T is

$$G = -RT \ln\left[(T/T_0)^{5/2}\right] \text{ J}$$

where $R = 8.314\,41$ J/K is the gas constant and $T_0 = 4.444\,18$ K. Determine the temperature at which $G = -10^5$ J.

22. ■ The chemical equilibrium equation in the production of methanol from CO and H_2 is[10]

$$\frac{\xi(3-2\xi)^2}{(1-\xi)^3} = 249.2$$

where ξ is the *equilibrium extent of the reaction*. Determine ξ.

23. ■ Determine the coordinates of the two points where the circles $(x-2)^2 + y^2 = 4$ and $x^2 + (y-3)^2 = 4$ intersect. Start by estimating the locations of the points from a sketch of the circles, and then use the Newton–Raphson method to compute the coordinates.

24. ■ The equations

$$\sin x + 3 \cos x - 2 = 0$$

$$\cos x - \sin y + 0.2 = 0$$

have a solution in the vicinity of the point (1, 1). Use the Newton–Raphson method to refine the solution.

25. ■ Use any method to find *all* real solutions in $0 < x < 1.5$ of the simultaneous equations

$$\tan x - y = 1$$

$$\cos x - 3 \sin y = 0$$

26. ■ The equation of a circle is

$$(x-a)^2 + (y-b)^2 = R^2$$

where R is the radius and (a, b) are the coordinates of the center. If the coordinates of three points on the circle are

x	8.21	0.34	5.96
y	0.00	6.62	-1.12

determine R, a and b.

27. ■

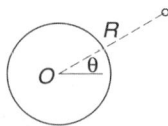

---

[10] From Alberty, R. A., *Physical Chemistry*, 7th ed., Wiley, 1987.

The trajectory of a satellite orbiting the earth is

$$R = \frac{C}{1 + e\sin(\theta + \alpha)}$$

where (R, θ) are the polar coordinates of the satellite, and C, e and α are constants (e is known as the eccentricity of the orbit). If the satellite was observed at the following three positions

θ	$-30°$	$0°$	$30°$
R (km)	6870	6728	6615

determine the smallest R of the trajectory and the corresponding value of θ.

28. ■

A projectile is launched at O with the velocity v at the angle θ to the horizontal. The parametric equations of the trajectory are

$$x = (v\cos\theta)t$$

$$y = -\frac{1}{2}gt^2 + (v\sin\theta)t$$

where t is the time measured from the instant of launch, and $g = 9.81$ m/s² represents the gravitational acceleration. If the projectile is to hit the target at the 45° angle shown in the figure, determine v, θ and the time of flight.

29. ■

The three angles shown in the figure of the four-bar linkage are related by

$$150\cos\theta_1 + 180\cos\theta_2 - 200\cos\theta_3 = 200$$

$$150\sin\theta_1 + 180\sin\theta_2 - 200\sin\theta_3 = 0$$

Determine θ_1 and θ_2 when $\theta_3 = 75°$. Note that there are two solutions.

*4.7 Zeroes of Polynomials

Introduction

A polynomial of degree n has the form

$$P_n(x) = a_0 + a_1 x + a_2 x^2 + \cdots + a_n x^n \tag{4.9}$$

where the coefficients a_i may be real or complex. We will concentrate on polynomials with real coefficients, but the algorithms presented in this chapter also work with complex coefficients.

The polynomial equation $P_n(x) = 0$ has exactly n roots, which may be real or complex. If the coefficients are real, the complex roots always occur in conjugate pairs $(x_r + ix_i, \ x_r - ix_i)$, where x_r and x_i are the real and imaginary parts, respectively. For real coefficients, the number of real roots can be estimated from the *rule of Descartes*:

- The number of positive, real roots equals the number of sign changes in the expression for $P_n(x)$, or less by an even number.
- The number of negative, real roots is equal to the number of sign changes in $P_n(-x)$, or less by an even number.

As an example, consider $P_3(x) = x^3 - 2x^2 - 8x + 27$. Since the sign changes twice, $P_3(x) = 0$ has either two or zero positive real roots. On the other hand, $P_3(-x) = -x^3 - 2x^2 + 8x + 27$ contains a single sign change; hence $P_3(x)$ possesses one negative real zero.

The real zeros of polynomials with real coefficients can always be computed by one of the methods already described. But if complex roots are to be computed, it is best to use a method that specializes in polynomials. Here we present a method due to Laguerre, which is reliable and simple to implement. Before proceeding to Laguerre's method, we must first develop two numerical tools that are needed in any method capable of determining the zeroes of a polynomial. The first of these is an efficient algorithm for evaluating a polynomial and its derivatives. The second algorithm we need is for the *deflation* of a polynomial, i.e., for dividing the $P_n(x)$ by $x - r$, where r is a root of $P_n(x) = 0$.

Evaluation of Polynomials

It is tempting to evaluate the polynomial in Eq. (4.9) from left to right by the following algorithm (we assume that the coefficients are stored in the array **a**):

```
p = 0.0
for i in range(n+1):
    p = p + a[i]*x**i
```

4.7 Zeroes of Polynomials

Since x^k is evaluated as $x \times x \times \cdots \times x$ ($k-1$ multiplications), we deduce that the number of multiplications in this algorithm is

$$1 + 2 + 3 + \cdots + n - 1 = \frac{1}{2}n(n-1)$$

If n is large, the number of multiplications can be reduced considerably if we evaluate the polynomial from right to left. For an example, take

$$P_4(x) = a_0 + a_1 x + a_2 x^2 + a_3 x^3 + a_4 x^4$$

After rewriting the polynomial as

$$P_4(x) = a_0 + x\{a_1 + x[a_2 + x(a_3 + xa_4)]\}$$

the preferred computational sequence becomes obvious:

$$P_0(x) = a_4$$
$$P_1(x) = a_3 + xP_0(x)$$
$$P_2(x) = a_2 + xP_1(x)$$
$$P_3(x) = a_1 + xP_2(x)$$
$$P_4(x) = a_0 + xP_3(x)$$

For a polynomial of degree n, the procedure can be summarized as

$$P_0(x) = a_n$$
$$P_i(x) = a_{n-i} + xP_{i-1}(x), \quad i = 1, 2, \ldots, n \quad (4.10)$$

leading to the algorithm

```
p = a[n]
for i in range(1,n+1):
    p = a[n-i] + p*x
```

The last algorithm involves only n multiplications, making it more efficient for $n > 3$. But computational economy is not the prime reason why this algorithm should be used. Because the result of each multiplication is rounded off, the procedure with the least number of multiplications invariably accumulates the smallest roundoff error.

Some root-finding algorithms, including Laguerre's method, also require evaluation of the first and second derivatives of $P_n(x)$. From Eq. (4.10) we obtain by differentiation

$$P'_0(x) = 0 \quad P'_i(x) = P_{i-1}(x) + xP'_{i-1}(x), \quad i = 1, 2, \ldots, n \quad (4.11\text{a})$$

$$P''_0(x) = 0 \quad P''_i(x) = 2P'_{i-1}(x) + xP''_{i-1}(x), \quad i = 1, 2, \ldots, n \quad (4.11\text{b})$$

evalPoly

Here is the function that evaluates a polynomial and its derivatives:

```
## module evalPoly
''' p,dp,ddp = evalPoly(a,x).
    Evaluates the polynomial
    p = a[0] + a[1]*x + a[2]*x^2 +...+ a[n]*x^n
    with its derivatives dp = p' and ddp = p''
    at x.
'''
def evalPoly(a,x):
    n = len(a) - 1
    p = a[n]
    dp = 0.0 + 0.0j
    ddp = 0.0 + 0.0j
    for i in range(1,n+1):
        ddp = ddp*x + 2.0*dp
        dp = dp*x + p
        p = p*x + a[n-i]
    return p,dp,ddp
```

Deflation of Polynomials

After a root r of $P_n(x) = 0$ has been computed, it is desirable to factor the polynomial as follows:

$$P_n(x) = (x - r) P_{n-1}(x) \qquad (4.12)$$

This procedure, known as deflation or *synthetic division*, involves nothing more than computing the coefficients of $P_{n-1}(x)$. Since the remaining zeros of $P_n(x)$ are also the zeros of $P_{n-1}(x)$, the root-finding procedure can now be applied to $P_{n-1}(x)$ rather than $P_n(x)$. Deflation thus makes it progressively easier to find successive roots, because the degree of the polynomial is reduced every time a root is found. Moreover, by eliminating the roots that have already been found, the chances of computing the same root more than once are eliminated.

If we let

$$P_{n-1}(x) = b_0 + b_1 x + b_2 x^2 + \cdots + b_{n-1} x^{n-1}$$

then Eq. (4.12) becomes

$$a_0 + a_1 x + a_2 x^2 + \cdots + a_{n-1} x^{n-1} + a_n x^n$$
$$= (x - r)(b_0 + b_1 x + b_2 x^2 + \cdots + b_{n-1} x^{n-1})$$

4.7 Zeroes of Polynomials

Equating the coefficients of like powers of x, we obtain

$$b_{n-1} = a_n \quad b_{n-2} = a_{n-1} + rb_{n-1} \quad \cdots \quad b_0 = a_1 + rb_1 \tag{4.13}$$

which leads to the *Horner's deflation algorithm*:

```
b[n-1] = a[n]
for i in range(n-2,-1,-1):
    b[i] = a[i+1] + r*b[i+1]
```

Laguerre's Method

Laguerre's formulas are not easily derived for a general polynomial $P_n(x)$. However, the derivation is greatly simplified if we consider the special case where the polynomial has a zero at $x = r$ and $(n-1)$ zeros at $x = q$. Hence the polynomial can be written as

$$P_n(x) = (x-r)(x-q)^{n-1} \tag{a}$$

Our problem is now this: given the polynomial in Eq. (a) in the form

$$P_n(x) = a_0 + a_1 x + a_2 x^2 + \cdots + a_n x^n$$

determine r (note that q is also unknown). It turns out that the result, which is exact for the special case considered here, works well as an iterative formula with any polynomial.

Differentiating Eq. (a) with respect to x, we get

$$P_n'(x) = (x-q)^{n-1} + (n-1)(x-r)(x-q)^{n-2}$$
$$= P_n(x)\left(\frac{1}{x-r} + \frac{n-1}{x-q}\right)$$

Thus

$$\frac{P_n'(x)}{P_n(x)} = \frac{1}{x-r} + \frac{n-1}{x-q} \tag{b}$$

which upon differentiation yields

$$\frac{P_n''(x)}{P_n(x)} - \left[\frac{P_n'(x)}{P_n(x)}\right]^2 = -\frac{1}{(x-r)^2} - \frac{n-1}{(x-q)^2} \tag{c}$$

It is convenient to introduce the notation

$$G(x) = \frac{P_n'(x)}{P_n(x)} \quad H(x) = G^2(x) - \frac{P_n''(x)}{P_n(x)} \tag{4.14}$$

so that Eqs. (b) and (c) become

$$G(x) = \frac{1}{x-r} + \frac{n-1}{x-q} \quad (4.15a)$$

$$H(x) = \frac{1}{(x-r)^2} + \frac{n-1}{(x-q)^2} \quad (4.15b)$$

If we solve Eq. (4.15a) for $x - q$ and substitute the result into Eq. (4.15b), we obtain a quadratic equation for $x - r$. The solution of this equation is the *Laguerre's formula*

$$x - r = \frac{n}{G(x) \pm \sqrt{(n-1)\left[nH(x) - G^2(x)\right]}} \quad (4.16)$$

The procedure for finding a zero of a general polynomial by Laguerre's formula is:

1. Let x be a guess for the root of $P_n(x) = 0$ (any value will do).
2. Evaluate $P_n(x)$, $P_n'(x)$ and $P_n''(x)$ using the procedure outlined in Eqs. (4.10) and (4.11).
3. Compute $G(x)$ and $H(x)$ from Eqs. (4.14).
4. Determine the improved root r from Eq. (4.16) choosing the sign that results in the *larger magnitude of the denominator* (this can be shown to improve convergence).
5. Let $x \leftarrow r$ and repeat steps 2–5 until $|P_n(x)| < \varepsilon$ or $|x - r| < \varepsilon$, where ε is the error tolerance.

One nice property of Laguerre's method is that it converges to a root, with very few exceptions, from any starting value of x.

■ polyRoots

The function `polyRoots` in this module computes all the roots of $P_n(x) = 0$, where the polynomial $P_n(x)$ defined by its coefficient array $\mathbf{a} = [a_0, a_1, \ldots, a_n]$. After the first root is computed by the nested function `laguerre`, the polynomial is deflated using `deflPoly` and the next zero computed by applying `laguerre` to the deflated polynomial. This process is repeated until all n roots have been found. If a computed root has a very small imaginary part, it is very likely that it represents roundoff error. Therefore, `polyRoots` replaces a tiny imaginary part by zero.

```
## module polyRoots
''' roots = polyRoots(a).
    Uses Laguerre's method to compute all the roots of
    a[0] + a[1]*x + a[2]*x^2 +...+ a[n]*x^n = 0.
    The roots are returned in the vector {roots},
```

4.7 Zeroes of Polynomials

```
'''
from evalPoly import *
from numarray import zeros,Complex64
from cmath import sqrt
from random import random

def polyRoots(a,tol=1.0e-12):

    def laguerre(a,tol):
        x = random()   # Starting value (random number)
        n = len(a) - 1
        for i in range(30):
            p,dp,ddp = evalPoly(a,x)
            if abs(p) < tol: return x
            g = dp/p
            h = g*g - ddp/p
            f = sqrt((n - 1)*(n*h - g*g))
            if abs(g + f) > abs(g - f): dx = n/(g + f)
            else: dx = n/(g - f)
            x = x - dx
            if abs(dx) < tol*max(abs(x),1.0): return x
        print 'Too many iterations in Laguerre'

    def deflPoly(a,root):  # Deflates a polynomial
        n = len(a)-1
        b = [(0.0 + 0.0j)]*n
        b[n-1] = a[n]
        for i in range(n-2,-1,-1):
            b[i] = a[i+1] + root*b[i+1]
        return b

    n = len(a) - 1
    roots = zeros((n),type=Complex64)
    for i in range(n):
        x = laguerre(a,tol)
        if abs(x.imag) < tol: x = x.real
        roots[i] = x
        a = deflPoly(a,x)
    return roots
    raw_input(''\nPress return to exit'')
```

Since the roots are computed with finite accuracy, each deflation introduces small errors in the coefficients of the deflated polynomial. The accumulated roundoff error increases with the degree of the polynomial and can become severe if the polynomial is ill-conditioned (small changes in the coefficients produce large changes in the roots). Hence the results should be viewed with caution when dealing with polynomials of high degree.

The errors caused by deflation can be reduced by recomputing each root using the original, undeflated polynomial. The roots obtained previously in conjunction with deflation are employed as the starting values.

EXAMPLE 4.10
A zero of the polynomial $P_4(x) = 3x^4 - 10x^3 - 48x^2 - 2x + 12$ is $x = 6$. Deflate the polynomial with Horner's algorithm, i.e., find $P_3(x)$ so that $(x - 6)P_3(x) = P_4(x)$.

Solution With $r = 6$ and $n = 4$, Eqs. (4.13) become

$$b_3 = a_4 = 3$$
$$b_2 = a_3 + 6b_3 = -10 + 6(3) = 8$$
$$b_1 = a_2 + 6b_2 = -48 + 6(8) = 0$$
$$b_0 = a_1 + 6b_1 = -2 + 6(0) = -2$$

Therefore,

$$P_3(x) = 3x^3 + 8x^2 - 2$$

EXAMPLE 4.11
A root of the equation $P_3(x) = x^3 - 4.0x^2 - 4.48x + 26.1$ is approximately $x = 3 - i$. Find a more accurate value of this root by one application of Laguerre's iterative formula.

Solution Use the given estimate of the root as the starting value. Thus

$$x = 3 - i \qquad x^2 = 8 - 6i \qquad x^3 = 18 - 26i$$

Substituting these values in $P_3(x)$ and its derivatives, we get

$$P_3(x) = x^3 - 4.0x^2 - 4.48x + 26.1$$
$$= (18 - 26i) - 4.0(8 - 6i) - 4.48(3 - i) + 26.1 = -1.34 + 2.48i$$
$$P_3'(x) = 3.0x^2 - 8.0x - 4.48$$
$$= 3.0(8 - 6i) - 8.0(3 - i) - 4.48 = -4.48 - 10.0i$$
$$P_3''(x) = 6.0x - 8.0 = 6.0(3 - i) - 8.0 = 10.0 - 6.0i$$

4.7 Zeroes of Polynomials

Equations (4.14) then yield

$$G(x) = \frac{P_3'(x)}{P_3(x)} = \frac{-4.48 - 10.0i}{-1.34 + 2.48i} = -2.36557 + 3.08462i$$

$$H(x) = G^2(x) - \frac{P_3''(x)}{P_3(x)} = (-2.36557 + 3.08462i)^2 - \frac{10.0 - 6.0i}{-1.34 + 2.48i}$$

$$= 0.35995 - 12.48452i$$

The term under the square root sign of the denominator in Eq. (4.16) becomes

$$F(x) = \sqrt{(n-1)\left[nH(x) - G^2(x)\right]}$$

$$= \sqrt{2\left[3(0.35995 - 12.48452i) - (-2.36557 + 3.08462i)^2\right]}$$

$$= \sqrt{5.67822 - 45.71946i} = 5.08670 - 4.49402i$$

Now we must find which sign in Eq. (4.16) produces the larger magnitude of the denominator:

$$|G(x) + F(x)| = |(-2.36557 + 3.08462i) + (5.08670 - 4.49402i)|$$

$$= |2.72113 - 1.40940i| = 3.06448$$

$$|G(x) - F(x)| = |(-2.36557 + 3.08462i) - (5.08670 - 4.49402i)|$$

$$= |-7.45227 + 7.57864i| = 10.62884$$

Using the minus sign, we obtain from Eq. (4.16) the following improved approximation for the root

$$r = x - \frac{n}{G(x) - F(x)} = (3 - i) - \frac{3}{-7.45227 + 7.57864i}$$

$$= 3.19790 - 0.79875i$$

Thanks to the good starting value, this approximation is already quite close to the exact value $r = 3.20 - 0.80i$.

EXAMPLE 4.12
Use polyRoots to compute *all* the roots of $x^4 - 5x^3 - 9x^2 + 155x - 250 = 0$.

Solution The commands

```
>>> from polyRoots import *
>>> print polyRoots([-250.0,155.0,-9.0,-5.0,1.0])
```

resulted in the output

```
[2.+0.j   4.-3.j   4.+3.j   -5.+0.j]
```

Roots of Equations

PROBLEM SET 4.2

Problems 1–5 A zero $x = r$ of $P_n(x)$ is given. Verify that r is indeed a zero, and then deflate the polynomial, i.e., find $P_{n-1}(x)$ so that $P_n(x) = (x - r)P_{n-1}(x)$.

1. $P_3(x) = 3x^3 + 7x^2 - 36x + 20, r = -5$.
2. $P_4(x) = x^4 - 3x^2 + 3x - 1, r = 1$.
3. $P_5(x) = x^5 - 30x^4 + 361x^3 - 2178x^2 + 6588x - 7992, r = 6$.
4. $P_4(x) = x^4 - 5x^3 - 2x^2 - 20x - 24, r = 2i$.
5. $P_3(x) = 3x^3 - 19x^2 + 45x - 13, r = 3 - 2i$.

Problems 6–9 A zero $x = r$ of $P_n(x)$ is given. Determine all the other zeros of $P_n(x)$ by using a calculator. You should need no tools other than deflation and the quadratic formula.

6. $P_3(x) = x^3 + 1.8x^2 - 9.01x - 13.398, r = -3.3$.
7. $P_3(x) = x^3 - 6.64x^2 + 16.84x - 8.32, r = 0.64$.
8. $P_3(x) = 2x^3 - 13x^2 + 32x - 13, r = 3 - 2i$.
9. $P_4(x) = x^4 - 3x^3 + 10x^2 - 6x - 20, r = 1 + 3i$.

Problems 10–15 Find all the zeros of the given $P_n(x)$.

10. ■$P_4(x) = x^4 + 2.1x^3 - 2.52x^2 + 2.1x - 3.52$.
11. ■$P_5(x) = x^5 - 156x^4 - 5x^3 + 780x^2 + 4x - 624$.
12. ■$P_6(x) = x^6 + 4x^5 - 8x^4 - 34x^3 + 57x^2 + 130x - 150$.
13. ■$P_7(x) = 8x^7 + 28x^6 + 34x^5 - 13x^4 - 124x^3 + 19x^2 + 220x - 100$.
14. ■$P_8(x) = x^8 - 7x^7 + 7x^6 + 25x^5 + 24x^4 - 98x^3 - 472x^2 + 440x + 800$.
15. ■$P_4(x) = x^4 + (5 + i)x^3 - (8 - 5i)x^2 + (30 - 14i)x - 84$.
16. ■

The two blocks of mass m each are connected by springs and a dashpot. The stiffness of each spring is k, and c is the coefficient of damping of the dashpot. When the system is displaced and released, the displacement of each block during the ensuing motion has the form

$$x_k(t) = A_k e^{\omega_r t} \cos(\omega_i t + \phi_k), \; k = 1, 2$$

where A_k and ϕ_k are constants, and $\omega = \omega_r \pm i\omega_i$ are the roots of

$$\omega^4 + 2\frac{c}{m}\omega^3 + 3\frac{k}{m}\omega^2 + \frac{c}{m}\frac{k}{m}\omega + \left(\frac{k}{m}\right)^2 = 0$$

Determine the two possible combinations of ω_r and ω_i if $c/m = 12 \text{ s}^{-1}$ and $k/m = 1500 \text{ s}^{-2}$.

4.8 Other Methods

The most prominent root-finding algorithms omitted from this chapter are the *secant method* and its close relative, the *false position method*. Both methods compute the improved value of the root by linear interpolation. They differ only by how they choose the points involved in the interpolation. The secant method always uses the two most recent estimates of the root, whereas the false position method employs the points that keep the root bracketed. The secant method is faster of the two, but the false position method is more stable. Since both are considerably slower than Brent's method, there is little reason to use them.

There are many methods for finding zeros of polynomials. Of these, the *Jenkins–Traub algorithm*[11] deserves special mention due to its robustness and widespread use in packaged software.

The zeros of a polynomial can also be obtained by calculating the eigenvalues of the $n \times n$ "companion matrix"

$$\mathbf{A} = \begin{bmatrix} -a_{n-1}/a_n & -a_2/a_n & \cdots & -a_1/a_n & -a_0/a_n \\ 1 & 0 & \cdots & 0 & 0 \\ 0 & 1 & & 0 & 0 \\ \vdots & \vdots & \ddots & \vdots & \vdots \\ 0 & 0 & \cdots & 1 & 0 \end{bmatrix}$$

[11] Jenkins, M., and Traub, J., *SIAM Journal on Numerical Analysis*, Vol. 7 (1970), p. 545.

where a_i are the coefficients of the polynomial. The characteristic equation (see Section 9.1) of this matrix is

$$x^n + \frac{a_{n-1}}{a_n}x^{n-1} + \frac{a_{n-2}}{a_n}x^{n-2} + \cdots + \frac{a_1}{a_n}x + \frac{a_0}{a_n} = 0$$

which is equivalent to $P_n(x) = 0$. Thus the eigenvalues of **A** are the zeroes of $P_n(x)$. The eigenvalue method is robust, but considerably slower than Laguerre's method. But it is worthy of consideration if a good program for eigenvalue problems is available.

5 Numerical Differentiation

> Given the function $f(x)$, compute $d^n f/dx^n$ at given x

5.1 Introduction

Numerical differentiation deals with the following problem: we are given the function $y = f(x)$ and wish to obtain one of its derivatives at the point $x = x_k$. The term "given" means that we either have an algorithm for computing the function, or possess a set of discrete data points (x_i, y_i), $i = 0, 1, \ldots, n$. In either case, we have access to a finite number of (x, y) data pairs from which to compute the derivative. If you suspect by now that numerical differentiation is related to interpolation, you are right—one means of finding the derivative is to approximate the function locally by a polynomial and then differentiate it. An equally effective tool is the Taylor series expansion of $f(x)$ about the point x_k, which has the advantage of providing us with information about the error involved in the approximation.

Numerical differentiation is not a particularly accurate process. It suffers from a conflict between roundoff errors (due to limited machine precision) and errors inherent in interpolation. For this reason, a derivative of a function can never be computed with the same precision as the function itself.

5.2 Finite Difference Approximations

The derivation of the finite difference approximations for the derivatives of $f(x)$ is based on forward and backward Taylor series expansions of $f(x)$ about x, such as

Numerical Differentiation

$$f(x+h) = f(x) + hf'(x) + \frac{h^2}{2!}f''(x) + \frac{h^3}{3!}f'''(x) + \frac{h^4}{4!}f^{(4)}(x) + \cdots \quad \text{(a)}$$

$$f(x-h) = f(x) - hf'(x) + \frac{h^2}{2!}f''(x) - \frac{h^3}{3!}f'''(x) + \frac{h^4}{4!}f^{(4)}(x) - \cdots \quad \text{(b)}$$

$$f(x+2h) = f(x) + 2hf'(x) + \frac{(2h)^2}{2!}f''(x) + \frac{(2h)^3}{3!}f'''(x) + \frac{(2h)^4}{4!}f^{(4)}(x) + \cdots \quad \text{(c)}$$

$$f(x-2h) = f(x) - 2hf'(x) + \frac{(2h)^2}{2!}f''(x) - \frac{(2h)^3}{3!}f'''(x) + \frac{(2h)^4}{4!}f^{(4)}(x) - \cdots \quad \text{(d)}$$

We also record the sums and differences of the series:

$$f(x+h) + f(x-h) = 2f(x) + h^2 f''(x) + \frac{h^4}{12} f^{(4)}(x) + \cdots \quad \text{(e)}$$

$$f(x+h) - f(x-h) = 2hf'(x) + \frac{h^3}{3} f'''(x) + \cdots \quad \text{(f)}$$

$$f(x+2h) + f(x-2h) = 2f(x) + 4h^2 f''(x) + \frac{4h^4}{3} f^{(4)}(x) + \cdots \quad \text{(g)}$$

$$f(x+2h) - f(x-2h) = 4hf'(x) + \frac{8h^3}{3} f'''(x) + \cdots \quad \text{(h)}$$

Note that the sums contain only even derivatives, whereas the differences retain just the odd derivatives. Equations (a)–(h) can be viewed as simultaneous equations that can be solved for various derivatives of $f(x)$. The number of equations involved and the number of terms kept in each equation depend on the order of the derivative and the desired degree of accuracy.

First Central Difference Approximations

The solution of Eq. (f) for $f'(x)$ is

$$f'(x) = \frac{f(x+h) - f(x-h)}{2h} - \frac{h^2}{6} f'''(x) - \cdots$$

or

$$f'(x) = \frac{f(x+h) - f(x-h)}{2h} + \mathcal{O}(h^2) \quad (5.1)$$

which is called the *first central difference approximation* for $f'(x)$. The term $\mathcal{O}(h^2)$ reminds us that the truncation error behaves as h^2.

5.2 Finite Difference Approximations

Similarly, Eq. (e) yields the first central difference approximation for $f''(x)$:

$$f''(x) = \frac{f(x+h) - 2f(x) + f(x-h)}{h^2} + \frac{h^2}{12}f^{(4)}(x) + \cdots$$

or

$$f''(x) = \frac{f(x+h) - 2f(x) + f(x-h)}{h^2} + \mathcal{O}(h^2) \quad (5.2)$$

Central difference approximations for other derivatives can be obtained from Eqs. (a)–(h) in the same manner. For example, eliminating $f'(x)$ from Eqs. (f) and (h) and solving for $f'''(x)$ yield

$$f'''(x) = \frac{f(x+2h) - 2f(x+h) + 2f(x-h) - f(x-2h)}{2h^3} + \mathcal{O}(h^2) \quad (5.3)$$

The approximation

$$f^{(4)}(x) = \frac{f(x+2h) - 4f(x+h) + 6f(x) - 4f(x-h) + f(x-2h)}{h^4} + \mathcal{O}(h^2) \quad (5.4)$$

is available from Eqs. (e) and (g) after eliminating $f''(x)$. Table 5.1 summarizes the results.

	$f(x-2h)$	$f(x-h)$	$f(x)$	$f(x+h)$	$f(x+2h)$
$2hf'(x)$		-1	0	1	
$h^2 f''(x)$		1	-2	1	
$2h^3 f'''(x)$	-1	2	0	-2	1
$h^4 f^{(4)}(x)$	1	-4	6	-4	1

Table 5.1. Coefficients of central finite difference approximations of $\mathcal{O}(h^2)$

First Noncentral Finite Difference Approximations

Central finite difference approximations are not always usable. For example, consider the situation where the function is given at the n discrete points $x_0, x_1, \ldots, x_n$. Since central differences use values of the function on each side of x, we would be unable to compute the derivatives at x_0 and x_n. Clearly, there is a need for finite difference expressions that require evaluations of the function only on one side of x. These expressions are called *forward* and *backward* finite difference approximations.

Numerical Differentiation

Noncentral finite differences can also be obtained from Eqs. (a)–(h). Solving Eq. (a) for $f'(x)$ we get

$$f'(x) = \frac{f(x+h) - f(x)}{h} - \frac{h}{2} f''(x) - \frac{h^2}{6} f'''(x) - \frac{h^3}{4!} f^{(4)}(x) - \cdots$$

Keeping only the first term on the right-hand side leads to the *first forward difference approximation*

$$f'(x) = \frac{f(x+h) - f(x)}{h} + \mathcal{O}(h) \tag{5.5}$$

Similarly, Eq. (b) yields the *first backward difference approximation*

$$f'(x) = \frac{f(x) - f(x-h)}{h} + \mathcal{O}(h) \tag{5.6}$$

Note that the truncation error is now $\mathcal{O}(h)$, which is not as good as $\mathcal{O}(h^2)$ in central difference approximations.

We can derive the approximations for higher derivatives in the same manner. For example, Eqs. (a) and (c) yield

$$f''(x) = \frac{f(x+2h) - 2f(x+h) + f(x)}{h^2} + \mathcal{O}(h) \tag{5.7}$$

The third and fourth derivatives can be derived in a similar fashion. The results are shown in Tables 5.2a and 5.2b.

	$f(x)$	$f(x+h)$	$f(x+2h)$	$f(x+3h)$	$f(x+4h)$
$hf'(x)$	-1	1			
$h^2 f''(x)$	1	-2	1		
$h^3 f'''(x)$	-1	3	-3	1	
$h^4 f^{(4)}(x)$	1	-4	6	-4	1

Table 5.2a. Coefficients of forward finite difference approximations of $\mathcal{O}(h)$

	$f(x-4h)$	$f(x-3h)$	$f(x-2h)$	$f(x-h)$	$f(x)$
$hf'(x)$				-1	1
$h^2 f''(x)$			1	-2	1
$h^3 f'''(x)$		-1	3	-3	1
$h^4 f^{(4)}(x)$	1	-4	6	-4	1

Table 5.2b. Coefficients of backward finite difference approximations of $\mathcal{O}(h)$

Second Noncentral Finite Difference Approximations

Finite difference approximations of $\mathcal{O}(h)$ are not popular due to reasons that will be explained shortly. The common practice is to use expressions of $\mathcal{O}(h^2)$. To obtain noncentral difference formulas of this order, we have to retain more terms in the Taylor series. As an illustration, we will derive the expression for $f'(x)$. We start with Eqs. (a) and (c), which are

$$f(x+h) = f(x) + hf'(x) + \frac{h^2}{2}f''(x) + \frac{h^3}{6}f'''(x) + \frac{h^4}{24}f^{(4)}(x) + \cdots$$

$$f(x+2h) = f(x) + 2hf'(x) + 2h^2 f''(x) + \frac{4h^3}{3}f'''(x) + \frac{2h^4}{3}f^{(4)}(x) + \cdots$$

We eliminate $f''(x)$ by multiplying the first equation by 4 and subtracting it from the second equation. The result is

$$f(x+2h) - 4f(x+h) = -3f(x) - 2hf'(x) + \frac{2h^3}{3}f'''(x) + \cdots$$

Therefore,

$$f'(x) = \frac{-f(x+2h) + 4f(x+h) - 3f(x)}{2h} + \frac{h^2}{3}f'''(x) + \cdots$$

or

$$f'(x) = \frac{-f(x+2h) + 4f(x+h) - 3f(x)}{2h} + \mathcal{O}(h^2) \tag{5.8}$$

Equation (5.8) is called the *second forward finite difference approximation*.

Derivation of finite difference approximations for higher derivatives involve additional Taylor series. Thus the forward difference approximation for $f''(x)$ utilizes series for $f(x+h)$, $f(x+2h)$ and $f(x+3h)$; the approximation for $f'''(x)$ involves Taylor expansions for $f(x+h)$, $f(x+2h)$, $f(x+3h)$ and $f(x+4h)$, etc. As you can see, the computations for high-order derivatives can become rather tedious. The results for both the forward and backward finite differences are summarized in Tables 5.3a and 5.3b.

	$f(x)$	$f(x+h)$	$f(x+2h)$	$f(x+3h)$	$f(x+4h)$	$f(x+5h)$
$2hf'(x)$	-3	4	-1			
$h^2 f''(x)$	2	-5	4	-1		
$2h^3 f'''(x)$	-5	18	-24	14	-3	
$h^4 f^{(4)}(x)$	3	-14	26	-24	11	-2

Table 5.3a. Coefficients of forward finite difference approximations of $\mathcal{O}(h^2)$

	$f(x-5h)$	$f(x-4h)$	$f(x-3h)$	$f(x-2h)$	$f(x-h)$	$f(x)$
$2hf'(x)$				1	−4	3
$h^2 f''(x)$			−1	4	−5	2
$2h^3 f'''(x)$		3	−14	24	−18	5
$h^4 f^{(4)}(x)$	−2	11	−24	26	−14	3

Table 5.3b. Coefficients of backward finite difference approximations of $\mathcal{O}(h^2)$

Errors in Finite Difference Approximations

Observe that in all finite difference expressions the sum of the coefficients is zero. The effect on the *roundoff error* can be profound. If h is very small, the values of $f(x)$, $f(x \pm h)$, $f(x \pm 2h)$ etc. will be approximately equal. When they are multiplied by the coefficients and added, several significant figures can be lost. On the other hand, we cannot make h too large, because then the *truncation error* would become excessive. This unfortunate situation has no remedy, but we can obtain some relief by taking the following precautions:

- Use double-precision arithmetic.
- Employ finite difference formulas that are accurate to at least $\mathcal{O}(h^2)$.

To illustrate the errors, let us compute the second derivative of $f(x) = e^{-x}$ at $x = 1$ from the central difference formula, Eq. (5.2). We carry out the calculations with six- and eight-digit precision, using different values of h. The results, shown in Table 5.4, should be compared with $f''(1) = e^{-1} = 0.367\,879\,44$.

h	6-digit precision	8-digit precision
0.64	0.380 610	0.380 609 11
0.32	0.371 035	0.371 029 39
0.16	0.368 711	0.368 664 84
0.08	0.368 281	0.368 076 56
0.04	0.368 75	0.367 831 25
0.02	0.37	0.3679
0.01	0.38	0.3679
0.005	0.40	0.3676
0.0025	0.48	0.3680
0.00125	1.28	0.3712

Table 5.4. $(e^{-x})''$ at $x = 1$ from central finite difference approximation

In the six-digit computations, the optimal value of h is 0.08, yielding a result accurate to three significant figures. Hence three significant figures are lost due to a combination of truncation and roundoff errors. Above optimal h, the dominant error is due to truncation; below it, the roundoff error becomes pronounced. The best result obtained with the eight-digit computation is accurate to four significant figures. Because the extra precision decreases the roundoff error, the optimal h is smaller (about 0.02) than in the six-figure calculations.

5.3 Richardson Extrapolation

Richardson extrapolation is a simple method for boosting the accuracy of certain numerical procedures, including finite difference approximations (we also use it later in other applications).

Suppose that we have an approximate means of computing some quantity G. Moreover, assume that the result depends on a parameter h. Denoting the approximation by $g(h)$, we have $G = g(h) + E(h)$, where $E(h)$ represents the error. Richardson extrapolation can remove the error, provided that it has the form $E(h) = ch^p$, c and p being constants. We start by computing $g(h)$ with some value of h, say $h = h_1$. In that case we have

$$G = g(h_1) + ch_1^p \tag{i}$$

Then we repeat the calculation with $h = h_2$, so that

$$G = g(h_2) + ch_2^p \tag{j}$$

Eliminating c and solving for G, we obtain from Eqs. (i) and (j)

$$G = \frac{(h_1/h_2)^p g(h_2) - g(h_1)}{(h_1/h_2)^p - 1} \tag{5.9a}$$

which is the *Richardson extrapolation formula*. It is common practice to use $h_2 = h_1/2$, in which case Eq. (5.9a) becomes

$$G = \frac{2^p g(h_1/2) - g(h_1)}{2^p - 1} \tag{5.9b}$$

Let us illustrate Richardson extrapolation by applying it to the finite difference approximation of $(e^{-x})''$ at $x = 1$. We work with six-digit precision and utilize the results in Table 5.4. Since the extrapolation works only on the truncation error, we must confine h to values that produce negligible roundoff. In Table 5.4 we have

$$g(0.64) = 0.380\,610 \qquad g(0.32) = 0.371\,035$$

Numerical Differentiation

The truncation error in the central difference approximation is $E(h) = \mathcal{O}(h^2) = c_1 h^2 + c_2 h^4 + c_3 h^6 + \cdots$. Therefore, we can eliminate the first (dominant) error term if we substitute $p = 2$ and $h_1 = 0.64$ in Eq. (5.9b). The result is

$$G = \frac{2^2 g(0.32) - g(0.64)}{2^2 - 1} = \frac{4(0.371\,035) - 0.380\,610}{3} = 0.367\,843$$

which is an approximation of $(e^{-x})''$ with the error $\mathcal{O}(h^4)$. Note that it is as accurate as the best result obtained with eight-digit computations in Table 5.4.

EXAMPLE 5.1
Given the evenly spaced data points

x	0	0.1	0.2	0.3	0.4
$f(x)$	0.0000	0.0819	0.1341	0.1646	0.1797

compute $f'(x)$ and $f''(x)$ at $x = 0$ and 0.2 using finite difference approximations of $\mathcal{O}(h^2)$.

Solution We will use finite difference approximations of $\mathcal{O}(h^2)$. From the forward difference tables in Table 5.3a we get

$$f'(0) = \frac{-3f(0) + 4f(0.1) - f(0.2)}{2(0.1)} = \frac{-3(0) + 4(0.0819) - 0.1341}{0.2} = 0.967$$

$$f''(0) = \frac{2f(0) - 5f(0.1) + 4f(0.2) - f(0.3)}{(0.1)^2}$$

$$= \frac{2(0) - 5(0.0819) + 4(0.1341) - 0.1646}{(0.1)^2} = -3.77$$

The central difference approximations in Table 5.1 yield

$$f'(0.2) = \frac{-f(0.1) + f(0.3)}{2(0.1)} = \frac{-0.0819 + 0.1646}{0.2} = 0.4135$$

$$f''(0.2) = \frac{f(0.1) - 2f(0.2) + f(0.3)}{(0.1)^2} = \frac{0.0819 - 2(0.1341) + 0.1646}{(0.1)^2} = -2.17$$

EXAMPLE 5.2
Use the data in Example 5.1 to compute $f'(0)$ as accurately as you can.

Solution One solution is to apply Richardson extrapolation to finite difference approximations. We start with two forward difference approximations of $\mathcal{O}(h^2)$ for $f'(0)$: one using $h = 0.2$ and the other one using $h = 0.1$. Referring to the formulas of $O(h^2)$

5.3 Richardson Extrapolation

in Table 5.3a, we get

$$g(0.2) = \frac{-3f(0) + 4f(0.2) - f(0.4)}{2(0.2)} = \frac{3(0) + 4(0.1341) - 0.1797}{0.4} = 0.8918$$

$$g(0.1) = \frac{-3f(0) + 4f(0.1) - f(0.2)}{2(0.1)} = \frac{-3(0) + 4(0.0819) - 0.1341}{0.2} = 0.9675$$

Recall that the error in both approximations is of the form $E(h) = c_1 h^2 + c_2 h^4 + c_3 h^6 + \cdots$. We can now use Richardson extrapolation, Eq. (5.9), to eliminate the dominant error term. With $p = 2$ we obtain

$$f'(0) \approx G = \frac{2^2 g(0.1) - g(0.2)}{2^2 - 1} = \frac{4(0.9675) - 0.8918}{3} = 0.9927$$

which is a finite difference approximation of $O(h^4)$.

EXAMPLE 5.3

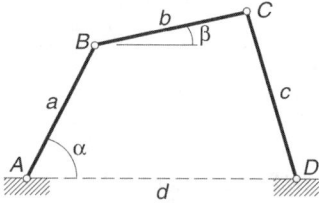

The linkage shown has the dimensions $a = 100$ mm, $b = 120$ mm, $c = 150$ mm and $d = 180$ mm. It can be shown by geometry that the relationship between the angles α and β is

$$(d - a\cos\alpha - b\cos\beta)^2 + (a\sin\alpha + b\sin\beta)^2 - c^2 = 0$$

For a given value of α, we can solve this transcendental equation for β by one of the root-finding methods in Chapter 4. This was done with $\alpha = 0°, 5°, 10°, \ldots, 30°$, the results being

α (deg)	0	5	10	15	20	25	30
β (rad)	1.6595	1.5434	1.4186	1.2925	1.1712	1.0585	0.9561

If link AB rotates with the constant angular velocity of 25 rad/s, use finite difference approximations of $\mathcal{O}(h^2)$ to tabulate the angular velocity $d\beta/dt$ of link BC against α.

Solution The angular speed of BC is

$$\frac{d\beta}{dt} = \frac{d\beta}{d\alpha}\frac{d\alpha}{dt} = 25\frac{d\beta}{d\alpha} \text{ rad/s}$$

where $d\beta/d\alpha$ can be computed from finite difference approximations using the data in the table. Forward and backward differences of $\mathcal{O}(h^2)$ are used at the endpoints, central differences elsewhere. Note that the increment of α is

$$h = (5 \text{ deg})\left(\frac{\pi}{180} \text{rad/deg}\right) = 0.087\,266 \text{ rad}$$

The computations yield

$$\dot{\beta}(0°) = 25\frac{-3\beta(0°) + 4\beta(5°) - \beta(10°)}{2h} = 25\frac{-3(1.6595) + 4(1.5434) - 1.4186}{2(0.087\,266)}$$
$$= -32.01 \text{ rad/s}$$
$$\dot{\beta}(5°) = 25\frac{\beta(10°) - \beta(0°)}{2h} = 25\frac{1.4186 - 1.6595}{2(0.087\,266)} = -34.51 \text{ rad/s}$$

etc.

The complete set of results is

α (deg)	0	5	10	15	20	25	30
$\dot{\beta}$ (rad/s)	-32.01	-34.51	-35.94	-35.44	-33.52	-30.81	-27.86

5.4 Derivatives by Interpolation

If $f(x)$ is given as a set of discrete data points, interpolation can be a very effective means of computing its derivatives. The idea is to approximate the derivative of $f(x)$ by the derivative of the interpolant. This method is particularly useful if the data points are located at uneven intervals of x, when the finite difference approximations listed in the last article are not applicable.[12]

Polynomial Interpolant

The idea here is simple: fit the polynomial of degree n

$$P_{n-1}(x) = a_0 + a_1 x + a_2 x^2 + \cdots + a_n x^n$$

through $n+1$ data points and then evaluate its derivatives at the given x. As pointed out in Section 3.2, it is generally advisable to limit the degree of the polynomial to less than six in order to avoid spurious oscillations of the interpolant. Since these oscillations are magnified with each differentiation, their effect can devastating. In

---

[12] It is possible to derive finite difference approximations for unevenly spaced data, but they would not be as accurate as the formulas derived in Section 5.2.

5.4 Derivatives by Interpolation

view of the above limitation, the interpolation is usually a local one, involving no more than a few nearest-neighbor data points.

For evenly spaced data points, polynomial interpolation and finite difference approximations produce identical results. In fact, the finite difference formulas are equivalent to polynomial interpolation.

Several methods of polynomial interpolation were introduced in Section 3.2. Unfortunately, none of them is suited for the computation of derivatives of the interpolant. The method that we need is one that determines the coefficients $a_0, a_1, \ldots, a_n$ of the polynomial. There is only one such method discussed in Chapter 3: the *least-squares fit*. Although this method is designed mainly for smoothing of data, it will carry out interpolation if we use $m = n$ in Eq. (3.22)—recall that m is the degree of the interpolating polynomial and $n+1$ represents the number of data points to be fitted. If the data contains noise, then the least-squares fit should be used in the smoothing mode, that is, with $m < n$. After the coefficients of the polynomial have been found, the polynomial and its first two derivatives can be evaluated efficiently by the function `evalPoly` listed in Section 4.7.

Cubic Spline Interpolant

Due to its stiffness, cubic spline is a good global interpolant; moreover, it is easy to differentiate. The first step is to determine the second derivatives k_i of the spline at the knots by solving Eqs. (3.11). This can be done with the function `curvatures` in the module `cubicSpline` listed in Section 3.3. The first and second derivatives are then computed from

$$f'_{i,i+1}(x) = \frac{k_i}{6}\left[\frac{3(x-x_{i+1})^2}{x_i - x_{i+1}} - (x_i - x_{i+1})\right]$$
$$- \frac{k_{i+1}}{6}\left[\frac{3(x-x_i)^2}{x_i - x_{i+1}} - (x_i - x_{i+1})\right] + \frac{y_i - y_{i+1}}{x_i - x_{i+1}} \quad (5.10)$$

$$f''_{i,i+1}(x) = k_i \frac{x - x_{i+1}}{x_i - x_{i+1}} - k_{i+1}\frac{x - x_i}{x_i - x_{i+1}} \quad (5.11)$$

which are obtained by differentiation of Eq. (3.10).

EXAMPLE 5.4
Given the data

x	1.5	1.9	2.1	2.4	2.6	3.1
$f(x)$	1.0628	1.3961	1.5432	1.7349	1.8423	2.0397

Numerical Differentiation

compute $f'(2)$ and $f''(2)$ using (1) polynomial interpolation over three nearest-neighbor points, and (2) natural cubic spline interpolant spanning all the data points.

Solution of Part (1) The interpolant is $P_2(x) = a_0 + a_1 x + a_2 x^2$ passing through the points at $x = 1.9, 2.1$ and 2.4. The normal equations, Eqs. (3.23), of the least-squares fit are

$$\begin{bmatrix} n & \sum x_i & \sum x_i^2 \\ \sum x_i & \sum x_i^2 & \sum x_i^3 \\ \sum x_i^2 & \sum x_i^3 & \sum x_i^4 \end{bmatrix} \begin{bmatrix} a_0 \\ a_1 \\ a_2 \end{bmatrix} = \begin{bmatrix} \sum y_i \\ \sum y_i x_i \\ \sum y_i x_i^2 \end{bmatrix}$$

After substituting the data, we get

$$\begin{bmatrix} 3 & 6.4 & 13.78 \\ 6.4 & 13.78 & 29.944 \\ 13.78 & 29.944 & 65.6578 \end{bmatrix} \begin{bmatrix} a_0 \\ a_1 \\ a_2 \end{bmatrix} = \begin{bmatrix} 4.6742 \\ 10.0571 \\ 21.8385 \end{bmatrix}$$

which yields $\mathbf{a} = \begin{bmatrix} -0.7714 & 1.5075 & -0.1930 \end{bmatrix}^T$.

The derivatives of the interpolant are $P_2'(x) = a_1 + 2a_2 x$ and $P_2''(x) = 2a_2$. Therefore,

$$f'(2) \approx P_2'(2) = 1.5075 + 2(-0.1930)(2) = 0.7355$$
$$f''(2) \approx P_2''(2) = 2(-0.1930) = -0.3860$$

Solution of Part (2) We must first determine the second derivatives k_i of the spline at its knots, after which the derivatives of $f(x)$ can be computed from Eqs. (5.10) and (5.11). The first part can be carried out by the following small program:

```
#!/usr/bin/python
## example5_4
from cubicSpline import curvatures
from LUdecomp3 import *
from numarray import array

xData = array([1.5, 1.9, 2.1, 2.4, 2.6, 3.1])
yData = array([1.0628, 1.3961, 1.5432, 1.7349, 1.8423, 2.0397])
print curvatures(LUdecomp3,LUsolve3,xData,yData)
raw_input(''Press return to exit'')
```

The output of the program, consisting of k_0 to k_5, is

```
[ 0.   -0.4258431 -0.37744139 -0.38796663 -0.55400477  0.  ]
Press return to exit
```

5.4 Derivatives by Interpolation

Since $x = 2$ lies between knots 1 and 2, we must use Eqs. (5.10) and (5.11) with $i = 1$. This yields

$$f'(2) \approx f'_{1,2}(2) = \frac{k_1}{6}\left[\frac{3(x-x_2)^2}{x_1-x_2} - (x_1-x_2)\right]$$

$$-\frac{k_2}{6}\left[\frac{3(x-x_1)^2}{x_1-x_2} - (x_1-x_2)\right] + \frac{y_1-y_2}{x_1-x_2}$$

$$= \frac{(-0.4258)}{6}\left[\frac{3(2-2.1)^2}{(-0.2)} - (-0.2)\right]$$

$$-\frac{(-0.3774)}{6}\left[\frac{3(2-1.9)^2}{(-0.2)} - (-0.2)\right] + \frac{1.3961-1.5432}{(-0.2)}$$

$$= 0.7351$$

$$f''(2) \approx f''_{1,2}(2) = k_1\frac{x-x_2}{x_1-x_2} - k_2\frac{x-x_1}{x_1-x_2}$$

$$= (-0.4258)\frac{2-2.1}{(-0.2)} - (-0.3774)\frac{2-1.9}{(-0.2)} = -0.4016$$

Note that the solutions for $f'(2)$ in parts (1) and (2) differ only in the fourth significant figure, but the values of $f''(2)$ are much farther apart. This is not unexpected, considering the general rule: the higher the order of the derivative, the lower the precision with which it can be computed. It is impossible to tell which of the two results is better without knowing the expression for $f(x)$. In this particular problem, the data points fall on the curve $f(x) = x^2 e^{-x/2}$, so that the "true" values of the derivatives are $f'(2) = 0.7358$ and $f''(2) = -0.3679$.

EXAMPLE 5.5
Determine $f'(0)$ and $f'(1)$ from the following noisy data

x	0	0.2	0.4	0.6
$f(x)$	1.9934	2.1465	2.2129	2.1790
x	0.8	1.0	1.2	1.4
$f(x)$	2.0683	1.9448	1.7655	1.5891

Solution We used the program listed in Example 3.10 to find the best polynomial fit (in the least-squares sense) to the data. The program was run three times with the following results:

```
Degree of polynomial ==> 2
Coefficients are:
[2.0261875   0.64703869 -0.70239583]
```

Numerical Differentiation

```
Std. deviation = 0.0360968935809
Degree of polynomial ==> 3
Coefficients are:
[1.99215      1.09276786  -1.55333333   0.40520833]
Std. deviation = 0.0082604082973

Degree of polynomial ==> 4
Coefficients are:
[1.99185568   1.10282373  -1.59056108   0.44812973  -0.01532907]
Std. deviation = 0.00951925073521

Degree of polynomial ==>
Finished. Press return to exit
```

Based on standard deviation, the cubic seems to be the best candidate for the interpolant. Before accepting the result, we compare the plots of the data points and the interpolant—see the figure. The fit does appear to be satisfactory

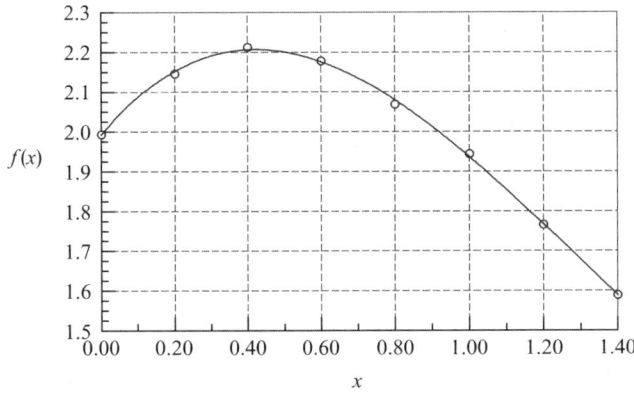

Approximating $f(x)$ by the interpolant, we have

$$f(x) \approx a_0 + a_1 x + a_2 x^2 + a_3 x^3$$

so that

$$f'(x) \approx a_1 + 2a_2 x + 3a_3 x^2$$

Therefore,

$$f'(0) \approx a_1 = 1.093$$

$$f'(1) = a_1 + 2a_2 + 3a_3 = 1.093 + 2(-1.553) + 3(0.405) = -0.798$$

5.4 Derivatives by Interpolation

In general, derivatives obtained from noisy data are at best rough approximations. In this problem, the data represent $f(x) = (x+2)/\cosh x$ with added random noise. Thus $f'(x) = [1 - (x+2)\tanh x]/\cosh x$, so that the "correct" derivatives are $f'(0) = 1.000$ and $f'(1) = -0.833$.

PROBLEM SET 5.1

1. Given the values of $f(x)$ at the points x, $x - h_1$ and $x + h_2$, where $h_1 \neq h_2$, determine the finite difference approximation for $f''(x)$. What is the order of the truncation error?

2. Given the first backward finite difference approximations for $f'(x)$ and $f''(x)$, derive the first backward finite difference approximation for $f'''(x)$ using the operation $f'''(x) = [f''(x)]'$.

3. Derive the central difference approximation for $f''(x)$ accurate to $\mathcal{O}(h^4)$ by applying Richardson extrapolation to the central difference approximation of $\mathcal{O}(h^2)$.

4. Derive the second forward finite difference approximation for $f'''(x)$ from the Taylor series.

5. Derive the first central difference approximation for $f^{(4)}(x)$ from the Taylor series.

6. Use finite difference approximations of $\mathcal{O}(h^2)$ to compute $f'(2.36)$ and $f''(2.36)$ from the data

x	2.36	2.37	2.38	2.39
$f(x)$	0.85866	0.86289	0.86710	0.87129

7. Estimate $f'(1)$ and $f''(1)$ from the following data:

x	0.97	1.00	1.05
$f(x)$	0.85040	0.84147	0.82612

8. Given the data

x	0.84	0.92	1.00	1.08	1.16
$f(x)$	0.431711	0.398519	0.367879	0.339596	0.313486

    calculate $f''(1)$ as accurately as you can.

9. Use the data in the table to compute $f'(0.2)$ as accurately as possible.

x	0	0.1	0.2	0.3	0.4
$f(x)$	0.000 000	0.078 348	0.138 910	0.192 916	0.244 981

10. Using five significant figures in the computations, determine $d(\sin x)/dx$ at $x = 0.8$ from (a) the first forward difference approximation and (b) the first central difference approximation. In each case, use h that gives the most accurate result (this requires experimentation).

11. ■ Use polynomial interpolation to compute f' and f'' at $x = 0$, using the data

x	-2.2	-0.3	0.8	1.9
$f(x)$	15.180	10.962	1.920	-2.040

12. ■

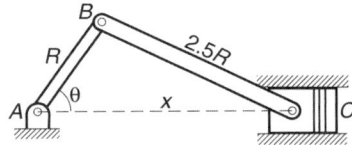

The crank AB of length $R = 90$ mm is rotating at the constant angular speed of $d\theta/dt = 5000$ rev/min. The position of the piston C can be shown to vary with the angle θ as

$$x = R\left(\cos\theta + \sqrt{2.5^2 - \sin^2\theta}\right)$$

Write a program that computes the acceleration of the piston at $\theta = 0°, 5°, 10°, \ldots, 180°$ by numerical differentiation.

13. ■

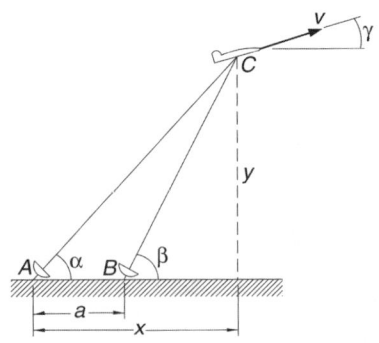

5.4 Derivatives by Interpolation

The radar stations A and B, separated by the distance $a = 500$ m, track the plane C by recording the angles α and β at one-second intervals. If three successive readings are

t (s)	9	10	11
α	54.80°	54.06°	53.34°
β	65.59°	64.59°	63.62°

calculate the speed v of the plane and the climb angle γ at $t = 10$ s. The coordinates of the plane can be shown to be

$$x = a\frac{\tan\beta}{\tan\beta - \tan\alpha} \qquad y = a\frac{\tan\alpha \tan\beta}{\tan\beta - \tan\alpha}$$

14. ■

Geometric analysis of the linkage shown resulted in the following table relating the angles θ and β:

θ (deg)	0	30	60	90	120	150
β (deg)	59.96	56.42	44.10	25.72	−0.27	−34.29

Assuming that member AB of the linkage rotates with the constant angular velocity $d\theta/dt = 1$ rad/s, compute $d\beta/dt$ in rad/s at the tabulated values of θ. Use cubic spline interpolation.

6 Numerical Integration

> Compute $\int_a^b f(x)\,dx$, where $f(x)$ is a given function

6.1 Introduction

Numerical integration, also known as *quadrature*, is intrinsically a much more accurate procedure than numerical differentiation. Quadrature approximates the definite integral

$$\int_a^b f(x)\,dx$$

by the sum

$$I = \sum_{i=0}^{n} A_i f(x_i)$$

where the *nodal abscissas* x_i and *weights* A_i depend on the particular rule used for the quadrature. All rules of quadrature are derived from polynomial interpolation of the integrand. Therefore, they work best if $f(x)$ can be approximated by a polynomial.

Methods of numerical integration can be divided into two groups: Newton–Cotes formulas and Gaussian quadrature. Newton–Cotes formulas are characterized by equally spaced abscissas, and include well-known methods such as the trapezoidal rule and Simpson's rule. They are most useful if $f(x)$ has already been computed at equal intervals, or can be computed at low cost. Since Newton–Cotes formulas are based on local interpolation, they require only a piecewise fit to a polynomial.

In Gaussian quadrature the locations of the abscissas are chosen to yield the best possible accuracy. Because Gaussian quadrature requires fewer evaluations of the integrand for a given level of precision, it is popular in cases where $f(x)$ is expensive

6.2 Newton–Cotes Formulas

to evaluate. Another advantage of Gaussian quadrature is its ability to handle integrable singularities, enabling us to evaluate expressions such as

$$\int_0^1 \frac{g(x)}{\sqrt{1-x^2}} dx$$

provided that $g(x)$ is a well-behaved function.

6.2 Newton–Cotes Formulas

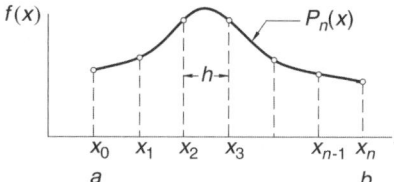

Figure 6.1. Polynomial approximation of $f(x)$.

Consider the definite integral

$$\int_a^b f(x)\, dx \qquad (6.1)$$

We divide the range of integration (a, b) into n equal intervals of length $h = (b-a)/n$, as shown in Fig. 6.1, and denote the abscissas of the resulting nodes by $x_0, x_1, \ldots, x_n$. Next we approximate $f(x)$ by a polynomial of degree n that intersects all the nodes. Lagrange's form of this polynomial, Eq. (3.1a), is

$$P_n(x) = \sum_{i=0}^n f(x_i)\ell_i(x)$$

where $\ell_i(x)$ are the cardinal functions defined in Eq. (3.1b). Therefore, an approximation to the integral in Eq. (6.1) is

$$I = \int_a^b P_n(x)dx = \sum_{i=0}^n \left[f(x_i) \int_a^b \ell_i(x)dx \right] = \sum_{i=0}^n A_i f(x_i) \qquad (6.2a)$$

where

$$A_i = \int_a^b \ell_i(x)dx, \quad i = 0, 1, \ldots, n \qquad (6.2b)$$

Equations (6.2) are the *Newton–Cotes formulas*. Classical examples of these formulas are the *trapezoidal rule* ($n = 1$), *Simpson's rule* ($n = 2$) and *Simpson's 3/8 rule* ($n = 3$). The most important of these is the trapezoidal rule. It can be combined with Richardson extrapolation into an efficient algorithm known as *Romberg integration*, which makes the other classical rules somewhat redundant.

Trapezoidal Rule

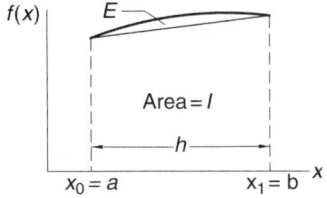

Figure 6.2. Trapezoidal rule.

If $n = 1$ (one panel), as illustrated in Fig. 6.2, we have $\ell_0 = (x - x_1)/(x_0 - x_1) = -(x - b)/h$. Therefore,

$$A_0 = \frac{1}{h}\int_a^b (x - b)\, dx = \frac{1}{2h}(b - a)^2 = \frac{h}{2}$$

Also $\ell_1 = (x - x_0)/(x_1 - x_0) = (x - a)/h$, so that

$$A_1 = \frac{1}{h}\int_a^b (x - a)\, dx = \frac{1}{2h}(b - a)^2 = \frac{h}{2}$$

Substitution in Eq. (6.2a) yields

$$I = [f(a) + f(b)]\frac{h}{2} \qquad (6.3)$$

which is known as the *trapezoidal rule*. It represents the area of the trapezoid in Fig. 6.2.

The error in the trapezoidal rule

$$E = \int_a^b f(x)\,dx - I$$

is the area of the region between $f(x)$ and the straight-line interpolant, as indicated in Fig. 6.2. It can be obtained by integrating the interpolation error in Eq. (3.3):

$$E = \frac{1}{2!}\int_a^b (x - x_0)(x - x_1) f''(\xi)\,dx = \frac{1}{2} f''(\xi) \int_a^b (x - a)(x - b)\,dx$$

$$= -\frac{1}{12}(b - a)^3 f''(\xi) = -\frac{h^3}{12} f''(\xi) \qquad (6.4)$$

Composite Trapezoidal Rule

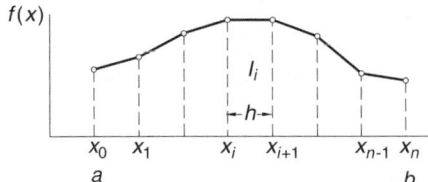

Figure 6.3. Composite trapezoidal rule.

6.2 Newton–Cotes Formulas

In practice the trapezoidal rule is applied in a piecewise fashion. Figure 6.3 shows the region (a, b) divided into n panels, each of width h. The function $f(x)$ to be integrated is approximated by a straight line in each panel. From the trapezoidal rule we obtain for the approximate area of a typical (ith) panel

$$I_i = [f(x_i) + f(x_{i+1})] \frac{h}{2}$$

Hence total area, representing $\int_a^b f(x)\,dx$, is

$$I = \sum_{i=0}^{n-1} I_i = [f(x_0) + 2f(x_1) + 2f(x_2) + \cdots + 2f(x_{n-1}) + f(x_n)] \frac{h}{2} \quad (6.5)$$

which is the *composite trapezoidal rule*.

The truncation error in the area of a panel is, from Eq. (6.4),

$$E_i = -\frac{h^3}{12} f''(\xi_i)$$

where ξ_i lies in (x_i, x_{i+1}). Hence the truncation error in Eq. (6.5) is

$$E = \sum_{i=0}^{n-1} E_i = -\frac{h^3}{12} \sum_{i=0}^{n-1} f''(\xi_i) \quad \text{(a)}$$

But

$$\sum_{i=0}^{n-1} f''(\xi_i) = n\bar{f}''$$

where $\bar{f}''$ is the arithmetic mean of the second derivatives. If $f''(x)$ is continuous, there must be a point ξ in (a, b) at which $f''(\xi) = \bar{f}''$, enabling us to write

$$\sum_{i=0}^{n-1} f''(\xi_i) = nf''(\xi) = \frac{b-a}{h} f''(\xi)$$

Therefore, Eq. (a) becomes

$$E = -\frac{(b-a)h^2}{12} f''(\xi) \quad (6.6)$$

It would be incorrect to conclude from Eq. (6.6) that $E = ch^2$ (c being a constant), because $f''(\xi)$ is not entirely independent of h. A deeper analysis of the error[13] shows that if $f(x)$ and its derivatives are finite in (a, b), then

$$E = c_1 h^2 + c_2 h^4 + c_3 h^6 + \cdots \quad (6.7)$$

---

[13] The analysis requires familiarity with the *Euler–Maclaurin summation formula*, which is covered in advanced texts.

Recursive Trapezoidal Rule

Let I_k be the integral evaluated with the composite trapezoidal rule using 2^{k-1} panels. Note that if k is increased by one, the number of panels is doubled. Using the notation

$$H = b - a$$

we obtain from Eq. (6.5) the following results for $k = 1, 2$ and 3.

$k = 1$ (1 panel):

$$I_1 = [f(a) + f(b)]\frac{H}{2} \tag{6.8}$$

$k = 2$ (2 panels):

$$I_2 = \left[f(a) + 2f\left(a + \frac{H}{2}\right) + f(b)\right]\frac{H}{4} = \frac{1}{2}I_1 + f\left(a + \frac{H}{2}\right)\frac{H}{2}$$

$k = 3$ (4 panels):

$$I_3 = \left[f(a) + 2f\left(a + \frac{H}{4}\right) + 2f\left(a + \frac{H}{2}\right) + 2f\left(a + \frac{3H}{4}\right) + f(b)\right]\frac{H}{8}$$

$$= \frac{1}{2}I_2 + \left[f\left(a + \frac{H}{4}\right) + f\left(a + \frac{3H}{4}\right)\right]\frac{H}{4}$$

We can now see that for arbitrary $k > 1$ we have

$$I_k = \frac{1}{2}I_{k-1} + \frac{H}{2^{k-1}}\sum_{i=1}^{2^{k-2}} f\left[a + \frac{(2i-1)H}{2^{k-1}}\right], \quad k = 2, 3, \ldots \tag{6.9a}$$

which is the *recursive trapezoidal rule*. Observe that the summation contains only the new nodes that were created when the number of panels was doubled. Therefore, the computation of the sequence $I_1, I_2, I_3, \ldots, I_k$ from Eqs. (6.8) and (6.9) involves the same amount of algebra as the calculation of I_k directly from Eq. (6.5). The advantage of using the recursive trapezoidal rule is that it allows us to monitor convergence and terminate the process when the difference between I_{k-1} and I_k becomes sufficiently small. A form of Eq. (6.9a) that is easier to remember is

$$I(h) = \frac{1}{2}I(2h) + h\sum f(x_{\text{new}}) \tag{6.9b}$$

where $h = H/n$ is the width of each panel.

■ trapezoid

The function `trapezoid` computes I_k (`Inew`), given I_{k-1} (`Iold`) from Eqs. (6.8) and (6.9). We can compute $\int_a^b f(x)\,dx$ by calling `trapezoid` with $k = 1, 2, \ldots$ until the desired precision is attained.

6.2 Newton–Cotes Formulas

```
## module trapezoid
''' Inew = trapezoid(f,a,b,Iold,k).
    Recursive trapezoidal rule:
    Iold = Integral of f(x) from x = a to b computed by
    trapezoidal rule with 2^(k-1) panels.
    Inew = Same integral computed with 2^k panels.
'''
def trapezoid(f,a,b,Iold,k):
    if k == 1:Inew = (f(a) + f(b))*(b - a)/2.0
    else:
        n = 2**(k -2 )       # Number of new points
        h = (b - a)/n        # Spacing of new points
        x = a + h/2.0        # Coord. of 1st new point
        sum = 0.0
        for i in range(n):
            sum = sum + f(x)
            x = x + h
        Inew = (Iold + h*sum)/2.0
    return Inew
```

Simpson's Rules

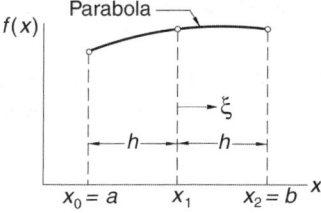

Figure 6.4. Simpson's 1/3 rule.

Simpson's 1/3 rule can be obtained from Newton–Cotes formulas with $n = 2$; that is, by passing a parabolic interpolant through three adjacent nodes, as shown in Fig. 6.4. The area under the parabola, which represents an approximation of $\int_a^b f(x)\,dx$, is (see derivation in Example 6.1)

$$I = \left[f(a) + 4f\left(\frac{a+b}{2}\right) + f(b) \right] \frac{h}{3} \tag{a}$$

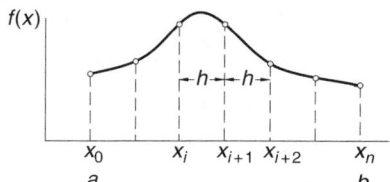

Figure 6.5. Composite Simpson's 1/3 rule.

Numerical Integration

To obtain the *composite Simpson's 1/3 rule*, the integration range (a, b) is divided into n panels (n even) of width $h = (b - a)/n$ each, as indicated in Fig. 6.5. Applying Eq. (a) to two adjacent panels, we have

$$\int_{x_i}^{x_{i+2}} f(x)\,dx \approx [f(x_i) + 4f(x_{i+1}) + f(x_{i+2})]\frac{h}{3} \qquad (b)$$

Substituting Eq. (b) into

$$\int_a^b f(x)\,dx = \int_{x_0}^{x_m} f(x)\,dx = \sum_{i=0,2,\ldots}^n \left[\int_{x_i}^{x_{i+2}} f(x)\,dx\right]$$

yields

$$\int_a^b f(x)\,dx \approx I = [f(x_0) + 4f(x_1) + 2f(x_2) + 4f(x_3) + \cdots \qquad (6.10)$$

$$\cdots + 2f(x_{n-2}) + 4f(x_{n-1}) + f(x_n)]\frac{h}{3}$$

The composite Simpson's 1/3 rule in Eq. (6.10) is perhaps the best-known method of numerical integration. Its reputation is somewhat undeserved, since the trapezoidal rule is more robust, and Romberg integration is more efficient.

The error in the composite Simpson's rule is

$$E = \frac{(b-a)h^4}{180} f^{(4)}(\xi) \qquad (6.11)$$

from which we conclude that Eq. (6.10) is exact if $f(x)$ is a polynomial of degree three or less.

Simpson's 1/3 rule requires the number of panels n to be even. If this condition is not satisfied, we can integrate over the first (or last) three panels with *Simpson's 3/8 rule*:

$$I = [f(x_0) + 3f(x_1) + 3f(x_2) + f(x_3)]\frac{3h}{8} \qquad (6.12)$$

and use Simpson's 1/3 rule for the remaining panels. The error in Eq. (6.12) is of the same order as in Eq. (6.10).

EXAMPLE 6.1
Derive Simpson's 1/3 rule from Newton–Cotes formulas.

Solution Referring to Fig. 6.4, we see that Simpson's 1/3 rule uses three nodes located at $x_0 = a$, $x_1 = (a+b)/2$ and $x_2 = b$. The spacing of the nodes is $h = (b-a)/2$. The cardinal functions of Lagrange's three-point interpolation are (see Section 3.2)

$$\ell_0(x) = \frac{(x-x_1)(x-x_2)}{(x_0-x_1)(x_0-x_2)} \qquad \ell_1(x) = \frac{(x-x_0)(x-x_2)}{(x_1-x_0)(x_1-x_2)}$$

$$\ell_2(x) = \frac{(x-x_0)(x-x_1)}{(x_2-x_0)(x_2-x_1)}$$

6.2 Newton–Cotes Formulas

The integration of these functions is easier if we introduce the variable ξ with origin at x_1. Then the coordinates of the nodes are $\xi_0 = -h$, $\xi_1 = 0$, $\xi_2 = h$, and Eq. (6.2b) becomes $A_i = \int_a^b \ell_i(x)dx = \int_{-h}^{h} \ell_i(\xi)d\xi$. Therefore,

$$A_0 = \int_{-h}^{h} \frac{(\xi - 0)(\xi - h)}{(-h)(-2h)} d\xi = \frac{1}{2h^2} \int_{-h}^{h} (\xi^2 - h\xi) d\xi = \frac{h}{3}$$

$$A_1 = \int_{-h}^{h} \frac{(\xi + h)(\xi - h)}{(h)(-h)} d\xi = -\frac{1}{h^2} \int_{-h}^{h} (\xi^2 - h^2) d\xi = \frac{4h}{3}$$

$$A_2 = \int_{-h}^{h} \frac{(\xi + h)(\xi - 0)}{(2h)(h)} d\xi = \frac{1}{2h^2} \int_{-h}^{h} (\xi^2 + h\xi) d\xi = \frac{h}{3}$$

Equation (6.2a) then yields

$$I = \sum_{i=0}^{2} A_i f(x_i) = \left[f(a) + 4f\left(\frac{a+b}{2}\right) + f(b) \right] \frac{h}{3}$$

which is Simpson's 1/3 rule.

EXAMPLE 6.2
Evaluate the bounds on $\int_0^{\pi} \sin(x)\, dx$ with the composite trapezoidal rule using (1) eight panels and (2) sixteen panels.

Solution of Part (1) With 8 panels there are 9 nodes spaced at $h = \pi/8$. The abscissas of the nodes are $x_i = i\pi/8$, $i = 0, 1, \ldots, 8$. From Eq. (6.5) we get

$$I = \left[\sin 0 + 2 \sum_{i=1}^{7} \sin \frac{i\pi}{8} + \sin \pi \right] \frac{\pi}{16} = 1.97423$$

The error is given by Eq. (6.6):

$$E = -\frac{(b-a)h^2}{12} f''(\xi) = -\frac{(\pi - 0)(\pi/8)^2}{12} (-\sin \xi) = \frac{\pi^3}{768} \sin \xi$$

where $0 < \xi < \pi$. Since we do not know the value of ξ, we cannot evaluate E, but we can determine its bounds:

$$E_{\min} = \frac{\pi^3}{768} \sin(0) = 0 \qquad E_{\max} = \frac{\pi^3}{768} \sin \frac{\pi}{2} = 0.040\,37$$

Therefore, $I + E_{\min} < \int_0^{\pi} \sin(x)\, dx < I + E_{\max}$, or

$$1.974\,23 < \int_0^{\pi} \sin(x)\, dx < 2.014\,60$$

The exact integral is, of course, 2.

Solution of Part (2) The new nodes created by the doubling of panels are located at midpoints of the old panels. Their abscissas are

$$x_j = \pi/16 + j\pi/8 = (1 + 2j)\pi/16, \quad j = 0, 1, \ldots, 7$$

Using the recursive trapezoidal rule in Eq. (6.9b), we get

$$I = \frac{1.974\,23}{2} + \frac{\pi}{16}\sum_{j=0}^{7}\sin\frac{(1+2j)\pi}{16} = 1.993\,58$$

and the bounds on the error become (note that E is quartered when h is halved) $E_{min} = 0$, $E_{max} = 0.040\,37/4 = 0.010\,09$. Hence

$$1.993\,58 < \int_{0}^{\pi}\sin(x)\,dx < 2.003\,67$$

EXAMPLE 6.3
Estimate $\int_{0}^{2.5} f(x)\,dx$ from the data

x	0	0.5	1.0	1.5	2.0	2.5
$f(x)$	1.5000	2.0000	2.0000	1.6364	1.2500	0.9565

Solution We will use Simpson's rules, since they are more accurate than the trapezoidal rule. Because the number of panels is odd, we compute the integral over the first three panels by Simpson's 3/8 rule, and use the 1/3 rule for the last two panels:

$$I = [f(0) + 3f(0.5) + 3f(1.0) + f(1.5)]\frac{3(0.5)}{8}$$

$$+ [f(1.5) + 4f(2.0) + f(2.5)]\frac{0.5}{3}$$

$$= 2.8381 + 1.2655 = 4.1036$$

EXAMPLE 6.4
Use the recursive trapezoidal rule to evaluate $\int_{0}^{\pi} \sqrt{x}\cos x\,dx$ to six decimal places. How many panels are needed to achieve this result?

Solution The program listed below utilizes the function `trapezoid`.

```
#!/usr/bin/python
## example6_4
from math import sqrt,cos,pi
from trapezoid import *

def f(x): return sqrt(x)*cos(x)

Iold = 0.0
for k in range(1,21):
    Inew = trapezoid(f,0.0,pi,Iold,k)
    if (k > 1) and (abs(Inew - Iold)) < 1.0e-6: break
    Iold = Inew
```

```
print ''Integral ='',Inew
print ''nPanels ='',2**(k-1)
raw_input(''\nPress return to exit'')
```

The output from the program is:

```
Integral = -0.894831664853
nPanels = 32768
```

Hence $\int_0^\pi \sqrt{x} \cos x \, dx = -0.894\,832$ requiring 32 768 panels. The slow convergence is the result of all the derivatives of $f(x)$ being singular at $x = 0$. Consequently, the error does not behave as shown in Eq. (6.7): $E = c_1 h^2 + c_2 h^4 + \cdots$, but is unpredictable. Difficulties of this nature can often be remedied by a change in variable. In this case, we introduce $t = \sqrt{x}$, so that $dt = dx/(2\sqrt{x}) = dx/(2t)$, or $dx = 2t\, dt$. Thus

$$\int_0^\pi \sqrt{x} \cos x \, dx = \int_0^{\sqrt{\pi}} 2t^2 \cos t^2 \, dt$$

Evaluation of the integral on the right-hand side was completed with 4096 panels.

6.3 Romberg Integration

Romberg integration combines the trapezoidal rule with Richardson extrapolation (see Section 5.3). Let us first introduce the notation

$$R_{i,1} = I_i$$

where, as before, I_i represents the approximate value of $\int_a^b f(x)dx$ computed by the recursive trapezoidal rule using 2^{i-1} panels. Recall that the error in this approximation is $E = c_1 h^2 + c_2 h^4 + \cdots$, where

$$h = \frac{b-a}{2^{i-1}}$$

is the width of a panel.

Romberg integration starts with the computation of $R_{1,1} = I_1$ (one panel) and $R_{2,1} = I_2$ (two panels) from the trapezoidal rule. The leading error term $c_1 h^2$ is then eliminated by Richardson extrapolation. Using $p = 2$ (the exponent in the leading error term) in Eq. (5.9) and denoting the result by $R_{2,2}$, we obtain

$$R_{2,2} = \frac{2^2 R_{2,1} - R_{1,1}}{2^2 - 1} = \frac{4}{3} R_{2,1} - \frac{1}{3} R_{1,1} \tag{a}$$

Numerical Integration

It is convenient to store the results in an array of the form

$$\begin{bmatrix} R_{1,1} & \\ R_{2,1} & R_{2,2} \end{bmatrix}$$

The next step is to calculate $R_{3,1} = I_3$ (four panels) and repeat Richardson extrapolation with $R_{2,1}$ and $R_{3,1}$, storing the result as $R_{3,2}$:

$$R_{3,2} = \frac{4}{3} R_{3,1} - \frac{1}{3} R_{2,1} \tag{b}$$

The elements of array **R** calculated so far are

$$\begin{bmatrix} R_{1,1} & & \\ R_{2,1} & R_{2,2} & \\ R_{3,1} & R_{3,2} & \end{bmatrix}$$

Both elements of the second column have an error of the form $c_2 h^4$, which can also be eliminated with Richardson extrapolation. Using $p = 4$ in Eq. (5.9), we get

$$R_{3,3} = \frac{2^4 R_{3,2} - R_{2,2}}{2^4 - 1} = \frac{16}{15} R_{3,2} - \frac{1}{15} R_{2,2} \tag{c}$$

This result has an error of $\mathcal{O}(h^6)$. The array has now expanded to

$$\begin{bmatrix} R_{1,1} & & \\ R_{2,1} & R_{2,2} & \\ R_{3,1} & R_{3,2} & R_{3,3} \end{bmatrix}$$

After another round of calculations we get

$$\begin{bmatrix} R_{1,1} & & & \\ R_{2,1} & R_{2,2} & & \\ R_{3,1} & R_{3,2} & R_{3,3} & \\ R_{4,1} & R_{4,2} & R_{4,3} & R_{4,4} \end{bmatrix}$$

where the error in $R_{4,4}$ is $\mathcal{O}(h^8)$. Note that the most accurate estimate of the integral is always the last diagonal term of the array. This process is continued until the difference between two successive diagonal terms becomes sufficiently small. The general extrapolation formula used in this scheme is

$$R_{i,j} = \frac{4^{j-1} R_{i,j-1} - R_{i-1,j-1}}{4^{j-1} - 1}, \quad i > 1, \quad j = 2, 3, \ldots, i \tag{6.13a}$$

A pictorial representation of Eq. (6.13a) is

6.3 Romberg Integration

$$R_{i-1,j-1}$$
$$\searrow$$
$$\alpha \qquad (6.13b)$$
$$\searrow$$
$$R_{i,j-1} \rightarrow \beta \rightarrow R_{i,j}$$

where the multipliers α and β depend on j in the following manner:

j	2	3	4	5	6
α	$-1/3$	$-1/15$	$-1/63$	$-1/255$	$-1/1023$
β	$4/3$	$16/15$	$64/63$	$256/255$	$1024/1023$

(6.13c)

The triangular array is convenient for hand computations, but computer implementation of the Romberg algorithm can be carried out within a one-dimensional array $\mathbf{R}'$. After the first extrapolation—see Eq. (a)—$R_{1,1}$ is never used again, so that it can be replaced with $R_{2,2}$. As a result, we have the array

$$\begin{bmatrix} R'_1 = R_{2,2} \\ R'_2 = R_{2,1} \end{bmatrix}$$

In the second extrapolation round, defined by Eqs. (b) and (c), $R_{3,2}$ overwrites $R_{2,1}$, and $R_{3,3}$ replaces $R_{2,2}$, so that the array contains

$$\begin{bmatrix} R'_1 = R_{3,3} \\ R'_2 = R_{3,2} \\ R'_3 = R_{3,1} \end{bmatrix}$$

and so on. In this manner, R'_1 always contains the best current result. The extrapolation formula for the kth round is

$$R'_j = \frac{4^{k-j} R'_{j+1} - R'_j}{4^{k-j} - 1}, \quad j = k-1, k-2, \ldots, 1 \qquad (6.14)$$

■ **romberg**

The algorithm for Romberg integration is implemented in the function romberg. It returns the integral and the number of panels used. Richardson's extrapolation is carried out by the subfunction richardson.

```
## module romberg
''' I,nPanels = romberg(f,a,b,tol=1.0e-6).
    Romberg integration of f(x) from x = a to b.
    Returns the integral and the number of panels used.
```

Numerical Integration

```
    '''
    from numarray import zeros,Float64
    from trapezoid import *

    def romberg(f,a,b,tol=1.0e-6):

        def richardson(r,k):
            for j in range(k-1,0,-1):
                const = 4.0**(k-j)
                r[j] = (const*r[j+1] - r[j])/(const - 1.0)
            return r

        r = zeros((21),type=Float64)
        r[1] = trapezoid(f,a,b,0.0,1)
        r_old = r[1]
        for k in range(2,21):
            r[k] = trapezoid(f,a,b,r[k-1],k)
            r = richardson(r,k)
            if abs(r[1]-r_old) < tol*max(abs(r[1]),1.0):
                return r[1],2**(k-1)
            r_old = r[1]
        print ''Romberg quadrature did not converge''
```

EXAMPLE 6.5
Show that $R_{k,2}$ in Romberg integration is identical to the composite Simpson's 1/3 rule in Eq. (6.10) with 2^{k-1} panels.

Solution Recall that in Romberg integration $R_{k,1} = I_k$ denoted the approximate integral obtained by the composite trapezoidal rule with $n = 2^{k-1}$ panels. Denoting the abscissas of the nodes by $x_0, x_1, \ldots, x_n$, we have from the composite trapezoidal rule in Eq. (6.5)

$$R_{k,1} = I_k = \left[f(x_0) + 2\sum_{i=1}^{n-1} f(x_i) + \frac{1}{2} f(x_n) \right] \frac{h}{2}$$

When we halve the number of panels (panel width $2h$), only the even-numbered abscissas enter the composite trapezoidal rule, yielding

$$R_{k-1,1} = I_{k-1} = \left[f(x_0) + 2 \sum_{i=2,4,\ldots}^{n-2} f(x_i) + f(x_n) \right] h$$

6.3 Romberg Integration

Applying Richardson extrapolation yields

$$R_{k,2} = \frac{4}{3} R_{k,1} - \frac{1}{3} R_{k-1,1}$$

$$= \left[\frac{1}{3} f(x_0) + \frac{4}{3} \sum_{i=1,3,\ldots}^{n-1} f(x_i) + \frac{2}{3} \sum_{i=2,4,\ldots}^{n-2} f(x_i) + \frac{1}{3} f(x_n) \right] h$$

which agrees with Eq. (6.10).

EXAMPLE 6.6
Use Romberg integration to evaluate $\int_0^\pi f(x)\,dx$, where $f(x) = \sin x$. Work with four decimal places.

Solution From the recursive trapezoidal rule in Eq. (6.9b) we get

$$R_{1,1} = I(\pi) = \frac{\pi}{2}[f(0) + f(\pi)] = 0$$

$$R_{2,1} = I(\pi/2) = \frac{1}{2} I(\pi) + \frac{\pi}{2} f(\pi/2) = 1.5708$$

$$R_{3,1} = I(\pi/4) = \frac{1}{2} I(\pi/2) + \frac{\pi}{4} [f(\pi/4) + f(3\pi/4)] = 1.8961$$

$$R_{4,1} = I(\pi/8) = \frac{1}{2} I(\pi/4) + \frac{\pi}{8} [f(\pi/8) + f(3\pi/8) + f(5\pi/8) + f(7\pi/8)]$$
$$= 1.9742$$

Using the extrapolation formulas in Eqs. (6.13), we can now construct the following table:

$$\begin{bmatrix} R_{1,1} & & & \\ R_{2,1} & R_{2,2} & & \\ R_{3,1} & R_{3,2} & R_{3,3} & \\ R_{4,1} & R_{4,2} & R_{4,3} & R_{4,4} \end{bmatrix} = \begin{bmatrix} 0 & & & \\ 1.5708 & 2.0944 & & \\ 1.8961 & 2.0046 & 1.9986 & \\ 1.9742 & 2.0003 & 2.0000 & 2.0000 \end{bmatrix}$$

It appears that the procedure has converged. Therefore, $\int_0^\pi \sin x\,dx = R_{4,4} = 2.0000$, which is, of course, the correct result.

EXAMPLE 6.7
Use Romberg integration to evaluate $\int_0^{\sqrt{\pi}} 2x^2 \cos x^2\,dx$ and compare the results with Example 6.4.

Solution
```
#!/usr/bin/python
## example6_7
from math import cos,sqrt,pi
```

Numerical Integration

```
from romberg import *

def f(x): return 2.0*(x**2)*cos(x**2)

I,n = romberg(f,0,sqrt(pi))
print ''Integral ='',I
print ''nPanels ='',n
raw_input(''\nPress return to exit'')
```

The results of running the program are:

```
Integral = -0.894831469504
nPanels = 64
```

It is clear that Romberg integration is considerably more efficient than the trapezoidal rule—it required 64 panels as compared to 4096 panels for the trapezoidal rule in Example 6.4.

PROBLEM SET 6.1

1. Use the recursive trapezoidal rule to evaluate $\int_0^{\pi/4} \ln(1 + \tan x)dx$. Explain the results.

2. The table shows the power P supplied to the driving wheels of a car as a function of the speed v. If the mass of the car is $m = 2000$ kg, determine the time Δt it takes for the car to accelerate from 1 m/s to 6 m/s. Use the trapezoidal rule for integration. *Hint*:

$$\Delta t = m \int_{1s}^{6s} (v/P)\, dv$$

which can be derived from Newton's law $F = m(dv/dt)$ and the definition of power $P = Fv$.

v (m/s)	0	1.0	1.8	2.4	3.5	4.4	5.1	6.0
P (kW)	0	4.7	12.2	19.0	31.8	40.1	43.8	43.2

3. Evaluate $\int_{-1}^{1} \cos(2\cos^{-1} x)dx$ with Simpson's 1/3 rule using 2, 4 and 6 panels. Explain the results.

4. Determine $\int_1^\infty (1 + x^4)^{-1} dx$ with the trapezoidal rule using five panels and compare the result with the "exact" integral 0.243 75. *Hint*: use the transformation $x^3 = 1/t$.

5.

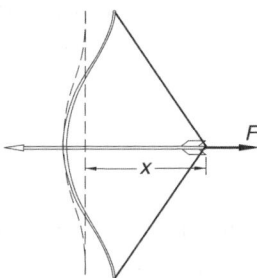

The table below gives the pull F of the bow as a function of the draw x. If the bow is drawn 0.5 m, determine the speed of the 0.075-kg arrow when it leaves the bow. *Hint*: the kinetic energy of arrow equals the work done in drawing the bow; that is, $mv^2/2 = \int_0^{0.5m} F\, dx$.

x (m)	0.00	0.05	0.10	0.15	0.20	0.25
F (N)	0	37	71	104	134	161
x (m)	0.30	0.35	0.40	0.45	0.50	
F (N)	185	207	225	239	250	

6. Evaluate $\int_0^2 (x^5 + 3x^3 - 2)\, dx$ by Romberg integration.

7. Estimate $\int_0^\pi f(x)\, dx$ as accurately as possible, where $f(x)$ is defined by the data

x	0	$\pi/4$	$\pi/2$	$3\pi/4$	π
$f(x)$	1.0000	0.3431	0.2500	0.3431	1.0000

8. Evaluate

$$\int_0^1 \frac{\sin x}{\sqrt{x}}\, dx$$

with Romberg integration. *Hint*: use transformation of variable to eliminate the indeterminacy at $x = 0$.

9. Newton–Cotes formulas for evaluating $\int_a^b f(x)\, dx$ were based on polynomial approximations of $f(x)$. Show that if $y = f(x)$ is approximated by a natural cubic spline with evenly spaced knots at $x_0, x_1, \ldots, x_n$, the quadrature formula becomes

$$I = \frac{h}{2}(y_0 + 2y_1 + 2y_2 + \cdots + 2y_{n-1} + y_n)$$
$$- \frac{h^3}{24}(k_0 + 2k_1 + k_2 + \cdots + 2k_{n-1} + k_n)$$

where h is the distance between the knots and $k_i = y_i''$. Note that the first part is the composite trapezoidal rule; the second part may be viewed as a "correction" for curvature.

10. ■ Evaluate

$$\int_0^{\pi/4} \frac{dx}{\sqrt{\sin x}}$$

with Romberg integration. *Hint*: use the transformation $\sin x = t^2$.

11. ■ The period of a simple pendulum of length L is $\tau = 4\sqrt{L/g}\, h(\theta_0)$, where g is the gravitational acceleration, θ_0 represents the angular amplitude and

$$h(\theta_0) = \int_0^{\pi/2} \frac{d\theta}{\sqrt{1 - \sin^2(\theta_0/2)\sin^2\theta}}$$

Compute $h(15°)$, $h(30°)$ and $h(45°)$, and compare these values with $h(0) = \pi/2$ (the approximation used for small amplitudes).

12. ■

The figure shows an elastic half-space that carries uniform loading of intensity q over a circular area of radius a. The vertical displacement of the surface at point P can be shown to be

$$w(r) = w_0 \int_0^{\pi/2} \frac{\cos^2\theta}{\sqrt{(r/a)^2 - \sin^2\theta}} d\theta \qquad r \geq a$$

where w_0 is the displacement at $r = a$. Use numerical integration to determine w/w_0 at $r = 2a$.

13. ■

6.4 Gaussian Integration

The mass m is attached to a spring of free length b and stiffness k. The coefficient of friction between the mass and the horizontal rod is μ. The acceleration of the mass can be shown to be (you may wish to prove this) $\ddot{x} = -f(x)$, where

$$f(x) = \mu g + \frac{k}{m}(\mu b + x)\left(1 - \frac{b}{\sqrt{b^2 + x^2}}\right)$$

If the mass is released from rest at $x = b$, its speed at $x = 0$ is given by

$$v_0 = \sqrt{2\int_0^b f(x)dx}$$

Compute v_0 by numerical integration using the data $m = 0.8$ kg, $b = 0.4$ m, $\mu = 0.3$, $k = 80$ N/m and $g = 9.81$ m/s^2.

14. ■ Debye's formula for the heat capacity C_V of a solid is $C_V = 9Nkg(u)$, where

$$g(u) = u^3 \int_0^{1/u} \frac{x^4 e^x}{(e^x - 1)^2} dx$$

The terms in this equation are

$$N = \text{number of particles in the solid}$$
$$k = \text{Boltzmann constant}$$
$$u = T/\Theta_D$$
$$T = \text{absolute temperature}$$
$$\Theta_D = \text{Debye temperature}$$

Compute $g(u)$ from $u = 0$ to 1.0 in intervals of 0.05 and plot the results.

15. ■ A power spike in an electric circuit results in the current

$$i(t) = i_0 e^{-t/t_0} \sin(2t/t_0)$$

across a resistor. The energy E dissipated by the resistor is

$$E = \int_0^\infty R\,[i(t)]^2\,dt$$

Find E using the data $i_0 = 100$ A, $R = 0.5\,\Omega$ and $t_0 = 0.01$ s.

6.4 Gaussian Integration

Gaussian Integration Formulas

We found that Newton–Cotes formulas for approximating $\int_a^b f(x)dx$ work best if $f(x)$ is a smooth function, such as a polynomial. This is also true for Gaussian quadrature. However, Gaussian formulas are also good at estimating integrals of the form

$$\int_a^b w(x) f(x) dx \qquad (6.15)$$

where $w(x)$, called the *weighting function*, can contain singularities, as long as they are integrable. An example of such an integral is $\int_0^1 (1 + x^2) \ln x \, dx$. Sometimes infinite limits, as in $\int_0^\infty e^{-x} \sin x \, dx$, can also be accommodated.

Gaussian integration formulas have the same form as Newton–Cotes rules:

$$I = \sum_{i=0}^n A_i f(x_i) \qquad (6.16)$$

where, as before, I represents the approximation to the integral in Eq. (6.15). The difference lies in the way that the weights A_i and nodal abscissas x_i are determined. In Newton–Cotes integration the nodes were evenly spaced in (a, b), i.e., their locations were predetermined. In Gaussian quadrature the nodes and weights are chosen so that Eq. (6.16) yields the exact integral if $f(x)$ is a polynomial of degree $2n + 1$ or less; that is,

$$\int_a^b w(x) P_m(x) dx = \sum_{i=0}^n A_i P_m(x_i), \quad m \leq 2n + 1 \qquad (6.17)$$

One way of determining the weights and abscissas is to substitute $P_0(x) = 1$, $P_1(x) = x, \ldots, P_{2n+1}(x) = x^{2n+1}$ in Eq. (6.17) and solve the resulting $2n + 2$ equations

$$\int_a^b w(x) x^j dx = \sum_{i=0}^n A_i x_i^j, \quad j = 0, 1, \ldots, 2n + 1$$

for the unknowns A_i and x_i.

As an illustration, let $w(x) = e^{-x}$, $a = 0$, $b = \infty$ and $n = 1$. The four equations determining x_0, x_1, A_0 and A_1 are

$$\int_0^\infty e^{-x} dx = A_0 + A_1$$

$$\int_0^\infty e^{-x} x \, dx = A_0 x_0 + A_1 x_1$$

$$\int_0^\infty e^{-x} x^2 dx = A_0 x_0^2 + A_1 x_1^2$$

$$\int_0^\infty e^{-x} x^3 dx = A_0 x_0^3 + A_1 x_1^3$$

6.4 Gaussian Integration

After evaluating the integrals, we get

$$A_0 + A_1 = 1$$
$$A_0 x_0 + A_1 x_1 = 1$$
$$A_0 x_0^2 + A_1 x_1^2 = 2$$
$$A_0 x_0^3 + A_1 x_1^3 = 6$$

The solution is

$$x_0 = 2 - \sqrt{2} \qquad A_0 = \frac{\sqrt{2}+1}{2\sqrt{2}}$$

$$x_1 = 2 + \sqrt{2} \qquad A_1 = \frac{\sqrt{2}-1}{2\sqrt{2}}$$

so that the integration formula becomes

$$\int_0^\infty e^{-x} f(x) dx \approx \frac{1}{2\sqrt{2}} \left[(\sqrt{2}+1) f\left(2-\sqrt{2}\right) + (\sqrt{2}-1) f\left(2+\sqrt{2}\right) \right]$$

Due to the nonlinearity of the equations, this approach will not work well for large n. Practical methods of finding x_i and A_i require some knowledge of orthogonal polynomials and their relationship to Gaussian quadrature. There are, however, several "classical" Gaussian integration formulas for which the abscissas and weights have been computed with great precision and tabulated. These formulas can be used without knowing the theory behind them, since all one needs for Gaussian integration are the values of x_i and A_i. If you do not intend to venture outside the classical formulas, you can skip the next two topics of this chapter.

*Orthogonal Polynomials

Orthogonal polynomials are employed in many areas of mathematics and numerical analysis. They have been studied thoroughly and many of their properties are known. What follows is a very small compendium of a large topic.

The polynomials $\varphi_n(x)$, $n = 0, 1, 2, \ldots$ (n is the degree of the polynomial) are said to form an *orthogonal set* in the interval (a, b) with respect to the *weighting function* $w(x)$ if

$$\int_a^b w(x) \varphi_m(x) \varphi_n(x) dx = 0, \quad m \neq n \tag{6.18}$$

The set is determined, except for a constant factor, by the choice of the weighting function and the limits of integration. That is, each set of orthogonal polynomials is associated with certain $w(x)$, a and b. The constant factor is specified by standardization. Some of the classical orthogonal polynomials, named after well-known mathematicians, are listed in Table 6.1. The last column in the table shows the standardization used.

Name	Symbol	a	b	w(x)	$\int_a^b w(x)\left[\varphi_n(x)\right]^2 dx$
Legendre	$p_n(x)$	-1	1	1	$2/(2n+1)$
Chebyshev	$T_n(x)$	-1	1	$(1-x^2)^{-1/2}$	$\pi/2$ $(n>0)$
Laguerre	$L_n(x)$	0	∞	e^{-x}	1
Hermite	$H_n(x)$	$-\infty$	∞	e^{-x^2}	$\sqrt{\pi}2^n n!$

Table 6.1

Orthogonal polynomials obey recurrence relations of the form

$$a_n\varphi_{n+1}(x) = (b_n + c_n x)\varphi_n(x) - d_n\varphi_{n-1}(x) \tag{6.19}$$

If the first two polynomials of the set are known, the other members of the set can be computed from Eq. (6.19). The coefficients in the recurrence formula, together with $\varphi_0(x)$ and $\varphi_1(x)$, are given in Table 6.2.

Name	$\varphi_0(x)$	$\varphi_1(x)$	a_n	b_n	c_n	d_n
Legendre	1	x	$n+1$	0	$2n+1$	n
Chebyshev	1	x	1	0	2	1
Laguerre	1	$1-x$	$n+1$	$2n+1$	-1	n
Hermite	1	$2x$	1	0	2	2

Table 6.2

The classical orthogonal polynomials are also obtainable from the formulas

$$p_n(x) = \frac{(-1)^n}{2^n n!}\frac{d^n}{dx^n}\left[(1-x^2)^n\right]$$
$$T_n(x) = \cos(n\cos^{-1} x), \quad n > 0$$
$$L_n(x) = \frac{e^x}{n!}\frac{d^n}{dx^n}(x^n e^{-x}) \tag{6.20}$$
$$H_n(x) = (-1)^n e^{x^2}\frac{d^n}{dx^n}(e^{-x^2})$$

and their derivatives can be calculated from

$$(1-x^2)p'_n(x) = n[-xp_n(x) + p_{n-1}(x)]$$
$$(1-x^2)T'_n(x) = n[-xT_n(x) + nT_{n-1}(x)]$$
$$xL'_n(x) = n[L_n(x) - L_{n-1}(x)] \tag{6.21}$$
$$H'_n(x) = 2nH_{n-1}(x)$$

6.4 Gaussian Integration

Other properties of orthogonal polynomials that have relevance to Gaussian integration are:

- $\varphi_n(x)$ has n real, distinct zeroes in the interval (a, b).
- The zeros of $\varphi_n(x)$ lie between the zeros of $\varphi_{n+1}(x)$.
- Any polynomial $P_n(x)$ of degree n can be expressed in the form

$$P_n(x) = \sum_{i=0}^{n} c_i \varphi_i(x) \tag{6.22}$$

- It follows from Eq. (6.22) and the orthogonality property in Eq. (6.18) that

$$\int_a^b w(x) P_n(x) \varphi_{n+m}(x) dx = 0, \quad m \geq 0 \tag{6.23}$$

*Determination of Nodal Abscissas and Weights

Theorem The nodal abscissas $x_0, x_1, \ldots, x_n$ are the zeros of the polynomial $\varphi_{n+1}(x)$ that belongs to the orthogonal set defined in Eq. (6.18).

Proof We start the proof by letting $f(x) = P_{2n+1}(x)$ be a polynomial of degree $2n + 1$. Since the Gaussian integration with $n + 1$ nodes is exact for this polynomial, we have

$$\int_a^b w(x) P_{2n+1}(x) dx = \sum_{i=0}^{n} A_i P_{2n+1}(x_i) \tag{a}$$

A polynomial of degree $2n + 1$ can always written in the form

$$P_{2n+1}(x) = Q_n(x) + R_n(x) \varphi_{n+1}(x) \tag{b}$$

where $Q_n(x)$, $R_n(x)$ and $\varphi_{n+1}(x)$ are polynomials of the degree indicated by the subscripts.[14] Therefore,

$$\int_a^b w(x) P_{2n+1}(x) dx = \int_a^b w(x) Q_n(x) dx + \int_a^b w(x) R_n(x) \varphi_{n+1}(x) dx$$

But according to Eq. (6.23) the second integral on the right-hand side vanishes, so that

$$\int_a^b w(x) P_{2n+1}(x) dx = \int_a^b w(x) Q_n(x) dx \tag{c}$$

Because a polynomial of degree n is uniquely defined by $n + 1$ points, it is always possible to find A_i such that

$$\int_a^b w(x) Q_n(x) dx = \sum_{i=0}^{n} A_i Q_n(x_i) \tag{d}$$

---

[14] It can be shown that $Q_n(x)$ and $R_n(x)$ are unique for given $P_{2n+1}(x)$ and $\varphi_{n+1}(x)$.

In order to arrive at Eq. (a), we must choose for the nodal abscissas x_i the roots of $\varphi_{n+1}(x) = 0$. According to Eq. (b) we then have

$$P_{2n+1}(x_i) = Q_n(x_i), \quad i = 0, 1, \ldots, n \tag{e}$$

which together with Eqs. (c) and (d) leads to

$$\int_a^b w(x) P_{2n+1}(x) dx = \int_a^b w(x) Q_n(x) dx = \sum_{i=0}^n A_i P_{2n+1}(x_i)$$

This completes the proof.

Theorem

$$A_i = \int_a^b w(x) \ell_i(x) dx, \quad i = 0, 1, \ldots, n \tag{6.24}$$

where $\ell_i(x)$ are the Lagrange's cardinal functions spanning the nodes at $x_0, x_1, \ldots x_n$. These functions were defined in Eq. (3.2).

Proof Applying Lagrange's formula, Eq. (3.1a), to $Q_n(x)$ yields

$$Q_n(x) = \sum_{i=0}^n Q_n(x_i) \ell_i(x)$$

which upon substitution in Eq. (d) gives us

$$\sum_{i=0}^n \left[Q_n(x_i) \int_a^b w(x) \ell_i(x) dx \right] = \sum_{i=0}^n A_i Q_n(x_i)$$

or

$$\sum_{i=0}^n Q_n(x_i) \left[A_i - \int_a^b w(x) \ell_i(x) dx \right] = 0$$

This equation can be satisfied for arbitrary $Q(x)$ of degree n only if

$$A_i - \int_a^b w(x) \ell_i(x) dx = 0, \quad i = 0, 1, \ldots, n$$

which is equivalent to Eq. (6.24).

It is not difficult to compute the zeros x_i, $i = 0, 1, \ldots, n$ of a polynomial $\varphi_{n+1}(x)$ belonging to an orthogonal set by one of the methods discussed in Chapter 4. Once the zeros are known, the weights A_i, $i = 0, 1, \ldots, n$ could be found from Eq. (6.24). However the following formulas (given without proof) are easier to compute

6.4 Gaussian Integration

$$\text{Gauss–Legendre} \quad A_i = \frac{2}{(1-x_i^2)\left[P'_{n+1}(x_i)\right]^2}$$

$$\text{Gauss–Laguerre} \quad A_i = \frac{1}{x_i\left[L'_{n+1}(x_i)\right]^2} \quad (6.25)$$

$$\text{Gauss–Hermite} \quad A_i = \frac{2^{n+2}(n+1)!\sqrt{\pi}}{\left[H'_{n+1}(x_i)\right]^2}$$

Abscissas and Weights for Classical Gaussian Quadratures

Here we list some classical Gaussian integration formulas. The tables of nodal abscissas and weights, covering $n = 1$ to 5, have been rounded off to six decimal places. These tables should be adequate for hand computation, but in programming you may need more precision or a larger number of nodes. In that case you should consult other references,[15] or use a subroutine to compute the abscissas and weights within the integration program.[16]

The truncation error in Gaussian quadrature

$$E = \int_a^b w(x)f(x)dx - \sum_{i=0}^{n} A_i f(x_i)$$

has the form $E = K(n) f^{(2n+2)}(c)$, where $a < c < b$ (the value of c is unknown; only its bounds are given). The expression for $K(n)$ depends on the particular quadrature being used. If the derivatives of $f(x)$ can be evaluated, the error formulas are useful in estimating the error bounds.

Gauss–Legendre Quadrature

$$\int_{-1}^{1} f(\xi)d\xi \approx \sum_{i=0}^{n} A_i f(\xi_i) \quad (6.26)$$

$\pm\xi_i$	A_i	$\pm\xi_i$	A_i
$n=1$		$n=4$	
0.577 350	1.000 000	0.000 000	0.568 889
$n=2$		0.538 469	0.478 629
0.000 000	0.888 889	0.906 180	0.236 927
0.774 597	0.555 556	$n=5$	
$n=3$		0.238 619	0.467 914
0.339 981	0.652 145	0.661 209	0.360 762
0.861 136	0.347 855	0.932 470	0.171 324

Table 6.3

---

[15] Abramowitz, M., and Stegun, I. A., *Handbook of Mathematical Functions*, Dover Publications, 1965; Stroud, A. H., and Secrest, D., *Gaussian Quadrature Formulas*, Prentice-Hall, 1966.

[16] Several such subroutines are listed in W. H. Press et al., *Numerical Recipes in Fortran 90*, Cambridge University Press, 1996.

This is the most often used Gaussian integration formula. The nodes are arranged symmetrically about $\xi = 0$, and the weights associated with a symmetric pair of nodes are equal. For example, for $n = 1$ we have $\xi_0 = -\xi_1$ and $A_0 = A_1$. The truncation error in Eq. (6.26) is

$$E = \frac{2^{2n+3}\,[(n+1)!]^4}{(2n+3)\,[(2n+2)!]^3} f^{(2n+2)}(c), \quad -1 < c < 1 \tag{6.27}$$

To apply Gauss–Legendre quadrature to the integral $\int_a^b f(x)\,dx$, we must first map the integration range (a, b) into the "standard" range $(-1, 1)$. We can accomplish this by the transformation

$$x = \frac{b+a}{2} + \frac{b-a}{2}\xi \tag{6.28}$$

Now $dx = d\xi\,(b-a)/2$, and the quadrature becomes

$$\int_a^b f(x)\,dx \approx \frac{b-a}{2} \sum_{i=0}^n A_i f(x_i) \tag{6.29}$$

where the abscissas x_i must be computed from Eq. (6.28). The truncation error here is

$$E = \frac{(b-a)^{2n+3}\,[(n+1)!]^4}{(2n+3)\,[(2n+2)!]^3} f^{(2n+2)}(c), \quad a < c < b \tag{6.30}$$

Gauss–Chebyshev Quadrature

$$\int_{-1}^1 (1-x^2)^{-1/2} f(x)\,dx \approx \frac{\pi}{n+1} \sum_{i=0}^n f(x_i) \tag{6.31}$$

Note that all the weights are equal: $A_i = \pi/(n+1)$. The abscissas of the nodes, which are symmetric about $x = 0$, are given by

$$x_i = \cos\frac{(2i+1)\pi}{2n+2} \tag{6.32}$$

The truncation error is

$$E = \frac{2\pi}{2^{2n+2}(2n+2)!} f^{(2n+2)}(c), \quad -1 < c < 1 \tag{6.33}$$

Gauss–Laguerre Quadrature

$$\int_0^\infty e^{-x} f(x)\,dx \approx \sum_{i=0}^n A_i f(x_i) \tag{6.34}$$

6.4 Gaussian Integration

x_i	A_i	x_i	A_i
n = 1		*n* = 4	
0.585 786	0.853 554	0.263 560	0.521 756
3.414 214	0.146 447	1.413 403	0.398 667
n = 2		3.596 426	(−1)0.759 424
0.415 775	0.711 093	7.085 810	(−2)0.361 175
2.294 280	0.278 517	12.640 801	(−4)0.233 670
6.289 945	(−1)0.103 892	*n* = 5	
n = 3		0.222 847	0.458 964
0.322 548	0.603 154	1.188 932	0.417 000
1.745 761	0.357 418	2.992 736	0.113 373
4.536 620	(−1)0.388 791	5.775 144	(−1)0.103 992
9.395 071	(−3)0.539 295	9.837 467	(−3)0.261 017
		15.982 874	(−6)0.898 548

Table 6.4. Multiply numbers by 10^k, where k is given in parentheses

$$E = \frac{[(n+1)!]^2}{(2n+2)!} f^{(2n+2)}(c), \quad 0 < c < \infty \tag{6.35}$$

Gauss–Hermite Quadrature

$$\int_{-\infty}^{\infty} e^{-x^2} f(x) dx \approx \sum_{i=0}^{n} A_i f(x_i) \tag{6.36}$$

The nodes are placed symmetrically about $x = 0$, each symmetric pair having the same weight.

$\pm x_i$	A_i	$\pm x_i$	A_i
n = 1		*n* = 4	
0.707 107	0.886 227	0.000 000	0.945 308
n = 2		0.958 572	0.393 619
0.000 000	1.181 636	2.020 183	(−1) 0.199 532
1.224 745	0.295 409	*n* = 5	
n = 3		0.436 077	0.724 629
0.524 648	0.804 914	1.335 849	0.157 067
1.650 680	(−1)0.813 128	2.350 605	(−2)0.453 001

Table 6.5. Multiply numbers by 10^k, where k is given in parentheses

$$E = \frac{\sqrt{\pi}(n+1)!}{2^2(2n+2)!} f^{(2n+2)}(c), \quad 0 < c < \infty \tag{6.37}$$

Gauss Quadrature with Logarithmic Singularity

$$\int_0^1 f(x)\ln(x)dx \approx -\sum_{i=0}^n A_i f(x_i) \qquad (6.38)$$

x_i	A_i	x_i	A_i
n = 1		**n = 4**	
0.112 009	0.718 539	(−1)0.291 345	0.297 893
0.602 277	0.281 461	0.173 977	0.349 776
n = 2		0.411 703	0.234 488
(−1)0.638 907	0.513 405	0.677 314	(−1)0.989 305
0.368 997	0.391 980	0.894 771	(−1)0.189 116
0.766 880	(−1)0.946 154	**n = 5**	
n = 3		(−1)0.216 344	0.238 764
(−1)0.414 485	0.383 464	0.129 583	0.308 287
0.245 275	0.386 875	0.314 020	0.245 317
0.556 165	0.190 435	0.538 657	0.142 009
0.848 982	(−1)0.392 255	0.756 916	(−1)0.554 546
		0.922 669	(−1)0.101 690

Table 6.6. Multiply numbers by 10^k, where k is given in parentheses

$$E = \frac{k(n)}{(2n+1)!} f^{(2n+1)}(c), \quad 0 < c < 1 \qquad (6.39)$$

where $k(1) = 0.00285$, $k(2) = 0.00017$, $k(3) = 0.00001$.

■ gaussNodes

The function gaussNodes listed below[17] computes the nodal abscissas x_i and the corresponding weights A_i used in Gauss–Legendre quadrature over the "standard" interval $(-1, 1)$. It can be shown that the approximate values of the abscissas are

$$x_i = \cos\frac{\pi(i + 0.75)}{m + 0.5}$$

where $m = n + 1$ is the number of nodes, also called the *integration order*. Using these approximations as the starting values, the nodal abscissas are computed by finding the nonnegative zeros of the Legendre polynomial $p_m(x)$ with Newton's method (the negative zeros are obtained from symmetry). Note that gaussNodes calls the subfunction legendre, which returns $p_m(t)$ and its derivative as the tuple (p,dp).

---

[17] This function is an adaptation of a routine in Press, W. H. et al., *Numerical Recipes in Fortran 90*, Cambridge University Press, 1996.

6.4 Gaussian Integration

```
## module gaussNodes
''' x,A = gaussNodes(m,tol=10e-9)
    Returns nodal abscissas {x} and weights {A} of
    Gauss-Legendre m-point quadrature.
'''
from math import cos,pi
from numarray import zeros,Float64

def gaussNodes(m,tol=10e-9):

    def legendre(t,m):
        p0 = 1.0; p1 = t
        for k in range(1,m):
            p = ((2.0*k + 1.0)*t*p1 - k*p0)/(1.0 + k)
            p0 = p1; p1 = p
        dp = m*(p0 - t*p1)/(1.0 - t**2)
        return p,dp

    A = zeros((m),type=Float64)
    x = zeros((m),type=Float64)
    nRoots = (m + 1)/2          # Number of non-neg. roots
    for i in range(nRoots):
        t = cos(pi*(i + 0.75)/(m + 0.5))  # Approx. root
        for j in range(30):
            p,dp = legendre(t,m)         # Newton-Raphson
            dt = -p/dp; t = t + dt       # method
            if abs(dt) < tol:
                x[i] = t; x[m-i-1] = -t
                A[i] = 2.0/(1.0 - t**2)/(dp**2) # Eq.(6.25)
                A[m-i-1] = A[i]
                break
    return x,A
```

■ **gaussQuad**

The function gaussQuad utilizes gaussNodes to evaluate $\int_a^b f(x)\,dx$ with Gauss–Legendre quadrature using m nodes. The function routine for $f(x)$ must be supplied by the user.

```
## module gaussQuad
''' I = gaussQuad(f,a,b,m).
    Computes the integral of f(x) from x = a to b
```

with Gauss-Legendre quadrature using m nodes.
'''
from gaussNodes import *

def gaussQuad(f,a,b,m):
    c1 = (b + a)/2.0
    c2 = (b - a)/2.0
    x,A = gaussNodes(m)
    sum = 0.0
    for i in range(len(x)):
        sum = sum + A[i]*f(c1 + c2*x[i])
    return c2*sum
```

**EXAMPLE 6.8**
Evaluate $\int_{-1}^{1} (1-x^2)^{3/2} dx$ as accurately as possible with Gaussian integration.

**Solution** As the integrand is smooth and free of singularities, we could use Gauss–Legendre quadrature. However, the exact integral can obtained with the Gauss–Chebyshev formula. We write

$$\int_{-1}^{1} (1-x^2)^{3/2} dx = \int_{-1}^{1} \frac{(1-x^2)^2}{\sqrt{1-x^2}} dx$$

The numerator $f(x) = (1-x^2)^2$ is a polynomial of degree four, so that Gauss–Chebyshev quadrature is exact with three nodes.

The abscissas of the nodes are obtained from Eq. (6.32). Substituting $n=2$, we get

$$x_i = \cos\frac{(2i+1)\pi}{6}, \quad i = 0, 1, 2$$

Therefore,

$$x_0 = \cos\frac{\pi}{6} = \frac{\sqrt{3}}{2}$$

$$x_1 = \cos\frac{\pi}{2} = 0$$

$$x_2 = \cos\frac{5\pi}{6} = \frac{\sqrt{3}}{2}$$

and Eq. (6.31) yields

$$\int_{-1}^{1} (1-x^2)^{3/2} dx \approx \frac{\pi}{3} \sum_{i=0}^{2} (1-x_i^2)^2$$

$$= \frac{\pi}{3}\left[\left(1-\frac{3}{4}\right)^2 + (1-0)^2 + \left(1-\frac{3}{4}\right)^2\right] = \frac{3\pi}{8}$$
```

6.4 Gaussian Integration

EXAMPLE 6.9
Use Gaussian integration to evaluate $\int_0^{0.5} \cos \pi x \ln x \, dx$.

Solution We split the integral into two parts:

$$\int_0^{0.5} \cos \pi x \ln x \, dx = \int_0^1 \cos \pi x \ln x \, dx - \int_{0.5}^1 \cos \pi x \ln x \, dx$$

The first integral on the right-hand side, which contains a logarithmic singularity at $x = 0$, can be computed with the special Gaussian quadrature in Eq. (6.38). Choosing $n = 3$, we have

$$\int_0^1 \cos \pi x \ln x \, dx \approx -\sum_{i=0}^3 A_i \cos \pi x_i$$

The sum is evaluated in the following table:

x_i	$\cos \pi x_i$	A_i	$A_i \cos \pi x_i$
0.041 448	0.991 534	0.383 464	0.380 218
0.245 275	0.717 525	0.386 875	0.277 592
0.556 165	−0.175 533	0.190 435	−0.033 428
0.848 982	−0.889 550	0.039 225	−0.034 892
			$\Sigma = 0.589\,490$

Thus

$$\int_0^1 \cos \pi x \ln x \, dx \approx -0.589\,490$$

The second integral is free of singularities, so that it can be evaluated with Gauss–Legendre quadrature. Choosing $n = 3$, we have

$$\int_{0.5}^1 \cos \pi x \ln x \, dx \approx 0.25 \sum_{i=0}^3 A_i \cos \pi x_i \ln x_i$$

where the nodal abscissas are (see Eq. (6.28))

$$x_i = \frac{1 + 0.5}{2} + \frac{1 - 0.5}{2} \xi_i = 0.75 + 0.25 \xi_i$$

Looking up ξ_i and A_i in Table 6.3 leads to the following computations:

ξ_i	x_i	$\cos \pi x_i \ln x_i$	A_i	$A_i \cos \pi x_i \ln x_i$
−0.861 136	0.534 716	0.068 141	0.347 855	0.023 703
−0.339 981	0.665 005	0.202 133	0.652 145	0.131 820
0.339 981	0.834 995	0.156 638	0.652 145	0.102 151
0.861 136	0.965 284	0.035 123	0.347 855	0.012 218
				$\Sigma = 0.269\,892$

from which

$$\int_{0.5}^{1} \cos \pi x \ln x \, dx \approx 0.25(0.269\,892) = 0.067\,473$$

Therefore,

$$\int_{0}^{1} \cos \pi x \ln x \, dx \approx -0.589\,490 - 0.067\,473 = -0.656\,963$$

which is correct to six decimal places.

EXAMPLE 6.10
Evaluate as accurately as possible

$$F = \int_{0}^{\infty} \frac{x+3}{\sqrt{x}} e^{-x} dx$$

Solution In its present form, the integral is not suited to any of the Gaussian quadratures listed in this chapter. But using the transformation

$$x = t^2 \qquad dx = 2t\,dt$$

we have

$$F = 2\int_{0}^{\infty} (t^2 + 3)e^{-t^2} dt = \int_{-\infty}^{\infty} (t^2 + 3)e^{-t^2} dt$$

which can be evaluated exactly with Gauss–Hermite formula using only two nodes ($n = 1$). Thus

$$F = A_0(t_0^2 + 3) + A_1(t_1^2 + 3)$$
$$= 0.886\,227\left[(0.707\,107)^2 + 3\right] + 0.886\,227\left[(-0.707\,107)^2 + 3\right]$$
$$= 6.203\,59$$

EXAMPLE 6.11
Determine how many nodes are required to evaluate

$$\int_{0}^{\pi} \left(\frac{\sin x}{x}\right)^2 dx$$

with Gauss–Legendre quadrature to six decimal places. The exact integral, rounded to six places, is 1.418 15.

Solution The integrand is a smooth function; hence it is suited for Gauss–Legendre integration. There is an indeterminacy at $x = 0$, but this does not bother the quadrature since the integrand is never evaluated at that point. We used the following program that computes the quadrature with 2, 3, ... nodes until the desired accuracy is reached:

6.4 Gaussian Integration

```
## example 6_11
from math import pi,sin
from gaussQuad import *

def f(x): return (sin(x)/x)**2

a = 0.0; b = pi;
Iexact = 1.41815
for m in range(2,12):
    I = gaussQuad(f,a,b,m)
    if abs(I - Iexact) < 0.00001:
        print ''Number of nodes ='',m
        print ''Integral ='', gaussQuad(f,a,b,m)
        break
raw_input(''\nPress return to exit'')
```

The program output is

```
Number of nodes = 5
Integral = 1.41815026778
```

EXAMPLE 6.12
Evaluate numerically $\int_{1.5}^{3} f(x)\,dx$, where $f(x)$ is represented by the unevenly spaced data

x	1.2	1.7	2.0	2.4	2.9	3.3
$f(x)$	$-0.362\,36$	$0.128\,84$	$0.416\,15$	$0.737\,39$	$0.970\,96$	$0.987\,48$

Knowing that the data points lie on the curve $f(x) = -\cos x$, evaluate the accuracy of the solution.

Solution We approximate $f(x)$ by the polynomial $P_5(x)$ that intersects all the data points, and then evaluate $\int_{1.5}^{3} f(x)dx \approx \int_{1.5}^{3} P_5(x)dx$ with the Gauss–Legendre formula. Since the polynomial is of degree five, only three nodes ($n = 2$) are required in the quadrature.

From Eq. (6.28) and Table 6.3, we obtain for the abscissas of the nodes

$$x_0 = \frac{3+1.5}{2} + \frac{3-1.5}{2}(-0.774597) = 1.6691$$

$$x_1 = \frac{3+1.5}{2} = 2.25$$

$$x_2 = \frac{3+1.5}{2} + \frac{3-1.5}{2}(0.774597) = 2.8309$$

Numerical Integration

We now compute the values of the interpolant $P_5(x)$ at the nodes. This can be done using the modules `newtonPoly` or `neville` listed in Section 3.2. The results are

$$P_5(x_0) = 0.098\,08 \qquad P_5(x_1) = 0.628\,16 \qquad P_5(x_2) = 0.952\,16$$

From Gauss–Legendre quadrature

$$I = \int_{1.5}^{3} P_5(x)\,dx = \frac{3-1.5}{2}\sum_{i=0}^{2} A_i P_5(x_i)$$

we get

$$I = 0.75\,[0.555\,556(0.098\,08) + 0.888\,889(0.628\,16) + 0.555\,556(0.952\,16)]$$
$$= 0.856\,37$$

Comparison with $\int_{1.5}^{3} \cos x\,dx = 0.856\,38$ shows that the discrepancy is within the roundoff error.

PROBLEM SET 6.2

1. Evaluate
$$\int_{1}^{\pi} \frac{\ln x}{x^2 - 2x + 2}\,dx$$
with Gauss–Legendre quadrature. Use (a) two nodes and (b) four nodes.

2. Use Gauss–Laguerre quadrature to evaluate $\int_{0}^{\infty}(1-x^2)^3 e^{-x}\,dx$.

3. Use Gauss–Chebyshev quadrature with six nodes to evaluate
$$\int_{0}^{\pi/2} \frac{dx}{\sqrt{\sin x}}$$
Compare the result with the "exact" value 2.62206. *Hint*: substitute $\sin x = t^2$.

4. The integral $\int_{0}^{\pi} \sin x\,dx$ is evaluated with Gauss–Legendre quadrature using four nodes. What are the bounds on the truncation error resulting from the quadrature?

5. How many nodes are required in Gauss–Laguerre quadrature to evaluate $\int_{0}^{\infty} e^{-x} \sin x\,dx$ to six decimal places?

6. Evaluate as accurately as possible
$$\int_{0}^{1} \frac{2x+1}{\sqrt{x(1-x)}}\,dx$$
Hint: substitute $x = (1+t)/2$.

7. Compute $\int_{0}^{\pi} \sin x \ln x\,dx$ to four decimal places.

8. Calculate the bounds on the truncation error if $\int_{0}^{\pi} x \sin x\,dx$ is evaluated with Gauss–Legendre quadrature using three nodes. What is the actual error?

6.4 Gaussian Integration

9. Evaluate $\int_0^2 (\sinh x/x)\, dx$ to four decimal places.

10. Evaluate the integral

$$\int_0^\infty \frac{x\, dx}{e^x + 1}$$

to six decimal places. *Hint*: substitute $e^x = 1/t$.

11. ■ The equation of an ellipse is $x^2/a^2 + y^2/b^2 = 1$. Write a program that computes the length

$$S = 2\int_{-a}^{a} \sqrt{1 + (dy/dx)^2}\, dx$$

of the circumference to five decimal places for given a and b. Test the program with $a = 2$ and $b = 1$.

12. ■ The error function, which is of importance in statistics, is defined as

$$\text{erf}(x) = \frac{2}{\sqrt{\pi}} \int_0^x e^{-t^2}\, dt$$

Write a program that uses Gauss–Legendre quadrature to evaluate erf(x) for a given x to six decimal places. Note that erf(x) = 1.000 000 (correct to 6 decimal places) when $x > 5$. Test the program by verifying that erf(1.0) = 0.842 701.

13. ■

The sliding weight of mass m is attached to a spring of stiffness k that has an undeformed length L. When the mass is released from rest at B, the time it takes to reach A can be shown to be $t = C\sqrt{m/k}$, where

$$C = \int_0^1 \left[\left(\sqrt{2} - 1\right)^2 - \left(\sqrt{1 + z^2} - 1\right)^2 \right]^{-1/2} dz$$

Compute C to six decimal places. *Hint*: the integrand has a singularity at $z = 1$ that behaves as $(1 - z^2)^{-1/2}$.

14. ■

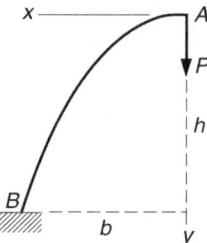

A uniform beam forms the semiparabolic cantilever arch AB. The vertical displacement of A due to the force P can be shown to be

$$\delta_A = \frac{Pb^3}{EI} C\left(\frac{h}{b}\right)$$

where EI is the bending rigidity of the beam and

$$C\left(\frac{h}{b}\right) = \int_0^1 z^2 \sqrt{1 + \left(\frac{2h}{b}z\right)^2}\, dz$$

Write a program that computes $C(h/b)$ for any given value of h/b to four decimal places. Use the program to compute $C(0.5)$, $C(1.0)$ and $C(2.0)$.

15. ■ There is no elegant way to compute $I = \int_0^{\pi/2} \ln(\sin x)\, dx$. A "brute force" method that works is to split the integral into several parts: from $x = 0$ to 0.01, from 0.01 to 0.2 and from $x = 0.2$ to $\pi/2$. In the first part we can use the approximation $\sin x \approx x$, which allows us to obtain the integral analytically. The other two parts can be evaluated with Gauss–Legendre quadrature. Use this method to evaluate I to six decimal places.

16. ■

The pressure of wind was measured at various heights on a vertical wall, as shown on the diagram. Find the height of the pressure center, which is defined as

$$\bar{h} = \frac{\int_0^{112\,\text{m}} h\, p(h)\, dh}{\int_0^{112\,\text{m}} p(h)\, dh}$$

Hint: fit a cubic polynomial to the data and then apply Gauss–Legendre quadrature.

*6.5 Multiple Integrals

Multiple integrals, such as the area integral $\int\int_A f(x, y)\, dx\, dy$, can also be evaluated by quadrature. The computations are straightforward if the region of integration has a simple geometric shape, such as a triangle or a quadrilateral. Due to complications in specifying the limits of integration on x and y, quadrature is not a practical means of evaluating integrals over irregular regions. However, an irregular region A can always be approximated as an assembly of triangular or quadrilateral subregions $A_1, A_2, \ldots$, called *finite elements*, as illustrated in Fig. 6.6. The integral over A can then be evaluated by summing the integrals over the finite elements:

$$\int\int_A f(x, y)\, dx\, dy \approx \sum_i \int\int_{A_i} f(x, y)\, dx\, dy$$

Volume integrals can computed in a similar manner, using tetrahedra or rectangular prisms for the finite elements.

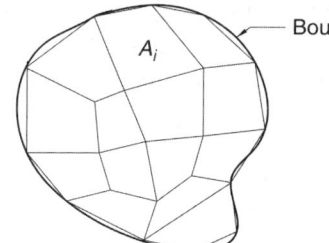

Figure 6.6. Finite element model of an irregular region.

Gauss–Legendre Quadrature over a Quadrilateral Element

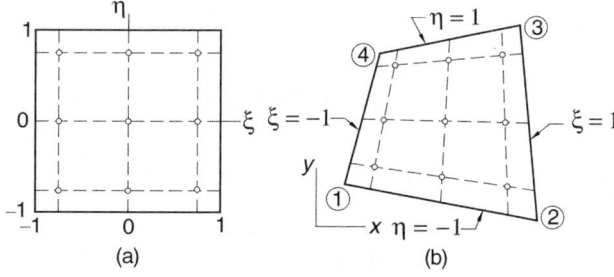

Figure 6.7. Mapping a quadrilateral into the standard rectangle.

Numerical Integration

Consider the double integral

$$I = \int_{-1}^{1}\int_{-1}^{1} f(\xi, \eta)\, d\xi\, d\eta$$

over the rectangular element shown in Fig. 6.7(a). Evaluating each integral in turn by Gauss–Legendre quadrature using $n+1$ integration points in each coordinate direction, we obtain

$$I = \int_{-1}^{1}\sum_{i=0}^{n} A_i f(\xi_i, \eta)\, d\eta = \sum_{j=0}^{n} A_j\left[\sum_{i=0}^{n} A_i f(\xi_i, \eta_i)\right]$$

or

$$I = \sum_{i=0}^{n}\sum_{j=0}^{n} A_i A_j f(\xi_i, \eta_j) \tag{6.40}$$

As noted previously, the number of integration points in each coordinate direction, $m = n+1$, is called the *integration order*. Figure 6.7(a) shows the locations of the integration points used in third-order integration ($m = 3$). Because the integration limits were the "standard" limits $(-1, 1)$ of Gauss–Legendre quadrature, the weights and the coordinates of the integration points are as listed Table 6.3.

In order to apply quadrature to the quadrilateral element in Fig. 6.7(b), we must first map the quadrilateral into the "standard" rectangle in Fig. 6.7(a). By mapping we mean a coordinate transformation $x = x(\xi, \eta)$, $y = y(\xi, \eta)$ that results in one-to-one correspondence between points in the quadrilateral and in the rectangle. The transformation that does the job is

$$x(\xi, \eta) = \sum_{k=1}^{4} N_k(\xi, \eta) x_k \qquad y(\xi, \eta) = \sum_{k=1}^{4} N_k(\xi, \eta) y_k \tag{6.41}$$

where (x_k, y_k) are the coordinates of corner k of the quadrilateral and

$$N_1(\xi, \eta) = \frac{1}{4}(1-\xi)(1-\eta)$$

$$N_2(\xi, \eta) = \frac{1}{4}(1+\xi)(1-\eta) \tag{6.42}$$

$$N_3(\xi, \eta) = \frac{1}{4}(1+\xi)(1+\eta)$$

$$N_4(\xi, \eta) = \frac{1}{4}(1-\xi)(1+\eta)$$

The functions $N_k(\xi, \eta)$, known as the *shape functions*, are bilinear (linear in each coordinate). Consequently, straight lines remain straight upon mapping. In particular, note that the sides of the quadrilateral are mapped into the lines $\xi = \pm 1$ and $\eta = \pm 1$.

Because mapping distorts areas, an infinitesimal area element $dA = dx\,dy$ of the quadrilateral is not equal to its counterpart $dA' = d\xi\,d\eta$ of the rectangle. It can be shown that the relationship between the areas is

$$dx\,dy = |\mathbf{J}(\xi, \eta)|\, d\xi\, d\eta \tag{6.43}$$

where

$$\mathbf{J}(\xi, \eta) = \begin{bmatrix} \dfrac{\partial x}{\partial \xi} & \dfrac{\partial y}{\partial \xi} \\ \dfrac{\partial x}{\partial \eta} & \dfrac{\partial y}{\partial \eta} \end{bmatrix} \tag{6.44a}$$

is known as the *Jacobian matrix* of the mapping. Substituting from Eqs. (6.41) and (6.42) and differentiating, we find that the components of the Jacobian matrix are

$$\begin{aligned}
J_{11} &= \frac{1}{4}[-(1-\eta)x_1 + (1-\eta)x_2 + (1+\eta)x_3 - (1-\eta)x_4] \\
J_{12} &= \frac{1}{4}[-(1-\eta)y_1 + (1-\eta)y_2 + (1+\eta)y_3 - (1-\eta)y_4] \\
J_{21} &= \frac{1}{4}[-(1-\xi)x_1 - (1+\xi)x_2 + (1+\xi)x_3 + (1-\xi)x_4] \\
J_{22} &= \frac{1}{4}[-(1-\xi)y_1 - (1+\xi)y_2 + (1+\xi)y_3 + (1-\xi)y_4]
\end{aligned} \tag{6.44b}$$

We can now write

$$\int\!\!\int_A f(x, y)\,dx\,dy = \int_{-1}^{1}\!\!\int_{-1}^{1} f[x(\xi,\eta), y(\xi,\eta)]\,|\mathbf{J}(\xi,\eta)|\,d\xi\,d\eta \tag{6.45}$$

Since the right-hand-side integral is taken over the "standard" rectangle, it can be evaluated using Eq. (6.40). Replacing $f(\xi, \eta)$ in Eq. (6.40) by the integrand in Eq. (6.45), we get the following formula for Gauss–Legendre quadrature over a quadrilateral region:

$$I = \sum_{i=0}^{n}\sum_{j=0}^{n} A_i A_j\, f\left[x(\xi_i, \eta_j), y(\xi_i, \eta_j)\right]\,|\mathbf{J}(\xi_i, \eta_j)| \tag{6.46}$$

The ξ and η-coordinates of the integration points and the weights can again be obtained from Table 6.3.

■ gaussQuad2

The function gaussQuad2 in this module computes $\int\!\int_A f(x, y)\,dx\,dy$ over a quadrilateral element with Gauss–Legendre quadrature of integration order m. The quadrilateral is defined by the arrays **x** and **y**, which contain the coordinates of the four corners ordered in a *counterclockwise direction* around the element. The determinant of the Jacobian matrix is obtained by calling the function jac; mapping is performed by map. The weights and the values of ξ and η at the integration points are computed by gaussNodes listed in the previous article (note that ξ and η appear as *s* and *t* in the listing).

```
## module gaussQuad2
''' I = gaussQuad2(f,xc,yc,m).
    Gauss-Legendre integration of f(x,y) over a
    quadrilateral using integration order m.
    {xc},{yc} are the corner coordinates of the quadrilateral.
'''
from gaussNodes import *
from numarray import zeros,Float64,dot

def gaussQuad2(f,x,y,m):

    def jac(x,y,s,t):
        J = zeros((2,2),type=Float64)
        J[0,0] = -(1.0 - t)*x[0] + (1.0 - t)*x[1]   \
                + (1.0 + t)*x[2] - (1.0 + t)*x[3]
        J[0,1] = -(1.0 - t)*y[0] + (1.0 - t)*y[1]   \
                + (1.0 + t)*y[2] - (1.0 + t)*y[3]
        J[1,0] = -(1.0 - s)*x[0] - (1.0 + s)*x[1]   \
                + (1.0 + s)*x[2] + (1.0 - s)*x[3]
        J[1,1] = -(1.0 - s)*y[0] - (1.0 + s)*y[1]   \
                + (1.0 + s)*y[2] + (1.0 - s)*y[3]
        return (J[0,0]*J[1,1] - J[0,1]*J[1,0])/16.0

    def map(x,y,s,t):
        N = zeros((4),type=Float64)
        N[0] = (1.0 - s)*(1.0 - t)/4.0
        N[1] = (1.0 + s)*(1.0 - t)/4.0
        N[2] = (1.0 + s)*(1.0 + t)/4.0
        N[3] = (1.0 - s)*(1.0 + t)/4.0
        xCoord = dot(N,x)
        yCoord = dot(N,y)
        return xCoord,yCoord

    s,A = gaussNodes(m)
    sum = 0.0
```

6.5 Multiple Integrals

```
for i in range(m):
    for j in range(m):
        xCoord,yCoord = map(x,y,s[i],s[j])
        sum = sum + A[i]*A[j]*jac(x,y,s[i],s[j]) \
              *f(xCoord,yCoord)
return sum
```

EXAMPLE 6.13

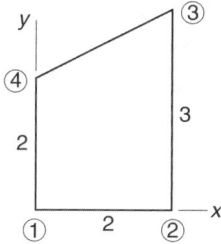

Evaluate the integral

$$I = \int\int_A (x^2 + y)\, dx\, dy$$

analytically by first transforming it from the quadrilateral region A shown to the "standard" rectangle.

Solution The corner coordinates of the quadrilateral are

$$\mathbf{x}^T = \begin{bmatrix} 0 & 2 & 2 & 0 \end{bmatrix} \qquad \mathbf{y}^T = \begin{bmatrix} 0 & 0 & 3 & 2 \end{bmatrix}$$

The mapping is

$$x(\xi, \eta) = \sum_{k=1}^{4} N_k(\xi, \eta) x_k$$

$$= 0 + \frac{(1+\xi)(1-\eta)}{4}(2) + \frac{(1+\xi)(1+\eta)}{4}(2) + 0$$

$$= 1 + \xi$$

$$y(\xi, \eta) = \sum_{k=1}^{4} N_k(\xi, \eta) y_k$$

$$= 0 + 0 + \frac{(1+\xi)(1+\eta)}{4}(3) + \frac{(1-\xi)(1+\eta)}{4}(2)$$

$$= \frac{(5+\xi)(1+\eta)}{4}$$

Numerical Integration

which yields for the Jacobian matrix

$$J(\xi, \eta) = \begin{bmatrix} \dfrac{\partial x}{\partial \xi} & \dfrac{\partial y}{\partial \xi} \\ \dfrac{\partial x}{\partial \eta} & \dfrac{\partial y}{\partial \eta} \end{bmatrix} = \begin{bmatrix} 1 & \dfrac{1+\eta}{4} \\ 0 & \dfrac{5+\xi}{4} \end{bmatrix}$$

Thus the area scale factor is

$$|J(\xi, \eta)| = \frac{5+\xi}{4}$$

Now we can map the integral from the quadrilateral to the standard rectangle. Referring to Eq. (6.45), we obtain

$$I = \int_{-1}^{1} \int_{-1}^{1} \left[(1+\xi)^2 + \frac{(5+\xi)(1+\eta)}{4} \right] \frac{5+\xi}{4} d\xi\, d\eta$$

$$= \int_{-1}^{1} \int_{-1}^{1} \left(\frac{45}{16} + \frac{27}{8}\xi + \frac{29}{16}\xi^2 + \frac{1}{4}\xi^3 + \frac{25}{16}\eta + \frac{5}{8}\xi\eta + \frac{1}{16}\xi^2\eta \right) d\xi\, d\eta$$

Noting that only even powers of ξ and η contribute to the integral, we can simplify the integral to

$$I = \int_{-1}^{1} \int_{-1}^{1} \left(\frac{45}{16} + \frac{29}{16}\xi^2 \right) d\xi\, d\eta = \frac{41}{3}$$

EXAMPLE 6.14

Evaluate the integral

$$\int_{-1}^{1} \int_{-1}^{1} \cos\frac{\pi x}{2} \cos\frac{\pi y}{2}\, dx\, dy$$

by Gauss–Legendre quadrature of order three.

Solution From the quadrature formula in Eq. (6.40), we have

$$I = \sum_{i=0}^{2} \sum_{j=0}^{2} A_i A_j \cos\frac{\pi x_i}{2} \cos\frac{\pi y_j}{2}$$

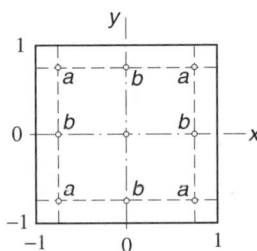

The integration points are shown in the figure; their coordinates and the corresponding weights are listed in Table 6.3. Note that the integrand, the integration points and

6.5 Multiple Integrals

the weights are all symmetric about the coordinate axes. It follows that the points labeled *a* contribute equal amounts to *I*; the same is true for the points labeled *b*. Therefore,

$$I = 4(0.555\,556)^2 \cos^2 \frac{\pi(0.774\,597)}{2}$$
$$+ 4(0.555\,556)(0.888\,889) \cos \frac{\pi(0.774\,597)}{2} \cos \frac{\pi(0)}{2}$$
$$+ (0.888\,889)^2 \cos^2 \frac{\pi(0)}{2}$$
$$= 1.623\,391$$

The exact value of the integral is $16/\pi^2 \approx 1.621\,139$.

EXAMPLE 6.15

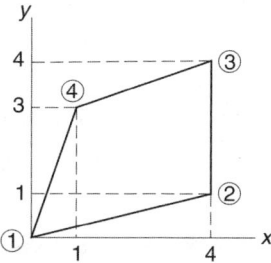

Utilize gaussQuad2 to evaluate $I = \int \int_A f(x, y)\, dx\, dy$ over the quadrilateral shown, where

$$f(x, y) = (x - 2)^2 (y - 2)^2$$

Use enough integration points for an "exact" answer.

Solution The required integration order is determined by the integrand in Eq. (6.45):

$$I = \int_{-1}^{1} \int_{-1}^{1} f[x(\xi, \eta), y(\xi, \eta)]\, |\mathbf{J}(\xi, \eta)|\, d\xi\, d\eta \qquad (a)$$

We note that $|J(\xi, \eta)|$, defined in Eqs. (6.44), is biquadratic. Since the specified $f(x, y)$ is also biquadratic, the integrand in Eq. (a) is a polynomial of degree 4 in both ξ and η. Thus third-order integration is sufficient for an "exact" result.

```
#!/usr/bin/python
## example 6_15
from gaussQuad2 import *
from numarray import array
```

```
def f(x,y): return ((x - 2.0)**2)*((y - 2.0)**2)

x = array([0.0, 4.0, 4.0, 1.0])
y = array([0.0, 1.0, 4.0, 3.0])
m = eval(raw_input(''Integration order ==> ''))
print ''Integral ='', gaussQuad2(f,x,y,m)
raw_input(''\nPress return to exit'')
```

Running the above program produced the following result:

```
Integration order ==> 3
Integral = 11.3777777778
```

Quadrature over a Triangular Element

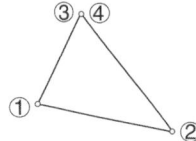

Figure 6.8. Degenerate quadrilateral.

A triangle may be viewed as a degenerate quadrilateral with two of its corners occupying the same location, as illustrated in Fig. 6.8. Therefore, the integration formulas over a quadrilateral region can also be used for a triangular element. However, it is computationally advantageous to use integration formulas specially developed for triangles, which we present without derivation.[18]

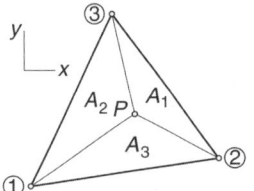

Figure 6.9. Triangular element.

Consider the triangular element in Fig. 6.9. Drawing straight lines from the point P in the triangle to each of the corners, we divide the triangle into three parts with areas A_1, A_2 and A_3. The so-called *area coordinates* of P are defined as

$$\alpha_i = \frac{A_i}{A}, \quad i = 1, 2, 3 \qquad (6.47)$$

---

[18] The triangle formulas are extensively used in the finite method analysis. See, for example, Zienkiewicz, O. C., and Taylor, R. L., *The Finite Element Method*, Vol. 1, 4th ed., McGraw-Hill, 1989.

6.5 Multiple Integrals

where A is the area of the element. Since $A_1 + A_2 + A_3 = A$, the area coordinated are related by

$$\alpha_1 + \alpha_2 + \alpha_3 = 1 \tag{6.48}$$

Note that α_i ranges from 0 (when P lies on the side opposite to corner i) to 1 (when P is at corner i).

A convenient formula of computing A from the corner coordinates (x_i, y_i) is

$$A = \frac{1}{2} \begin{vmatrix} 1 & 1 & 1 \\ x_1 & x_2 & x_3 \\ y_1 & y_2 & y_3 \end{vmatrix} \tag{6.49}$$

The area coordinates are mapped into the Cartesian coordinates by

$$x(\alpha_1, \alpha_2, \alpha_3) = \sum_{i=1}^{3} \alpha_i x_i \qquad y(\alpha_1, \alpha_2, \alpha_3) = \sum_{i=1}^{3} \alpha_i y_i \tag{6.50}$$

The integration formula over the element is

$$\int\int_A f[x(\alpha), y(\alpha)] \, dA = A \sum_k W_k f[x(\alpha_k), y(\alpha_k)] \tag{6.51}$$

where α_k represents the area coordinates of the integration point k, and W_k are the weights. The locations of the integration points are shown in Fig. 6.10, and the corresponding values of α_k and W_k are listed in Table 6.7. The quadrature in Eq. (6.51) is exact if $f(x, y)$ is a polynomial of the degree indicated.

(a) Linear

(b) Quadratic

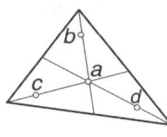
(c) Cubic

Figure 6.10. Integration points of triangular elements.

Degree of $f(x, y)$	Point	α_k	W_k
(a) Linear	a	1/3, 1/3, 1/3	1
(b) Quadratic	a	1/2, 0, 1/2	1/3
	b	1/2, 1/2, 0	1/3
	c	0, 1/2, 1/2	1/3
(c) Cubic	a	1/3, 1/3, 1/3	−27/48
	b	1/5, 1/5, 3/5	25/48
	c	3/5. 1/5, 1/5	25/48
	d	1/5, 3/5, 1/5	25/48

Table 6.7

242 Numerical Integration

■ triangleQuad

The function triangleQuad computes $\int \int_A f(x, y)\, dx\, dy$ over a triangular region using the cubic formula—case (c) in Fig. 6.10. The triangle is defined by its corner coordinate arrays xc and yc, where the coordinates are listed in a *counterclockwise* order around the triangle.

```
## module triangleQuad
''' I = triangleQuad(f,xc,yc).
    Integration of f(x,y) over a triangle using
    the cubic formula.
    {xc},{yc} are the corner coordinates of the triangle.
'''
from numarray import array,matrixmultiply

def triangleQuad(f,xc,yc):
    alpha = array([[1.0/3, 1.0/3.0, 1.0/3.0], \
                   [0.2, 0.2, 0.6],           \
                   [0.6, 0.2, 0.2],           \
                   [0.2, 0.6, 0.2]])
    W = array([-27.0/48.0 ,25.0/48.0, 25.0/48.0, 25.0/48.0])
    x = matrixmultiply(alpha,xc)
    y = matrixmultiply(alpha,yc)
    A = (xc[1]*yc[2] - xc[2]*yc[1]      \
      - xc[0]*yc[2] + xc[2]*yc[0]       \
      + xc[0]*yc[1] - xc[1]*yc[0])/2.0
    sum = 0.0
    for i in range(4):
        sum = sum + W[i] * f(x[i],y[i])
    return A*sum
```

EXAMPLE 6.16

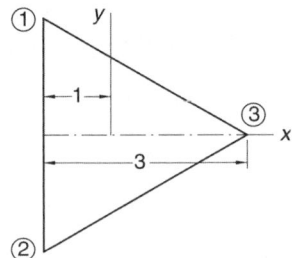

6.5 Multiple Integrals

Evaluate $I = \int \int_A f(x, y)\, dx\, dy$ over the equilateral triangle shown, where[19]

$$f(x, y) = \frac{1}{2}(x^2 + y^2) - \frac{1}{6}(x^3 - 3xy^2) - \frac{2}{3}$$

Use the quadrature formulas for (1) a quadrilateral; and (2) a triangle.

Solution of Part (1) Let the triangle be formed by collapsing corners 3 and 4 of a quadrilateral. The corner coordinates of this quadrilateral are $\mathbf{x} = [-1, -1, 2, 2]^T$ and $\mathbf{y} = \left[\sqrt{3}, -\sqrt{3}, 0, 0\right]^T$. To determine the minimum required integration order for an exact result, we must examine $f[x(\xi, \eta), y(\xi, \eta)]\,|\mathbf{J}(\xi, \eta)|$, the integrand in (6.45). Since $|\mathbf{J}(\xi, \eta)|$ is biquadratic, and $f(x, y)$ is cubic in x, the integrand is a polynomial of degree 5 in x. Therefore, third-order integration will suffice. The program used for the computations is similar to the one in Example 6.15:

```
#!/usr/bin/python
## example6_16a
from gaussQuad2 import *
from numarray import array
from math import sqrt

def f(x,y):
    return (x**2 + y**2)/2.0          \
         - (x**3 - 3.0*x*y**2)/6.0    \
         - 2.0/3.0

x = array([-1.0,-1.0,2.0,2.0])
y = array([sqrt(3.0),-sqrt(3.0),0.0,0.0])
m = eval(raw_input(''Integration order ==> ''))
print ''Integral ='', gaussQuad2(f,x,y,m)
raw_input(''\nPress return to exit'')
```

Here is the output:

```
Integration order ==> 3
Integral = -1.55884572681
```

Solution of Part (2) The following program utilizes `triangleQuad`:

```
#!/usr/bin/python
# example6_16b
```

---
[19] This function is identical to the Prandtl stress function for torsion of a bar with the cross section shown; the integral is related to the torsional stiffness of the bar. See, for example Timoshenko, S. P., and Goodier, J. N., *Theory of Elasticity*, 3rd ed., McGraw-Hill, 1970.

```
from numarray import array
from math import sqrt
from triangleQuad import *

def f(x,y):
    return (x**2 + y**2)/2.0           \
           - (x**3 - 3.0*x*y**2)/6.0   \
           - 2.0/3.0

xCorner = array([-1.0, -1.0, 2.0])
yCorner = array([sqrt(3.0), -sqrt(3.0), 0.0])
print ''Integral ='',triangleQuad(f,xCorner,yCorner)
raw_input(''Press return to  exit'')
```

Since the integrand is a cubic, this quadrature is also exact, the result being

```
Integral = -1.55884572681
```

Note that only four function evaluations were required when using the triangle formulas. In contrast, the function had to be evaluated at nine points in part (1).

EXAMPLE 6.17
The corner coordinates of a triangle are $(0, 0)$, $(16, 10)$ and $(12, 20)$. Compute $\int\int_A (x^2 - y^2)\, dx\, dy$ over this triangle.

Solution

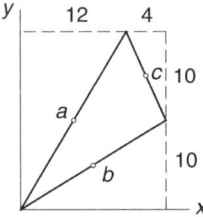

Because $f(x, y)$ is quadratic, quadrature over the three integration points shown in Fig. 6.10(b) will be sufficient for an "exact" result. Note that the integration points lie in the middle of each side; their coordinates are $(6, 10)$, $(8, 5)$ and $(14, 15)$. The area of the triangle is obtained from Eq. (6.49):

$$A = \frac{1}{2}\begin{vmatrix} 1 & 1 & 1 \\ x_1 & x_2 & x_3 \\ y_1 & y_2 & y_3 \end{vmatrix} = \frac{1}{2}\begin{vmatrix} 1 & 1 & 1 \\ 0 & 16 & 12 \\ 0 & 10 & 20 \end{vmatrix} = 100$$

6.5 Multiple Integrals

From Eq. (6.51) we get

$$I = A \sum_{k=a}^{c} W_k f(x_k, y_k)$$

$$= 100 \left[\frac{1}{3} f(6, 10) + \frac{1}{3} f(8, 5) + \frac{1}{3} f(14, 15) \right]$$

$$= \frac{100}{3} \left[(6^2 - 10^2) + (8^2 - 5^2) + (14^2 - 15^2) \right] = 1800$$

PROBLEM SET 6.3

1. Use Gauss–Legendre quadrature to compute

$$\int_{-1}^{1} \int_{-1}^{1} (1 - x^2)(1 - y^2) \, dx \, dy$$

2. Evaluate the following integral with Gauss–Legendre quadrature:

$$\int_{y=0}^{2} \int_{x=0}^{3} x^2 y^2 \, dx \, dy$$

3. Compute the approximate value of

$$\int_{-1}^{1} \int_{-1}^{1} e^{-(x^2 + y^2)} \, dx \, dy$$

with Gauss–Legendre quadrature. Use integration order (a) two and (b) three. (The "exact" value of the integral is 2.230 985.)

4. Use third-order Gauss–Legendre quadrature to obtain an approximate value of

$$\int_{-1}^{1} \int_{-1}^{1} \cos \frac{\pi (x - y)}{2} \, dx \, dy$$

(The "exact" value of the integral is 1.621 139.)

5.

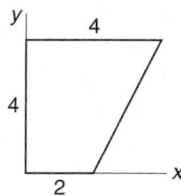

Map the integral $\int \int_A xy \, dx \, dy$ from the quadrilateral region shown to the "standard" rectangle and then evaluate it analytically.

6.

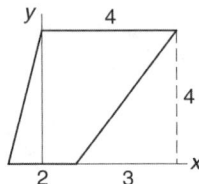

Compute $\int \int_A x \, dx \, dy$ over the quadrilateral region shown by first mapping it into the "standard" rectangle and then integrating analytically.

7.

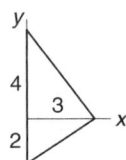

Use quadrature to compute $\int \int_A x^2 \, dx \, dy$ over the triangle shown.

8. Evaluate $\int \int_A x^3 \, dx \, dy$ over the triangle shown in Prob. 7.

9.

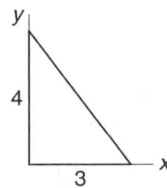

Evaluate $\int \int_A (3 - x) y \, dx \, dy$ over the region shown.

10. Evaluate $\int \int_A x^2 y \, dx \, dy$ over the triangle shown in Prob. 9.

11. ∎

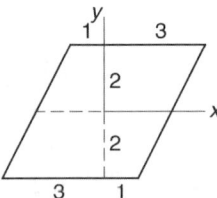

Evaluate $\int \int_A xy(2 - x^2)(2 - xy) \, dx \, dy$ over the region shown.

12. ∎ Compute $\int \int_A xy \exp(-x^2) \, dx \, dy$ over the region shown in Prob. 11 to four decimal places.

13. ■

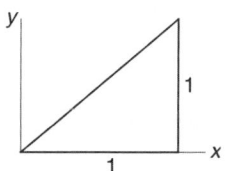

Evaluate $\int\int_A (1-x)(y-x)y\,dx\,dy$ over the triangle shown.

14. ■ Estimate $\int\int_A \sin \pi x\,dx\,dy$ over the region shown in Prob. 13. Use the cubic integration formula for a triangle. (The exact integral is $1/\pi$.)

15. ■ Compute $\int\int_A \sin \pi x \sin \pi (y-x)\,dx\,dy$ to six decimal places, where A is the triangular region shown in Prob. 13. Consider the triangle as a degenerate quadrilateral.

16. ■

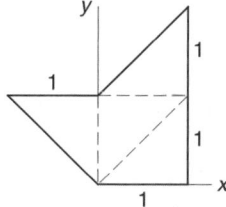

Write a program to evaluate $\int\int_A f(x,y)\,dx\,dy$ over an irregular region that has been divided into several triangular elements. Use the program to compute $\int\int_A xy(y-x)\,dx\,dy$ over the region shown.

7 Initial Value Problems

> Solve $\mathbf{y}' = \mathbf{F}(x, \mathbf{y})$ with the auxiliary conditions $\mathbf{y}(a) = \boldsymbol{\alpha}$

7.1 Introduction

The general form of a *first-order differential equation* is

$$y' = f(x, y) \tag{7.1a}$$

where $y' = dy/dx$ and $f(x, y)$ is a given function. The solution of this equation contains an arbitrary constant (the constant of integration). To find this constant, we must know a point on the solution curve; that is, y must be specified at some value of x, say at $x = a$. We write this auxiliary condition as

$$y(a) = \alpha \tag{7.1b}$$

An ordinary differential equation of order n

$$y^{(n)} = f\left(x, y, y', \ldots, y^{(n-1)}\right) \tag{7.2}$$

can always transformed into n first-order equations. Using the notation

$$y_0 = y \quad y_1 = y' \quad y_2 = y'' \quad \ldots \quad y_{n-1} = y^{(n-1)} \tag{7.3}$$

the equivalent first-order equations are

$$y'_0 = y_1 \quad y'_1 = y_2 \quad y'_2 = y_3 \quad \ldots \quad y'_n = f(x, y_0, y_1, \ldots, y_{n-1}) \tag{7.4a}$$

The solution now requires the knowledge n auxiliary conditions. If these conditions are specified at the same value of x, the problem is said to be an *initial value problem*. Then the auxiliary conditions, called *initial conditions*, have the form

$$y_0(a) = \alpha_0 \quad y_1(a) = \alpha_1 \quad \ldots \quad y_{n-1}(a) = \alpha_{n-1} \tag{7.4b}$$

If y_i are specified at different values of x, the problem is called a *boundary value problem*.

For example,
$$y'' = -y \qquad y(0) = 1 \qquad y'(0) = 0$$
is an initial value problem since both auxiliary conditions imposed on the solution are given at $x = 0$. On the other hand,
$$y'' = -y \qquad y(0) = 1 \qquad y(\pi) = 0$$
is a boundary value problem because the two conditions are specified at different values of x.

In this chapter we consider only initial value problems. The more difficult boundary value problems are discussed in the next chapter. We also make extensive use of vector notation, which allows us manipulate sets of first-order equations in a concise form. For example, Eqs. (7.4) are written as
$$\mathbf{y}' = \mathbf{F}(x, \mathbf{y}) \qquad \mathbf{y}(a) = \boldsymbol{\alpha} \tag{7.5a}$$
where
$$\mathbf{F}(x, \mathbf{y}) = \begin{bmatrix} y_1 \\ y_2 \\ \vdots \\ f(x, \mathbf{y}) \end{bmatrix} \tag{7.5b}$$

A numerical solution of differential equations is essentially a table of x- and $\mathbf{y}$-values listed at discrete intervals of x.

7.2 Taylor Series Method

The Taylor series method is conceptually simple and capable of high accuracy. Its basis is the truncated Taylor series for $\mathbf{y}$ about x:
$$\mathbf{y}(x+h) \approx \mathbf{y}(x) + \mathbf{y}'(x)h + \frac{1}{2!}\mathbf{y}''(x)h^2 + \frac{1}{3!}\mathbf{y}'''(x)h^3 + \cdots + \frac{1}{m!}\mathbf{y}^{(m)}(x)h^m \tag{7.6}$$

Because Eq. (7.6) predicts $\mathbf{y}$ at $x + h$ from the information available at x, it is also a formula for numerical integration. The last term kept in the series determines the *order of integration*. For the series in Eq. (7.6) the integration order is m.

The *truncation error*, due to the terms omitted from the series, is
$$\mathbf{E} = \frac{1}{(m+1)!}\mathbf{y}^{(m+1)}(\xi)h^{m+1}, \qquad x < \xi < x + h$$

Initial Value Problems

Using the finite difference approximation

$$\mathbf{y}^{(m+1)}(\xi) \approx \frac{\mathbf{y}^{(m)}(x+h) - \mathbf{y}^{(m)}(x)}{h}$$

we obtain the more usable form

$$\mathbf{E} \approx \frac{h^m}{(m+1)!} \left[\mathbf{y}^{(m)}(x+h) - \mathbf{y}^{(m)}(x) \right] \tag{7.7}$$

which could be incorporated in the algorithm to monitor the error in each integration step.

■ taylor

The function `taylor` implements the Taylor series method of integration of order four. It can handle any number of first-order differential equations $y'_i = f_i(x, y_0, y_1, \ldots)$, $i = 0, 1, \ldots$. The user is required to supply the function `deriv` that returns the $4 \times n$ array

$$\mathbf{D} = \begin{bmatrix} (\mathbf{y}')^T \\ (\mathbf{y}'')^T \\ (\mathbf{y}''')^T \\ (\mathbf{y}^{(4)})^T \end{bmatrix} = \begin{bmatrix} y'_0 & y'_1 & \cdots & y'_{n-1} \\ y''_0 & y''_1 & \cdots & y''_{n-1} \\ y'''_0 & y'''_1 & \cdots & y'''_{n-1} \\ y^{(4)}_0 & y^{(4)}_1 & \cdots & y^{(4)}_{n-1} \end{bmatrix}$$

The function returns the arrays X and Y that contain the values of x and $\mathbf{y}$ at intervals h.

```
## module taylor
''' X,Y = taylor(deriv,x,y,xStop,h).
    4th-order Taylor series method for solving the initial
    value problem {y}' = {F(x,{y})}, where
    {y} = {y[0],y[1],...y[n-1]}.
    x,y    = initial conditions
    xStop  = terminal value of x
    h      = increment of x used in integration
    deriv  = user-supplied function that returns the 4 x n array
             [y'[0]    y'[1]    y'[2]    ...  y'[n-1]
              y''[0]   y''[1]   y''[2]   ...  y''[n-1]
              y'''[0]  y'''[1]  y'''[2]  ...  y'''[n-1]
              y''''[0] y''''[1] y''''[2] ...  y''''[n-1]]
'''
from numarray import array
def taylor(deriv,x,y,xStop,h):
    X = []
```

7.2 Taylor Series Method

```
        Y = []
        X.append(x)
        Y.append(y)
        while x < xStop:          # Loop over integration steps
            h = min(h,xStop - x)
            D = deriv(x,y)        # Derivatives of y
            H = 1.0
            for j in range(4):    # Build Taylor series
                H = H*h/(j + 1)
                y = y + D[j]*H    # H = h^j/j!
            x = x + h
            X.append(x)           # Append results to
            Y.append(y)           # lists X and Y
        return array(X),array(Y)  # Convert lists into arrays
```

■ printSoln

We use this function to print X and Y obtained from numerical integration. The amount of data is controlled by the parameter freq. For example, if freq = 5, every 5th integration step would be displayed. If freq = 0, only the initial and final values will be shown.

```
## module printSoln
''' printSoln(X,Y,freq).
    Prints X and Y returned from the differential
    equation solvers using printput frequency 'freq'.
        freq = n prints every nth step.
        freq = 0 prints initial and final values only.
'''
def printSoln(X,Y,freq):

    def printHead(n):
        print ''\n          x  '',
        for i in range (n):
            print ''      y['',i,''] '',
        print

    def printLine(x,y,n):
        print ''%13.4e''% x,
        for i in range (n):
```

```
            print ''%13.4e''% y[i],
    print

m = len(Y)
try: n = len(Y[0])
except TypeError: n = 1
if freq == 0: freq = m
printHead(n)
for i in range(0,m,freq):
    printLine(X[i],Y[i],n)
if i != m - 1: printLine(X[m - 1],Y[m - 1],n)
```

EXAMPLE 7.1
Given that

$$y' + 4y = x^2 \qquad y(0) = 1$$

determine $y(0.1)$ with the fourth-order Taylor series method using a single integration step. Also compute the estimated error from Eq. (7.7) and compare it with the actual error. The analytical solution of the differential equation is

$$y = \frac{31}{32}e^{-4x} + \frac{1}{4}x^2 - \frac{1}{8}x + \frac{1}{32}$$

Solution The Taylor series up to and including the term with h^4 is

$$y(h) = y(0) + y'(0)h + \frac{1}{2!}y''(0)h^2 + \frac{1}{3!}y'''(0)h^3 + \frac{1}{4!}y^{(4)}(0)h^4 \qquad \text{(a)}$$

Differentiation of the differential equation yields

$$y' = -4y + x^2$$
$$y'' = -4y' + 2x = 16y - 4x^2 + 2x$$
$$y''' = 16y' - 8x + 2 = -64y + 16x^2 - 8x + 2$$
$$y^{(4)} = -64y' + 32x - 8 = 256y - 64x^2 + 32x - 8$$

Thus at $x = 0$ we have

$$y'(0) = -4(1) = -4$$
$$y''(0) = 16(1) = 16$$
$$y'''(0) = -64(1) + 2 = -62$$
$$y^{(4)}(0) = 256(1) - 8 = 248$$

7.2 Taylor Series Method

With $h = 0.1$ Eq. (a) becomes

$$y(0.1) = 1 + (-4)(0.1) + \frac{1}{2!}(16)(0.1)^2 + \frac{1}{3!}(-62)(0.1)^3 + \frac{1}{4!}(248)(0.1)^4$$
$$= 0.670700$$

According to Eq. (7.7) the approximate truncation error is

$$E = \frac{h^4}{5!}\left[y^{(4)}(0.1) - y^{(4)}(0)\right]$$

where

$$y^{(4)}(0) = 248$$
$$y^{(4)}(0.1) = 256(0.6707) - 64(0.1)^2 + 32(0.1) - 8 = 166.259$$

Therefore,

$$E = \frac{(0.1)^4}{5!}(166.259 - 248) = -6.8 \times 10^{-5}$$

The analytical solution yields

$$y(0.1) = \frac{31}{32}e^{-4(0.1)} + \frac{1}{4}(0.1)^2 - \frac{1}{8}(0.1) + \frac{1}{32} = 0.670623$$

so that the actual error is $0.670623 - 0.670700 = -7.7 \times 10^{-5}$.

EXAMPLE 7.2
Solve

$$y'' = -0.1y' - x \qquad y(0) = 0 \qquad y'(0) = 1$$

from $x = 0$ to 2 with the Taylor series method of order four. Use $h = 0.25$ and utilize the functions `taylor` and `printSoln`.

Solution With the notation $y_0 = y$ and $y_1 = y'$ the equivalent first-order equations and the initial conditions are

$$\mathbf{y}' = \begin{bmatrix} y_0' \\ y_1' \end{bmatrix} = \begin{bmatrix} y_1 \\ -0.1y_1 - x \end{bmatrix} \qquad \mathbf{y}(0) = \begin{bmatrix} 0 \\ 1 \end{bmatrix}$$

Repeated differentiation of the differential equations yields

$$\mathbf{y}'' = \begin{bmatrix} y_1' \\ -0.1y_1' - 1 \end{bmatrix} = \begin{bmatrix} -0.1y_1 - x \\ 0.01y_1 + 0.1x - 1 \end{bmatrix}$$

$$\mathbf{y}''' = \begin{bmatrix} -0.1y_1' - 1 \\ 0.01y_1' + 0.1 \end{bmatrix} = \begin{bmatrix} 0.01y_1 + 0.1x - 1 \\ -0.001y_1 - 0.01x + 0.1 \end{bmatrix}$$

$$\mathbf{y}^{(4)} = \begin{bmatrix} 0.01y_1' + 0.1 \\ -0.001y_1' - 0.01 \end{bmatrix} = \begin{bmatrix} -0.001y_1 - 0.01x + 0.1 \\ 0.0001y_1 + 0.001x - 0.01 \end{bmatrix}$$

Initial Value Problems

Thus the derivative array that has to be computed by the function `deriv` is

$$\mathbf{D} = \begin{bmatrix} y_1 & -0.1y_1 - x \\ -0.1y_1 - x & 0.01y_1 + 0.1x - 1 \\ 0.01y_1 + 0.1x - 1 & -0.001y_1 - 0.01x + 0.1 \\ -0.001y_1 - 0.01x + 0.1 & 0.0001y_1 + 0.001x - 0.01 \end{bmatrix}$$

Here is the program that performs the integration:

```
#!/usr/bin/python
## example7_2
from printSoln import *
from taylor import *

def deriv(x,y):
    D = zeros((4,2),type=Float64)
    D[0] = [y[1]   , -0.1*y[1] - x]
    D[1] = [D[0,1],  0.01*y[1] + 0.1*x - 1.0]
    D[2] = [D[1,1], -0.001*y[1] - 0.01*x + 0.1]
    D[3] = [D[2,1],  0.0001*y[1] + 0.001*x - 0.01]
    return D

x = 0.0                 # Start of integration
xStop = 2.0             # End of integration
y = array([0.0, 1.0])   # Initial values of {y}
h = 0.25                # Step size
freq = 1                # Printout frequency
X,Y = taylor(deriv,x,y,xStop,h)
printSoln(X,Y,freq)
raw_input(''\nPress return to exit'')
```

The results are:

```
       x              y[ 0 ]          y[ 1 ]
 0.0000e+000      0.0000e+000     1.0000e+000
 2.5000e-001      2.4431e-001     9.4432e-001
 5.0000e-001      4.6713e-001     8.2829e-001
 7.5000e-001      6.5355e-001     6.5339e-001
 1.0000e+000      7.8904e-001     4.2110e-001
 1.2500e+000      8.5943e-001     1.3281e-001
 1.5000e+000      8.5090e-001    -2.1009e-001
 1.7500e+000      7.4995e-001    -6.0625e-001
 2.0000e+000      5.4345e-001    -1.0543e+000
```

7.3 Runge–Kutta Methods

The analytical solution of the problem is

$$y = 100x - 5x^2 + 990(e^{-0.1x} - 1)$$

from which we obtain $y(2) = 0.543\,446$, which agrees well with the numerical solution.

7.3 Runge–Kutta Methods

The main drawback of the Taylor series method is that it requires repeated differentiation of the dependent variables. These expressions may become very long and are, therefore, error-prone and tedious to compute. Moreover, there is the extra work of coding each of the derivatives. The aim of Runge–Kutta methods is to eliminate the need for repeated differentiation of the differential equations. Since no such differentiation is involved in the first-order Taylor series integration formula

$$\mathbf{y}(x + h) = \mathbf{y}(x) + \mathbf{y}'(x)h = \mathbf{y}(x) + \mathbf{F}(x, \mathbf{y})h \tag{7.8}$$

it can also be considered as the first-order Runge–Kutta method; it is also called *Euler's method*. Due to excessive truncation error, this method is rarely used in practice.

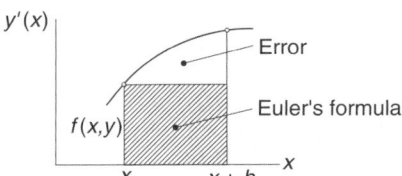

Figure 7.1. Graphical representation of Euler's formula.

Let us now take a look at the graphical interpretation of Euler's formula. For the sake of simplicity, we assume that there is a single dependent variable y, so that the differential equation is $y' = f(x, y)$. The change in the solution y between x and $x + h$ is

$$y(x + h) - y(h) = \int_x^{x+h} y'\,dx = \int_x^{x+h} f(x, y)\,dx$$

which is the area of the panel under the $y'(x)$ plot, shown in Fig. 7.1. Euler's formula approximates this area by the area of the cross-hatched rectangle. The area between the rectangle and the plot represents the truncation error. Clearly, the truncation error is proportional to the slope of the plot; that is, proportional to $y''(x)$.

Second-Order Runge–Kutta Method

To arrive at the second-order method, we assume an integration formula of the form

$$\mathbf{y}(x + h) = \mathbf{y}(x) + c_0 \mathbf{F}(x, \mathbf{y})h + c_1 \mathbf{F}\left[x + ph, \mathbf{y} + qh\mathbf{F}(x, \mathbf{y})\right]h \tag{a}$$

Initial Value Problems

and attempt to find the parameters c_0, c_1, p and q by matching Eq. (a) to the Taylor series

$$\mathbf{y}(x+h) = \mathbf{y}(x) + \mathbf{y}'(x)h + \frac{1}{2!}\mathbf{y}''(x)h^2 + \mathcal{O}(h^3)$$
$$= \mathbf{y}(x) + \mathbf{F}(x,\mathbf{y})h + \frac{1}{2}\mathbf{F}'(x,\mathbf{y})h^2 + \mathcal{O}(h^3) \quad \text{(b)}$$

Noting that

$$\mathbf{F}'(x,\mathbf{y}) = \frac{\partial \mathbf{F}}{\partial x} + \sum_{i=0}^{n-1}\frac{\partial \mathbf{F}}{\partial y_i}y_i' = \frac{\partial \mathbf{F}}{\partial x} + \sum_{i=0}^{n-1}\frac{\partial \mathbf{F}}{\partial y_i}F_i(x,\mathbf{y})$$

where n is the number of first-order equations, we can write Eq. (b) as

$$\mathbf{y}(x+h) = \mathbf{y}(x) + \mathbf{F}(x,\mathbf{y})h + \frac{1}{2}\left(\frac{\partial \mathbf{F}}{\partial x} + \sum_{i=0}^{n-1}\frac{\partial \mathbf{F}}{\partial y_i}F_i(x,\mathbf{y})\right)h^2 + \mathcal{O}(h^3) \quad \text{(c)}$$

Returning to Eq. (a), we can rewrite the last term by applying a Taylor series in several variables:

$$\mathbf{F}\left[x+ph,\mathbf{y}+qh\mathbf{F}(x,\mathbf{y})\right] = \mathbf{F}(x,\mathbf{y}) + \frac{\partial \mathbf{F}}{\partial x}ph + qh\sum_{i=1}^{n-1}\frac{\partial \mathbf{F}}{\partial y_i}F_i(x,\mathbf{y}) + \mathcal{O}(h^2)$$

so that Eq. (a) becomes

$$\mathbf{y}(x+h) = \mathbf{y}(x) + (c_0 + c_1)\mathbf{F}(x,\mathbf{y})h + c_1\left[\frac{\partial \mathbf{F}}{\partial x}ph + qh\sum_{i=1}^{n-1}\frac{\partial \mathbf{F}}{\partial y_i}F_i(x,\mathbf{y})\right]h + \mathcal{O}(h^3) \quad \text{(d)}$$

Comparing Eqs. (c) and (d), we find that they are identical if

$$c_0 + c_1 = 1 \qquad c_1 p = \frac{1}{2} \qquad c_1 q = \frac{1}{2} \quad \text{(e)}$$

Because Eqs. (e) represent three equations in four unknown parameters, we can assign any value to one of the parameters. Some of the popular choices and the names associated with the resulting formulas are:

$c_0 = 0$ $\quad c_1 = 1 \quad$ $p = 1/2 \quad q = 1/2 \quad$ Modified Euler's method
$c_0 = 1/2$ $\quad c_1 = 1/2 \quad$ $p = 1 \quad\quad q = 1 \quad\quad$ Heun's method
$c_0 = 1/3$ $\quad c_1 = 2/3 \quad$ $p = 3/4 \quad q = 3/4 \quad$ Ralston's method

All these formulas are classified as second-order Runge–Kutta methods, with no formula having a numerical superiority over the others. Choosing the *modified Euler's method*, we substitute the corresponding parameters into Eq. (a) to yield

$$\mathbf{y}(x+h) = \mathbf{y}(x) + \mathbf{F}\left[x+\frac{h}{2},\mathbf{y}+\frac{h}{2}\mathbf{F}(x,\mathbf{y})\right]h \quad \text{(f)}$$

7.3 Runge–Kutta Methods

This integration formula can be conveniently evaluated by the following sequence of operations

$$\mathbf{K}_0 = h\mathbf{F}(x, \mathbf{y})$$
$$\mathbf{K}_1 = h\mathbf{F}\left(x + \frac{h}{2}, \mathbf{y} + \frac{1}{2}\mathbf{K}_0\right) \quad (7.9)$$
$$\mathbf{y}(x + h) = \mathbf{y}(x) + \mathbf{K}_1$$

Second-order methods are not popular in computer application. Most programmers prefer integration formulas of order four, which achieve a given accuracy with less computational effort.

Figure 7.2. Graphical representation of modified Euler formula.

Figure 7.2 displays the graphical interpretation of modified Euler's formula for a single differential equation $y' = f(x, y)$. The first of Eqs. (7.9) yields an estimate of y at the midpoint of the panel by Euler's formula: $y(x + h/2) = y(x) + f(x, y)h/2 = y(x) + K_0/2$. The second equation then approximates the area of the panel by the area K_1 of the cross-hatched rectangle. The error here is proportional to the curvature y''' of the plot.

Fourth-Order Runge–Kutta Method

The fourth-order Runge–Kutta method is obtained from the Taylor series along the same lines as the second-order method. Since the derivation is rather long and not very instructive, we skip it. The final form of the integration formula again depends on the choice of the parameters; that is, there is no unique Runge–Kutta fourth-order formula. The most popular version, which is known simply as the *Runge–Kutta method*, entails the following sequence of operations:

$$\mathbf{K}_0 = h\mathbf{F}(x, \mathbf{y})$$
$$\mathbf{K}_1 = h\mathbf{F}\left(x + \frac{h}{2}, \mathbf{y} + \frac{\mathbf{K}_0}{2}\right)$$
$$\mathbf{K}_2 = h\mathbf{F}\left(x + \frac{h}{2}, \mathbf{y} + \frac{\mathbf{K}_1}{2}\right) \quad (7.10)$$
$$\mathbf{K}_3 = h\mathbf{F}(x + h, \mathbf{y} + \mathbf{K}_2)$$
$$\mathbf{y}(x + h) = \mathbf{y}(x) + \frac{1}{6}(\mathbf{K}_0 + 2\mathbf{K}_1 + 2\mathbf{K}_2 + \mathbf{K}_3)$$

Initial Value Problems

The main drawback of this method is that it does not lend itself to an estimate of the truncation error. Therefore, we must guess the integration step size *h*, or determine it by trial and error. In contrast, the so-called *adaptive methods* can evaluate the truncation error in each integration step and adjust the value of *h* accordingly (but at a higher cost of computation). One such adaptive method is introduced in the next article.

■ run_kut4

The function `integrate` in this module implements the Runge–Kutta method of order four. The user must provide `integrate` with the function F(x,y) that defines the first-order differential equations $\mathbf{y}' = \mathbf{F}(x, \mathbf{y})$.

```
## module run_kut4
''' X,Y = integrate(F,x,y,xStop,h).
    4th-order Runge-Kutta method for solving the
    initial value problem {y}' = {F(x,{y})}, where
    {y} = {y[0],y[1],...y[n-1]}.
    x,y   = initial conditions.
    xStop = terminal value of x.
    h     = increment of x used in integration.
    F     = user-supplied function that returns the
            array F(x,y) = {y'[0],y'[1],...,y'[n-1]}.
'''
from numarray import array
def integrate(F,x,y,xStop,h):

    def run_kut4(F,x,y,h):
   # Computes increment of y from Eqs. (7.10)
        K0 = h*F(x,y)
        K1 = h*F(x + h/2.0, y + K0/2.0)
        K2 = h*F(x + h/2.0, y + K1/2.0)
        K3 = h*F(x + h, y + K2)
        return (K0 + 2.0*K1 + 2.0*K2 + K3)/6.0

    X = []
    Y = []
    X.append(x)
    Y.append(y)
    while x < xStop:
        h = min(h,xStop - x)
        y = y + run_kut4(F,x,y,h)
```

7.3 Runge–Kutta Methods

```
    x = x + h
    X.append(x)
    Y.append(y)
return array(X),array(Y)
```

EXAMPLE 7.3
Use the second-order Runge–Kutta method to integrate
$$y' = \sin y \qquad y(0) = 1$$
from $x = 0$ to 0.5 in steps of $h = 0.1$. Keep four decimal places in the computations.

Solution In this problem we have
$$F(x, y) = \sin y$$
so that the integration formulas in Eqs. (7.9) are
$$K_0 = hF(x, y) = 0.1 \sin y$$
$$K_1 = hF\left(x + \frac{h}{2}, y + \frac{1}{2}K_0\right) = 0.1 \sin\left(y + \frac{1}{2}K_0\right)$$
$$y(x + h) = y(x) + K_1$$

Noting that $y(0) = 1$, we may proceed with the integration as follows:
$$K_0 = 0.1 \sin 1.0000 = 0.0841$$
$$K_1 = 0.1 \sin\left(1.0000 + \frac{0.0841}{2}\right) = 0.0863$$
$$y(0.1) = 1.0 + 0.0863 = 1.0863$$
$$K_0 = 0.1 \sin 1.0863 = 0.0885$$
$$K_1 = 0.1 \sin\left(1.0863 + \frac{0.0885}{2}\right) = 0.0905$$
$$y(0.2) = 1.0863 + 0.0905 = 1.1768$$

and so on. A summary of the computations is shown in the table below.

x	y	K_0	K_1
0.0	1.0000	0.0841	0.0863
0.1	1.0863	0.0885	0.0905
0.2	1.1768	0.0923	0.0940
0.3	1.2708	0.0955	0.0968
0.4	1.3676	0.0979	0.0988
0.5	1.4664		

The exact solution can be shown to be

$$x(y) = \ln(\csc y - \cot y) + 0.604582$$

which yields $x(1.4664) = 0.5000$. Therefore, up to this point the numerical solution is accurate to four decimal places. However, it is unlikely that this precision would be maintained if we were to continue the integration. Since the errors (due to truncation and roundoff) tend to accumulate, longer integration ranges require better integration formulas and more significant figures in the computations.

EXAMPLE 7.4
Solve

$$y'' = -0.1y' - x \qquad y(0) = 0 \qquad y'(0) = 1$$

from $x = 0$ to 2 in increments of $h = 0.25$ with the Runge–Kutta method of order four. (This problem was solved by the Taylor series method in Example 7.2.)

Solution Letting $y_0 = y$ and $y_1 = y'$, we write the equivalent first-order equations as

$$\mathbf{y}' = \mathbf{F}(x, \mathbf{y}) = \begin{bmatrix} y_0' \\ y_1' \end{bmatrix} = \begin{bmatrix} y_1 \\ -0.1y_1 - x \end{bmatrix}$$

Comparing the function F(x,y) here with deriv(x,y) in Example 7.2 we note that it is much simpler to input the differential equations in the Runge–Kutta method than in the Taylor series method.

```
#!/usr/bin/python
## example7_4
from numarray import array,zeros,Float64
from printSoln import *
from run_kut4 import *

def F(x,y):
    F = zeros((2),type=Float64)
    F[0] = y[1]
    F[1] = -0.1*y[1] - x
    return F

x = 0.0                    # Start of integration
xStop = 2.0                # End of integration
y = array([0.0, 1.0])      # Initial values of {y}
h = 0.25                   # Step size
freq = 1                   # Printout frequency
```

7.3 Runge–Kutta Methods

```
X,Y = integrate(F,x,y,xStop,h)
printSoln(X,Y,freq)
raw_input(''Press return to exit'')
```

The output from the fourth-order method is shown below. The results are the same as obtained by the Taylor series method in Example 7.2. This was expected, since both methods are of the same order.

```
       x            y[ 0 ]        y[ 1 ]
 0.0000e+000   0.0000e+000   1.0000e+000
 2.5000e-001   2.4431e-001   9.4432e-001
 5.0000e-001   4.6713e-001   8.2829e-001
 7.5000e-001   6.5355e-001   6.5339e-001
 1.0000e+000   7.8904e-001   4.2110e-001
 1.2500e+000   8.5943e-001   1.3281e-001
 1.5000e+000   8.5090e-001  -2.1009e-001
 1.7500e+000   7.4995e-001  -6.0625e-001
 2.0000e+000   5.4345e-001  -1.0543e+000
```

EXAMPLE 7.5

Use the fourth-order Runge–Kutta method to integrate

$$y' = 3y - 4e^{-x} \qquad y(0) = 1$$

from $x = 0$ to 10 in steps of $h = 0.1$. Compare the result with the analytical solution $y = e^{-x}$.

Solution We used the program shown below. Recalling that `run_kut4` assumes y to be an array, we specified the initial value as `y = array([1.0])` rather than `y = 1.0`.

```
#!/usr/bin/python
## example7_5
from numarray import zeros,Float64,array
from run_kut4 import *
from printSoln import *
from math import exp

def F(x,y):
    F = zeros((1),type=Float64)
    F[0] = 3.0*y[0] - 4.0*exp(-x)
    return F
```

```
x = 0.0                  # Start of integration
xStop = 10.0             # End of integration
y = array([1.0])         # Initial values of {y}
h = 0.1                  # Step size
freq = 20                # Printout frequency

X,Y = integrate(F,x,y,xStop,h)
printSoln(X,Y,freq)
raw_input(''\nPress return to exit'')
```

Running the program produced the following output:

```
      x             y[ 0 ]
0.0000e+000    1.0000e+000
2.0000e+000    1.3250e-001
4.0000e+000   -1.1237e+000
6.0000e+000   -4.6056e+002
8.0000e+000   -1.8575e+005
1.0000e+001   -7.4912e+007
```

It is clear that something went wrong. According to the analytical solution, y should approach zero with increasing x, but the output shows the opposite trend: after an initial decrease, the magnitude of y increases dramatically. The explanation is found by taking a closer look at the analytical solution. The general solution of the given differential equation is

$$y = Ce^{3x} + e^{-x}$$

which can be verified by substitution. The initial condition $y(0) = 1$ yields $C = 0$, so that the solution to the problem is indeed $y = e^{-x}$.

The cause of trouble in the numerical solution is the dormant term Ce^{3x}. Suppose that the initial condition contains a small error ε, so that we have $y(0) = 1 + \varepsilon$. This changes the analytical solution to

$$y = \varepsilon e^{3x} + e^{-x}$$

We now see that the term containing the error ε becomes dominant as x is increased. Since errors inherent in the numerical solution have the same effect as small changes in initial conditions, we conclude that our numerical solution is the victim of *numerical instability* due to sensitivity of the solution to initial conditions. The lesson is: do not blindly trust the results of numerical integration.

7.3 Runge–Kutta Methods

EXAMPLE 7.6

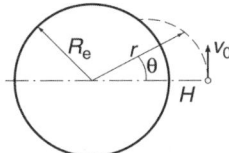

A spacecraft is launched at altitude $H = 772$ km above sea level with the speed $v_0 = 6700$ m/s in the direction shown. The differential equations describing the motion of the spacecraft are

$$\ddot{r} = r\dot{\theta}^2 - \frac{GM_e}{r^2} \qquad \ddot{\theta} = -\frac{2\dot{r}\dot{\theta}}{r}$$

where r and θ are the polar coordinates of the spacecraft. The constants involved in the motion are

$$G = 6.672 \times 10^{-11} \text{ m}^3\text{kg}^{-1}\text{s}^{-2} = \text{universal gravitational constant}$$

$$M_e = 5.9742 \times 10^{24} \text{ kg} = \text{mass of the earth}$$

$$R_e = 6378.14 \text{ km} = \text{radius of the earth at sea level}$$

(1) Derive the first-order differential equations and the initial conditions of the form $\dot{\mathbf{y}} = \mathbf{F}(t, \mathbf{y})$, $\mathbf{y}(0) = \mathbf{b}$. (2) Use the fourth-order Runge–Kutta method to integrate the equations from the time of launch until the spacecraft hits the earth. Determine θ at the impact site.

Solution of Part (1) We have

$$GM_e = \left(6.672 \times 10^{-11}\right)\left(5.9742 \times 10^{24}\right) = 3.9860 \times 10^{14} \text{ m}^3\text{s}^{-2}$$

Letting

$$\mathbf{y} = \begin{bmatrix} y_0 \\ y_1 \\ y_2 \\ y_3 \end{bmatrix} = \begin{bmatrix} r \\ \dot{r} \\ \theta \\ \dot{\theta} \end{bmatrix}$$

the equivalent first-order equations become

$$\dot{\mathbf{y}} = \begin{bmatrix} \dot{y}_0 \\ \dot{y}_1 \\ \dot{y}_2 \\ \dot{y}_3 \end{bmatrix} = \begin{bmatrix} y_1 \\ y_0 y_3^2 - 3.9860 \times 10^{14}/y_0^2 \\ y_3 \\ -2y_1 y_3/y_0 \end{bmatrix}$$

and the initial conditions are

$$r(0) = R_e + H = R_e = (6378.14 + 772) \times 10^3 = 7.15014 \times 10^6 \text{ m}$$

$$\dot{r}(0) = 0$$

$$\theta(0) = 0$$

$$\dot{\theta}(0) = v_0/r(0) = (6700)/(7.15014 \times 10^6) = 0.937045 \times 10^{-3} \text{ rad/s}$$

Therefore,

$$\mathbf{y}(0) = \begin{bmatrix} 7.15014 \times 10^6 \\ 0 \\ 0 \\ 0.937045 \times 10^{-3} \end{bmatrix}$$

Solution of Part (2) The program used for numerical integration is listed below. Note that the independent variable t is denoted by x. The period of integration xStop (the time when the spacecraft hits) was estimated from a previous run of the program.

```
#!/usr/bin/python
## example7_6
from numarray import zeros,Float64,array
from run_kut4 import *
from printSoln import *

def F(x,y):
    F = zeros((4),type=Float64)
    F[0] = y[1]
    F[1] = y[0]*(y[3]**2) - 3.9860e14/(y[0]**2)
    F[2] = y[3]
    F[3] = -2.0*y[1]*y[3]/y[0]
    return F

x = 0.0
xStop = 1200.0
y = array([7.15014e6, 0.0, 0.0, 0.937045e-3])
h = 50.0
freq = 2

X,Y = integrate(F,x,y,xStop,h)
printSoln(X,Y,freq)
raw_input(''\nPress return to exit'')
```

7.3 Runge–Kutta Methods

Here is the output:

```
     x           y[ 0 ]        y[ 1 ]        y[ 2 ]        y[ 3 ]
0.0000e+000    7.1501e+006    0.0000e+000   0.0000e+000   9.3704e-004
1.0000e+002    7.1426e+006   -1.5173e+002   9.3771e-002   9.3904e-004
2.0000e+002    7.1198e+006   -3.0276e+002   1.8794e-001   9.4504e-004
3.0000e+002    7.0820e+006   -4.5236e+002   2.8292e-001   9.5515e-004
4.0000e+002    7.0294e+006   -5.9973e+002   3.7911e-001   9.6951e-004
5.0000e+002    6.9622e+006   -7.4393e+002   4.7697e-001   9.8832e-004
6.0000e+002    6.8808e+006   -8.8389e+002   5.7693e-001   1.0118e-003
7.0000e+002    6.7856e+006   -1.0183e+003   6.7950e-001   1.0404e-003
8.0000e+002    6.6773e+006   -1.1456e+003   7.8520e-001   1.0744e-003
9.0000e+002    6.5568e+006   -1.2639e+003   8.9459e-001   1.1143e-003
1.0000e+003    6.4250e+006   -1.3708e+003   1.0083e+000   1.1605e-003
1.1000e+003    6.2831e+006   -1.4634e+003   1.1269e+000   1.2135e-003
1.2000e+003    6.1329e+006   -1.5384e+003   1.2512e+000   1.2737e-003
```

The spacecraft hits the earth when r equals $R_e = 6.378\,14 \times 10^6$ m. This occurs between $t = 1000$ and 1100 s. A more accurate value of t can be obtained by polynomial interpolation. If no great precision is needed, linear interpolation will do. Letting $1000 + \Delta t$ be the time of impact, we can write

$$r(1000 + \Delta t) = R_e$$

Expanding r in a two-term Taylor series, we get

$$r(1000) + \dot{r}(1000)\Delta t = R_e$$

$$6.4250 \times 10^6 + \left(-1.3708 \times 10^3\right)\Delta t = 6378.14 \times 10^3$$

from which

$$\Delta t = 34.184 \text{ s}$$

Thus the time of impact is 1034.25.

The coordinate θ of the impact site can be estimated in a similar manner. Using again two terms of the Taylor series, we have

$$\theta(1000 + \Delta t) = \theta(1000) + \dot{\theta}(1000)\Delta t$$
$$= 1.0083 + \left(1.1605 \times 10^{-3}\right)(34.184)$$
$$= 1.0480 \text{ rad} = 60.00°$$

PROBLEM SET 7.1

1. Given
$$y' + 4y = x^2 \qquad y(0) = 1$$
compute $y(0.1)$ using two steps of the Taylor series method of order two. Compare the result with Example 7.1.

2. Solve Prob. 1 with one step of the Runge–Kutta method of order (a) two and (b) four.

3. Integrate
$$y' = \sin y \qquad y(0) = 1$$
from $x = 0$ to 0.5 with the second-order Taylor series method using $h = 0.1$. Compare the result with Example 7.3.

4. Verify that the problem
$$y' = y^{1/3} \qquad y(0) = 0$$
has two solutions: $y = 0$ and $y = (2x/3)^{3/2}$. Which of the solutions would be reproduced by numerical integration if the initial condition is set at (a) $y = 0$ and (b) $y = 10^{-16}$? Verify your conclusions by integrating with any numerical method.

5. Convert the following differential equations into first-order equations of the form $\mathbf{y}' = \mathbf{F}(x, \mathbf{y})$:

   (a) $\ln y' + y = \sin x$
   (b) $y'' y - x y' - 2 y^2 = 0$
   (c) $y^{(4)} - 4 y'' \sqrt{1 - y^2} = 0$
   (d) $(y'')^2 = |32 y' x - y^2|$

6. In the following sets of coupled differential equations t is the independent variable. Convert these equations into first-order equations of the form $\dot{\mathbf{y}} = \mathbf{F}(t, \mathbf{y})$:

   (a) $\ddot{y} = x - 2y \qquad \ddot{x} = y - x$
   (b) $\ddot{y} = -y(\dot{y}^2 + \dot{x}^2)^{1/4} \qquad \ddot{x} = -x(\dot{y}^2 + \dot{x})^{1/4} - 32$
   (c) $\dot{y}^2 + t \sin y = 4 \dot{x} \qquad x \ddot{x} + t \cos y = 4 \dot{y}$

7. ■ The differential equation for the motion of a simple pendulum is
$$\frac{d^2 \theta}{dt^2} = -\frac{g}{L} \sin \theta$$
where

$\theta =$ angular displacement from the vertical
$g =$ gravitational acceleration
$L =$ length of the pendulum

With the transformation $\tau = t\sqrt{g/L}$ the equation becomes

$$\frac{d^2\theta}{d\tau^2} = -\sin\theta$$

Use numerical integration to determine the period of the pendulum if the amplitude is $\theta_0 = 1$ rad. Note that for small amplitudes ($\sin\theta \approx \theta$) the period is $2\pi\sqrt{L/g}$.

8. ■ A skydiver of mass m in a vertical free fall experiences an aerodynamic drag force $F_D = c_D \dot{y}^2$, where y is measured downward from the start of the fall. The differential equation describing the fall is

$$\ddot{y} = g - \frac{c_D}{m}\dot{y}^2$$

Determine the time of a 500 m fall. Use $g = 9.80665$ m/s², $c_D = 0.2028$ kg/m and $m = 80$ kg.

9. ■

The spring–mass system is at rest when the force $P(t)$ is applied, where

$$P(t) = \begin{cases} 10t \text{ N} & \text{when } t < 2 \text{ s} \\ 20 \text{ N} & \text{when } t \geq 2 \text{ s} \end{cases}$$

The differential equation of the ensuing motion is

$$\ddot{y} = \frac{P(t)}{m} - \frac{k}{m}y$$

Determine the maximum displacement of the mass. Use $m = 2.5$ kg and $k = 75$ N/m.

10. ■

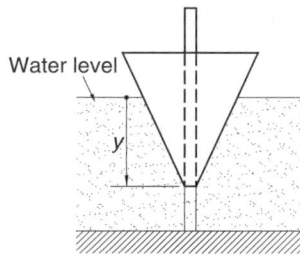

The conical float is free to slide on a vertical rod. When the float is disturbed from its equilibrium position, it undergoes oscillating motion described by the differential equation

$$\ddot{y} = g\left(1 - ay^3\right)$$

where $a = 16$ m^{-3} (determined by the density and dimensions of the float) and $g = 9.80665$ m/s^2. If the float is raised to the position $y = 0.1$ m and released, determine the period and the amplitude of the oscillations.

11. ■

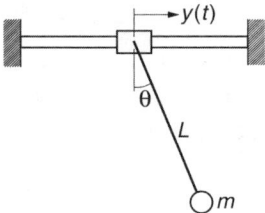

The pendulum is suspended from a sliding collar. The system is at rest when the oscillating motion $y(t) = Y \sin \omega t$ is imposed on the collar, starting at $t = 0$. The differential equation describing the motion of the pendulum is

$$\ddot{\theta} = -\frac{g}{L} \sin \theta + \frac{\omega^2}{L} Y \cos \theta \sin \omega t$$

Plot θ vs. t from $t = 0$ to 10 s and determine the largest θ during this period. Use $g = 9.80665$ m/s^2, $L = 1.0$ m, $Y = 0.25$ m and $\omega = 2.5$ rad/s.

12. ■

The system consisting of a sliding mass and a guide rod is at rest with the mass at $r = 0.75$ m. At time $t = 0$ a motor is turned on that imposes the motion $\theta(t) = (\pi/12) \cos \pi t$ on the rod. The differential equation describing the resulting motion of the slider is

$$\ddot{r} = \left(\frac{\pi^2}{12}\right)^2 r \sin^2 \pi t - g \sin\left(\frac{\pi}{12} \cos \pi t\right)$$

Determine the time when the slider reaches the tip of the rod. Use $g = 9.80665$ m/s^2.

13. ■

A ball of mass $m = 0.25$ kg is launched with the velocity $v_0 = 50$ m/s in the direction shown. If the aerodynamic drag force acting on the ball is $F_D = C_D v^{3/2}$, the differential equations describing the motion are

$$\ddot{x} = -\frac{C_D}{m}\dot{x}v^{1/2} \qquad \ddot{y} = -\frac{C_D}{m}\dot{y}v^{1/2} - g$$

where $v = \sqrt{\dot{x}^2 + \dot{y}^2}$. Determine the time of flight and the range R. Use $C_D = 0.03$ kg/(m·s)$^{1/2}$ and $g = 9.80665$ m/s^2.

14. ■ The differential equation describing the angular position θ of a mechanical arm is

$$\ddot{\theta} = \frac{a(b - \theta) - \theta\dot{\theta}^2}{1 + \theta^2}$$

where $a = 100$ s^{-2} and $b = 15$. If $\theta(0) = 2\pi$ and $\dot{\theta}(0) = 0$, compute θ and $\dot{\theta}$ when $t = 0.5$ s.

15. ■

The mass m is suspended from an elastic cord with an extensional stiffness k and undeformed length L. If the mass is released from rest at $\theta = 60°$ with the cord unstretched, find the length r of the cord when the position $\theta = 0$ is reached for the first time. The differential equations describing the motion are

$$\ddot{r} = r\dot{\theta}^2 + g\cos\theta - \frac{k}{m}(r - L)$$

$$\ddot{\theta} = \frac{-2\dot{r}\dot{\theta} - g\sin\theta}{r}$$

Use $g = 9.80665$ m/s^2, $k = 40$ N/m, $L = 0.5$ m and $m = 0.25$ kg.

16. ■ Solve Prob. 15 if the mass is released from the position $\theta = 60°$ with the cord stretched by 0.075 m.

17. ■

Consider the mass–spring system where dry friction is present between the block and the horizontal surface. The frictional force has a constant magnitude μmg (μ is the coefficient of friction) and always opposes the motion. The differential equation for the motion of the block can be expressed as

$$\ddot{y} = -\frac{k}{m}y - \mu g \frac{\dot{y}}{|\dot{y}|}$$

where y is measured from the position where the spring is unstretched. If the block is released from rest at $y = y_0$, verify by numerical integration that the next positive peak value of y is $y_0 - 4\mu\,mg/k$ (this relationship can be derived analytically). Use $k = 3000$ N/m, $m = 6$ kg, $\mu = 0.5$, $g = 9.80665$ m/s² and $y_0 = 0.1$ m.

18. ■ Integrate the following problems from $x = 0$ to 20 and plot y vs. x:

    (a) $y'' + 0.5(y^2 - 1)y' + y = 0 \qquad y(0) = 1 \qquad y'(0) = 0$
    (b) $y'' = y\cos 2x \qquad y(0) = 0 \qquad y'(0) = 1$

    These differential equations arise in nonlinear vibration analysis.

19. ■ The solution of the problem

$$y'' + \frac{1}{x}y' + y \qquad y(0) = 1 \qquad y'(0) = 0$$

is the Bessel function $J_0(x)$. Use numerical integration to compute $J_0(5)$ and compare the result with -0.17760, the value listed in mathematical tables. *Hint*: to avoid singularity at $x = 0$, start the integration at $x = 10^{-12}$.

20. ■ Consider the initial value problem

$$y'' = 16.81y \qquad y(0) = 1.0 \qquad y'(0) = -4.1$$

    (a) Derive the analytical solution. (b) Do you anticipate difficulties in numerical solution of this problem? (c) Try numerical integration from $x = 0$ to 8 to see if your concerns were justified.

21. ■

Kirchoff's equations for the circuit shown are

$$L\frac{di_1}{dt} + Ri_1 + 2R(i_1 + i_2) = E(t) \qquad \text{(a)}$$

$$\frac{q_2}{C} + Ri_2 + 2R(i_2 + i_1) = E(t) \qquad \text{(b)}$$

where i_1 and i_2 are the loop currents, and q_2 is the charge of the condenser. Differentiating Eq. (b) and substituting the charge–current relationship $dq_2/dt = i_2$, we get

$$\frac{di_1}{dt} = \frac{-3Ri_1 - 2Ri_2 + E(t)}{L} \tag{c}$$

$$\frac{di_2}{dt} = -\frac{2}{3}\frac{di_1}{dt} - \frac{i_2}{3RC} + \frac{1}{3R}\frac{dE}{dt} \tag{d}$$

We could substitute di_1/dt from Eq. (c) into Eq. (d), so that the latter would assume the usual form $di_2/dt = f(t, i_1, i_2)$, but it is more convenient to leave the equations as they are. Assuming that the voltage source is turned on at time $t = 0$, plot the loop currents i_1 and i_2 from $t = 0$ to 0.05 s. Use $E(t) = 240 \sin(120\pi t)$ V, $R = 1.0\,\Omega$, $L = 0.2 \times 10^{-3}$ H and $C = 3.5 \times 10^{-3}$ F.

22. ■

The constant voltage source of the circuit shown is turned on at $t = 0$, causing transient currents i_1 and i_2 in the two loops that last about 0.05 s. Plot these currents from $t = 0$ to 0.05 s, using the following data: $E = 9$ V, $R = 0.25\,\Omega$, $L = 1.2 \times 10^{-3}$ H and $C = 5 \times 10^{-3}$ F. Kirchoff's equations for the two loops are

$$L\frac{di_1}{dt} + Ri_1 + \frac{q_1 - q_2}{C} = E$$

$$L\frac{di_2}{dt} + Ri_2 + \frac{q_2 - q_1}{C} + \frac{q_2}{C} = 0$$

Additional two equations are the current–charge relationships

$$\frac{dq_1}{dt} = i_1 \qquad \frac{dq_2}{dt} = i_2$$

7.4 Stability and Stiffness

Loosely speaking, a method of numerical integration is said to be stable if the effects of local errors do not accumulate catastrophically; that is, if the global error remains bounded. If the method is unstable, the global error will increase exponentially, eventually causing numerical overflow. Stability has nothing to do with accuracy; in fact, an inaccurate method can be very stable.

Stability is determined by three factors: the differential equations, the method of solution and the value of the increment h. Unfortunately, it is not easy to determine stability beforehand, unless the differential equation is linear.

Stability of Euler's Method

As a simple illustration of stability, consider the linear problem

$$y' = -\lambda y \qquad y(0) = \beta \tag{7.11}$$

where λ is a positive constant. The exact solution of this problem is

$$y(x) = \beta e^{-\lambda x}$$

Let us now investigate what happens when we attempt to solve Eq. (7.11) numerically with Euler's formula

$$y(x+h) = y(x) + hy'(x) \tag{7.12}$$

Substituting $y'(x) = -\lambda y(x)$, we get

$$y(x+h) = (1 - \lambda h)y(x)$$

If $|1 - \lambda h| > 1$, the method is clearly unstable since $|y|$ increases in every integration step. Thus Euler's method is stable only if $|1 - \lambda h| \leq 1$, or

$$h \leq 2/\lambda \tag{7.13}$$

The results can be extended to a system of n differential equations of the form

$$\mathbf{y}' = -\mathbf{\Lambda} \mathbf{y} \tag{7.14}$$

where $\mathbf{\Lambda}$ is a constant matrix with the positive eigenvalues λ_i, $i = 1, 2, \ldots, n$. It can be shown that Euler's method of integration is stable only if

$$h < 2/\lambda_{\max} \tag{7.15}$$

where $\lambda_{\max}$ is the largest eigenvalue of $\mathbf{\Lambda}$.

Stiffness

An initial value problem is called *stiff* if some terms in the solution vector $\mathbf{y}(x)$ vary much more rapidly with x than others. Stiffness can be easily predicted for the differential equations $\mathbf{y}' = -\mathbf{\Lambda} \mathbf{y}$ with constant coefficient matrix $\mathbf{\Lambda}$. The solution of these equations is $\mathbf{y}(x) = \sum_i C_i \mathbf{v}_i \exp(-\lambda_i x)$, where λ_i are the eigenvalues of $\mathbf{\Lambda}$ and $\mathbf{v}_i$ are the corresponding eigenvectors. It is evident that the problem is stiff if there is a large disparity in the magnitudes of the positive eigenvalues.

7.4 Stability and Stiffness

Numerical integration of stiff equations requires special care. The step size h needed for stability is determined by the largest eigenvalue λ_{max}, even if the terms $\exp(-\lambda_{max} x)$ in the solution decay very rapidly and become insignificant as we move away from the origin.

For example, consider the differential equation[20]

$$y'' + 1001 y' + 1000 y = 0 \tag{7.16}$$

Using $y_0 = y$ and $y_1 = y_1$, the equivalent first-order equations are

$$\mathbf{y}' = \begin{bmatrix} y_1 \\ -1000 y_0 - 1001 y_1 \end{bmatrix}$$

In this case

$$\mathbf{\Lambda} = \begin{bmatrix} 0 & -1 \\ 1000 & 1001 \end{bmatrix}$$

The eigenvalues of $\mathbf{\Lambda}$ are the roots of

$$|\mathbf{\Lambda} - \lambda \mathbf{I}| = \begin{vmatrix} -\lambda & -1 \\ 1000 & 1001 - \lambda \end{vmatrix} = 0$$

Expanding the determinant we get

$$-\lambda(1001 - \lambda) + 1000 = 0$$

which has the solutions $\lambda_1 = 1$ and $\lambda_2 = 1000$. These equation are clearly stiff. According to Eq. (7.15) we would need $h < 2/\lambda_2 = 0.002$ for Euler's method to be stable. The Runge–Kutta method would have approximately the same limitation on the step size.

When the problem is very stiff, the usual methods of solution, such as the Runge–Kutta formulas, become impractical due to the very small h required for stability. These problems are best solved with methods that are specially designed for stiff equations. Stiff problem solvers, which are outside the scope of this text, have much better stability characteristics; some of them are even unconditionally stable. However, the higher degree of stability comes at a cost—the general rule is that stability can be improved only by reducing the order of the method (and thus increasing the truncation error).

EXAMPLE 7.7
(1) Show that the problem

$$y'' = -\frac{19}{4} y - 10 y' \qquad y(0) = -9 \qquad y'(0) = 0$$

---

[20] This example is taken from C. E. Pearson, *Numerical Methods in Engineering and Science*, van Nostrand and Reinhold (1986).

is moderately stiff and estimate h_{max}, the largest value of h for which the Runge–Kutta method would be stable. (2) Confirm the estimate by computing $y(10)$ with $h \approx h_{max}/2$ and $h \approx 2h_{max}$.

Solution of Part (1) With the notation $y = y_0$ and $y' = y_1$ the equivalent first-order differential equations are

$$\mathbf{y}' = \begin{bmatrix} y_1 \\ -\dfrac{19}{4} y_0 - 10 y_1 \end{bmatrix} = -\mathbf{\Lambda} \begin{bmatrix} y_0 \\ y_1 \end{bmatrix}$$

where

$$\mathbf{\Lambda} = \begin{bmatrix} 0 & -1 \\ \dfrac{19}{4} & 10 \end{bmatrix}$$

The eigenvalues of $\mathbf{\Lambda}$ are given by

$$|\mathbf{\Lambda} - \lambda \mathbf{I}| = \begin{vmatrix} -\lambda & -1 \\ \dfrac{19}{4} & 10 - \lambda \end{vmatrix} = 0$$

which yields $\lambda_1 = 1/2$ and $\lambda_2 = 19/2$. Because λ_2 is quite a bit larger than λ_1, the equations are moderately stiff.

Solution of Part (2) An estimate for the upper limit of the stable range of h can be obtained from Eq. (7.15):

$$h_{max} = \frac{2}{\lambda_{max}} = \frac{2}{19/2} = 0.2153$$

Although this formula is strictly valid for Euler's method, it is usually not too far off for higher-order integration formulas.

Here are the results from the Runge–Kutta method with $h = 0.1$ (by specifying `freq = 0` in `printSoln`, only the initial and final values were printed):

```
     x              y[ 0 ]           y[ 1 ]
0.0000e+000    -9.0000e+000     0.0000e+000
1.0000e+001    -6.4011e-002     3.2005e-002
```

The analytical solution is

$$y(x) = -\frac{19}{2} e^{-x/2} + \frac{1}{2} e^{-19x/2}$$

yielding $y(10) = -0.064011$, which agrees with the value obtained numerically.

With $h = 0.5$ we encountered instability, as expected:

```
     x              y[ 0 ]          y[ 1 ]
0.0000e+000    -9.0000e+000     0.0000e+000
1.0000e+001     2.7030e+020    -2.5678e+021
```

7.5 Adaptive Runge–Kutta Method

Determination of a suitable step size h can be a major headache in numerical integration. If h is too large, the truncation error may be unacceptable; if h is too small, we are squandering computational resources. Moreover, a constant step size may not be appropriate for the entire range of integration. For example, if the solution curve starts off with rapid changes before becoming smooth (as in a stiff problem), we should use a small h at the beginning and increase it as we reach the smooth region. This is where *adaptive methods* come in. They estimate the truncation error at each integration step and automatically adjust the step size to keep the error within prescribed limits.

The adaptive Runge–Kutta methods use so-called *embedded integration formulas*. These formulas come in pairs: one formula has the integration order m, the other one is of order $m + 1$. The idea is to use both formulas to advance the solution from x to $x + h$. Denoting the results by $\mathbf{y}_m(x + h)$ and $\mathbf{y}_{m+1}(x + h)$, we may estimate the truncation error in the formula of order m:

$$\mathbf{E}(h) = \mathbf{y}_{m+1}(x + h) - \mathbf{y}_m(x + h) \tag{7.17}$$

What makes the embedded formulas attractive is that they share the points where $\mathbf{F}(x, \mathbf{y})$ is evaluated. This means that once $\mathbf{y}_m(x + h)$ has been computed, relatively small additional effort is required to calculate $\mathbf{y}_{m+1}(x + h)$.

Here are the Runge–Kutta embedded formulas of orders 5 and 4 that were originally derived by Fehlberg; hence they are known as *Runge–Kutta–Fehlberg formulas*:

$$\mathbf{K}_0 = h\mathbf{F}(x, \mathbf{y})$$

$$\mathbf{K}_i = h\mathbf{F}\left(x + A_i h, \mathbf{y} + \sum_{j=0}^{i-1} B_{ij} \mathbf{K}_j\right), \quad i = 1, 2, \ldots, 5 \tag{7.18}$$

$$\mathbf{y}_5(x + h) = \mathbf{y}(x) + \sum_{i=0}^{5} C_i \mathbf{K}_i \quad \text{(fifth-order formula)} \tag{7.19a}$$

$$\mathbf{y}_4(x + h) = \mathbf{y}(x) + \sum_{i=0}^{5} D_i \mathbf{K}_i \quad \text{(fourth-order formula)} \tag{7.19b}$$

Initial Value Problems

The coefficients appearing in these formulas are not unique. Table 6.1 gives the coefficients proposed by Cash and Karp[21] which are claimed to be an improvement over Fehlberg's original values.

i	A_i	B_{ij}					C_i	D_i
0	—	—	—	—	—	—	$\dfrac{37}{378}$	$\dfrac{2825}{27\,648}$
1	$\dfrac{1}{5}$	$\dfrac{1}{5}$	—	—	—	—	0	0
2	$\dfrac{3}{10}$	$\dfrac{3}{40}$	$\dfrac{9}{40}$	—	—	—	$\dfrac{250}{621}$	$\dfrac{18\,575}{48\,384}$
3	$\dfrac{3}{5}$	$\dfrac{3}{10}$	$-\dfrac{9}{10}$	$\dfrac{6}{5}$	—	—	$\dfrac{125}{594}$	$\dfrac{13\,525}{55\,296}$
4	1	$-\dfrac{11}{54}$	$\dfrac{5}{2}$	$-\dfrac{70}{27}$	$\dfrac{35}{27}$	—	0	$\dfrac{277}{14\,336}$
5	$\dfrac{7}{8}$	$\dfrac{1631}{55296}$	$\dfrac{175}{512}$	$\dfrac{575}{13824}$	$\dfrac{44275}{110592}$	$\dfrac{253}{4096}$	$\dfrac{512}{1771}$	$\dfrac{1}{4}$

Table 6.1. Cash–Karp coefficients for Runge–Kutta-Fehlberg formulas

The solution is advanced with the fifth-order formula in Eq. (7.19a). The fourth-order formula is used only implicitly in estimating the truncation error

$$\mathbf{E}(h) = \mathbf{y}_5(x+h) - \mathbf{y}_4(x+h) = \sum_{i=0}^{5}(C_i - D_i)\mathbf{K}_i \qquad (7.20)$$

Since Eq. (7.20) actually applies to the fourth-order formula, it tends to overestimate the error in the fifth-order formula.

Note that $\mathbf{E}(h)$ is a vector, its components $E_i(h)$ representing the errors in the dependent variables y_i. This brings up the question: what is the error measure $e(h)$ that we wish to control? There is no single choice that works well in all problems. If we want to control the largest component of $\mathbf{E}(h)$, the error measure would be

$$e(h) = \max_i |E_i(h)| \qquad (7.21)$$

We could also control some gross measure of the error, such as the root-mean-square

---

[21] Cash, J. R., and Carp, A. H., *ACM Transactions on Mathematical Software*, Vol. 16, p. 201 (1990).

7.5 Adaptive Runge–Kutta Method

error defined by

$$\bar{E}(h) = \sqrt{\frac{1}{n}\sum_{i=0}^{n-1} E_i^2(h)} \tag{7.22}$$

where n is the number of first-order equations. Then we would use

$$e(h) = \bar{E}(h) \tag{7.23}$$

for the error measure. Since the root-mean-square error is easier to handle, we adopt it for our program.

Error control is achieved by adjusting the increment h so that the per-step error $e(h)$ is approximately equal to a prescribed tolerance ε. Noting that the truncation error in the fourth-order formula is $\mathcal{O}(h^5)$, we conclude that

$$\frac{e(h_1)}{e(h_2)} \approx \left(\frac{h_1}{h_2}\right)^5 \tag{a}$$

Let us suppose that we performed an integration step with h_1 that resulted in the error $e(h_1)$. The step size h_2 that we should have used can now be obtained from Eq. (a) by setting $e(h_2) = \varepsilon$:

$$h_2 = h_1 \left[\frac{\varepsilon}{e(h_1)}\right]^{1/5} \tag{b}$$

If $h_2 \geq h_1$, we could repeat the integration step with h_2, but since the error was below the tolerance, that would be a waste of a perfectly good result. So we accept the current step and try h_2 in the next step. On the other hand, if $h_2 < h_1$, we must scrap the current step and repeat it with h_2. As Eq. (b) is only an approximation, it is prudent to incorporate a small margin of safety. In our program we use the formula

$$h_2 = 0.9 h_1 \left[\frac{\varepsilon}{e(h_1)}\right]^{1/5} \tag{7.24}$$

Recall that $e(h)$ applies to a single integration step; that is, it is a measure of the local truncation error. The all-important global truncation error is due to the accumulation of the local errors. What should ε be set at in order to achieve a global error tolerance $\varepsilon_{\text{global}}$? Since $e(h)$ is a conservative estimate of the actual error, setting $\varepsilon = \varepsilon_{\text{global}}$ will usually be adequate. If the number integration steps is large, it is advisable to decrease ε accordingly.

Is there any reason to use the nonadaptive methods at all? Usually no; however, there are special cases where adaptive methods break down. For example, adaptive methods generally do not work if $\mathbf{F}(x, \mathbf{y})$ contains discontinuities. Because the error behaves erratically at the point of discontinuity, the program can get stuck in an infinite loop trying to find the appropriate value of h. We would also use a nonadaptive method if the output is to have evenly spaced values of x.

■ run_kut5

This module is compatible with run_kut4 listed in the previous chapter. Any program that calls integrate can choose between the adaptive and the nonadaptive methods by importing either run_kut5 or run_kut4. The input argument *h* is the trial value of the increment for the first integration step.

```
## module run_kut5
''' X,Y = integrate(F,x,y,xStop,h,tol=1.0e-6).
    Adaptive Runge-Kutta method for solving the
    initial value problem {y}' = {F(x,{y})}, where
    {y} = {y[0],y[1],...y[n-1]}.
    x,y    = initial conditions
    xStop  = terminal value of x
    h      = initial increment of x used in integration
    tol    = per-step error tolerance
    F      = user-supplied function that returns the
             array F(x,y) = {y'[0],y'[1],...,y'[n-1]}.
'''
from numarray import array,sum,zeros,Float64
from math import sqrt

def integrate(F,x,y,xStop,h,tol=1.0e-6):

    def run_kut5(F,x,y,h):
      # Runge-Kutta-Fehlberg formulas
        C = array([37./378, 0., 250./621, 125./594,         \
                   0., 512./1771])
        D = array([2825./27648, 0., 18575./48384,           \
                   13525./55296, 277./14336, 1./4])
        n = len(y)
        K = zeros((6,n),type=Float64)
        K[0] = h*F(x,y)
        K[1] = h*F(x + 1./5*h, y + 1./5*K[0])
        K[2] = h*F(x + 3./10*h, y + 3./40*K[0] + 9./40*K[1])
        K[3] = h*F(x + 3./5*h, y + 3./10*K[0]- 9./10*K[1]   \
               + 6./5*K[2])
        K[4] = h*F(x + h, y - 11./54*K[0] + 5./2*K[1]       \
               - 70./27*K[2] + 35./27*K[3])
        K[5] = h*F(x + 7./8*h, y + 1631./55296*K[0]         \
```

7.5 Adaptive Runge–Kutta Method

```
                + 175./512*K[1] + 575./13824*K[2]           \
                + 44275./110592*K[3] + 253./4096*K[4])
    # Initialize arrays {dy} and {E}
      E = zeros((n),type=Float64)
      dy = zeros((n),type=Float64)
    # Compute solution increment {dy} and per-step error {E}
      for i in range(6):
          dy = dy + C[i]*K[i]
          E = E + (C[i] - D[i])*K[i]
    # Compute RMS error e
      e = sqrt(sum(E**2)/n)
      return dy,e

X = []
Y = []
X.append(x)
Y.append(y)
stopper = 0  # Integration stopper(0 = off, 1 = on)

for i in range(10000):
    dy,e = run_kut5(F,x,y,h)
  # Accept integration step if error e is within tolerance
    if  e <= tol:
        y = y + dy
        x = x + h
        X.append(x)
        Y.append(y)
      # Stop if end of integration range is reached
        if stopper == 1: break
  # Compute next step size from Eq. (7.24)
    if e != 0.0:
        hNext = 0.9*h*(tol/e)**0.2
    else: hNext = h
  # Check if next step is the last one; is so, adjust h
    if (h > 0.0) == ((x + hNext) >= xStop):
        hNext = xStop - x
        stopper = 1
    h = hNext
return array(X),array(Y)
```

EXAMPLE 7.8
The aerodynamic drag force acting on a certain object in free fall can be approximated by
$$F_D = av^2 e^{-by}$$
where

v = velocity of the object in m/s
y = elevation of the object in meters
$a = 7.45$ kg/m
$b = 10.53 \times 10^{-5}$ m^{-1}

The exponential term accounts for the change of air density with elevation. The differential equation describing the fall is
$$m\ddot{y} = -mg + F_D$$
where $g = 9.80665$ m/s² and $m = 114$ kg is the mass of the object. If the object is released at an elevation of 9 km, determine its elevation and speed after a 10-s fall with the adaptive Runge–Kutta method.

Solution The differential equation and the initial conditions are
$$\ddot{y} = -g + \frac{a}{m}\dot{y}^2 \exp(-by)$$
$$= -9.80665 + \frac{7.45}{114}\dot{y}^2 \exp(-10.53 \times 10^{-5} y)$$
$$y(0) = 9000 \text{ m} \qquad \dot{y}(0) = 0$$

Letting $y_0 = y$ and $y_1 = \dot{y}$, we obtain the equivalent first-order equations as
$$\dot{\mathbf{y}} = \begin{bmatrix} \dot{y}_0 \\ \dot{y}_1 \end{bmatrix} = \begin{bmatrix} y_1 \\ -9.80665 + (65.351 \times 10^{-3}) y_1^2 \exp(-10.53 \times 10^{-5} y_0) \end{bmatrix}$$

$$\mathbf{y}(0) = \begin{bmatrix} 9000 \text{ m} \\ 0 \end{bmatrix}$$

The driver program for `run_kut5` is listed below. We specified a per-step error tolerance of 10^{-2} in `integrate`. Considering the magnitude of **y**, this should be enough for five decimal point accuracy in the solution.

```
#!/usr/bin/python
## example7_8
from numarray import array,zeros,Float64
from run_kut5 import *
```

7.5 Adaptive Runge–Kutta Method

```
from printSoln import *
from math import exp

def F(x,y):
    F = zeros((2),type=Float64)
    F[0] = y[1]
    F[1] = -9.80665 + 65.351e-3 * y[1]**2 * exp(-10.53e-5*y[0])
    return F

x = 0.0
xStop = 10.0
y = array([9000, 0.0])
h = 0.5
freq = 1
X,Y = integrate(F,x,y,xStop,h,1.0e-2)
printSoln(X,Y,freq)
raw_input(''\nPress return to exit'')
```

Running the program resulted in the following output:

x	y[0]	y[1]
0.0000e+000	9.0000e+003	0.0000e+000
5.0000e-001	8.9988e+003	-4.8043e+000
2.0584e+000	8.9821e+003	-1.5186e+001
3.4602e+000	8.9581e+003	-1.8439e+001
4.8756e+000	8.9312e+003	-1.9322e+001
6.5347e+000	8.8989e+003	-1.9533e+001
8.6276e+000	8.8580e+003	-1.9541e+001
1.0000e+001	8.8312e+003	-1.9519e+001

The first step was carried out with the prescribed trial value $h = 0.5$ s. Apparently the error was well within the tolerance, so that the step was accepted. Subsequent step sizes, determined from Eq. (7.24), were considerably larger.

Inspecting the output, we see that at $t = 10$ s the object is moving with the speed $v = -\dot{y} = 19.52$ m/s at an elevation of $y = 8831$ m.

EXAMPLE 7.9
Integrate the moderately stiff problem

$$y'' = -\frac{19}{4}y - 10y' \qquad y(0) = -9 \qquad y'(0) = 0$$

from $x = 0$ to 10 with the adaptive Runge–Kutta method and plot the results (this problem also appeared in Example 7.7).

Initial Value Problems

Solution Since we use an adaptive method, there is no need to worry about the stable range of h, as we did in Example 7.7. As long as we specify a reasonable tolerance for the per-step error (in this case the default value 10^{-6} is fine), the algorithm will find the appropriate step size. Here is the program and its output:

```
#!/usr/bin/python
## example7_9
from numarray import array,zeros,Float64
from run_kut5 import *
from printSoln import *

def F(x,y):
    F = zeros((2),type=Float64)
    F[0] = y[1]
    F[1] = -4.75*y[0] - 10.0*y[1]
    return F

x = 0.0
xStop = 10.0
y = array([-9.0, 0.0])
h = 0.1
freq = 4
X,Y = integrate(F,x,y,xStop,h)
printSoln(X,Y,freq)
raw_input(''\nPress return to exit'')
```

x	y[0]	y[1]
0.0000e+000	-9.0000e+000	0.0000e+000
9.8941e-002	-8.8461e+000	2.6651e+000
2.1932e-001	-8.4511e+000	3.6653e+000
3.7058e-001	-7.8784e+000	3.8061e+000
5.7229e-001	-7.1338e+000	3.5473e+000
8.6922e-001	-6.1513e+000	3.0745e+000
1.4009e+000	-4.7153e+000	2.3577e+000
2.8558e+000	-2.2783e+000	1.1391e+000
4.3990e+000	-1.0531e+000	5.2656e-001
5.9545e+000	-4.8385e-001	2.4193e-001
7.5596e+000	-2.1685e-001	1.0843e-001
9.1159e+000	-9.9591e-002	4.9794e-002
1.0000e+001	-6.4010e-002	3.2005e-002

7.6 Bulirsch–Stoer Method

The results are in agreement with the analytical solution.

The plots of y and y' show every fourth integration step. Note the high density of points near $x = 0$ where y' changes rapidly. As the y'-curve becomes smoother, the distance between the points increases.

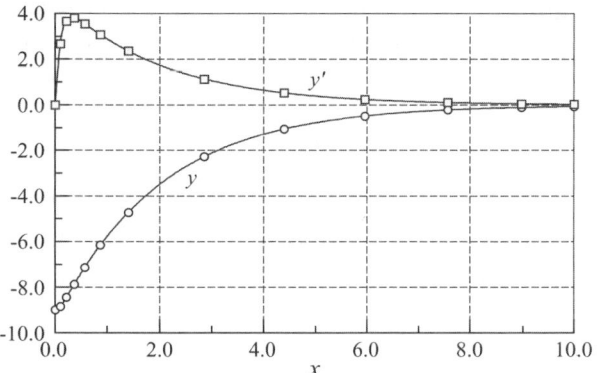

7.6 Bulirsch–Stoer Method

Midpoint Method

The midpoint formula of numerical integration of $\mathbf{y}' = \mathbf{F}(x, \mathbf{y})$ is

$$\mathbf{y}(x + h) = \mathbf{y}(x - h) + 2h\mathbf{F}\left[x, \mathbf{y}(x)\right] \qquad (7.25)$$

It is a second-order formula, like the modified Euler's formula. We discuss it here because it is the basis of the powerful *Bulirsch–Stoer method*, which is the technique of choice in problems where high accuracy is required.

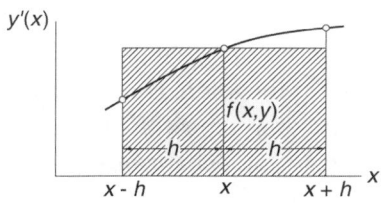

Figure 7.3. Graphical repesentation of the midpoint formula.

Figure 7.3 illustrates the midpoint formula for a single differential equation $y' = f(x, y)$. The change in y over the two panels shown is

$$y(x + h) - y(x - h) = \int_{x-h}^{x+h} y'(x)dx$$

which equals the area under the $y'(x)$ curve. The midpoint method approximates this area by the area $2hf(x, y)$ of the cross-hatched rectangle.

Initial Value Problems

Figure 7.4. Mesh used in the midpoint method.

Consider now advancing the solution of $\mathbf{y}'(x) = \mathbf{F}(x, \mathbf{y})$ from $x = x_0$ to $x_0 + H$ with the midpoint formula. We divide the interval of integration into n steps of length $h = H/n$ each, as shown in Fig. 7.4, and carry out the computations

$$\mathbf{y}_1 = \mathbf{y}_0 + h\mathbf{F}_0$$
$$\mathbf{y}_2 = \mathbf{y}_0 + 2h\mathbf{F}_1$$
$$\mathbf{y}_3 = \mathbf{y}_1 + 2h\mathbf{F}_2 \qquad (7.26)$$
$$\vdots$$
$$\mathbf{y}_n = \mathbf{y}_{n-2} + 2h\mathbf{F}_{n-1}$$

Here we used the notation $\mathbf{y}_i = \mathbf{y}(x_i)$ and $\mathbf{F}_i = \mathbf{F}(x_i, \mathbf{y}_i)$. The first of Eqs. (7.26) uses the Euler formula to "seed" the midpoint method; the other equations are midpoint formulas. The final result is obtained by averaging $\mathbf{y}_n$ in Eq. (7.26) and the estimate $\mathbf{y}_n \approx \mathbf{y}_{n-1} + h\mathbf{F}_n$ available from Euler formula:

$$\mathbf{y}(x_0 + H) = \frac{1}{2}\left[\mathbf{y}_n + \left(\mathbf{y}_{n-1} + h\mathbf{F}_n\right)\right] \qquad (7.27)$$

Richardson Extrapolation

It can be shown that the error in Eq. (7.27) is

$$\mathbf{E} = \mathbf{c}_1 h^2 + \mathbf{c}_2 h^4 + \mathbf{c}_3 h^6 + \cdots$$

Herein lies the great utility of the midpoint method: we can eliminate as many of the leading error terms as we wish by Richardson's extrapolation. For example, we could compute $\mathbf{y}(x_0 + H)$ with a certain value of h and then repeat the process with $h/2$. Denoting the corresponding results by $\mathbf{g}(h)$ and $\mathbf{g}(h/2)$, Richardson's extrapolation—see Eq. (5.9)—then yields the improved result

$$\mathbf{y}_{\text{better}}(x_0 + H) = \frac{4\mathbf{g}(h/2) - \mathbf{g}(h)}{3}$$

which is fourth-order accurate. Another round of integration with $h/4$ followed by Richardson's extrapolation get us sixth-order accuracy, etc.

The $\mathbf{y}$'s in Eqs. (7.26) should be viewed as a intermediate variables, because unlike $\mathbf{y}(x_0 + H)$, they cannot be refined by Richardson's extrapolation.

7.6 Bulirsch–Stoer Method

■ midpoint

The function midpoint in this module combines the midpoint method with Richardson extrapolation. The first application of the midpoint method uses two integration steps. The number of steps is doubled in successive integrations, each integration being followed by Richardson extrapolation. The procedure is stopped when two successive solutions differ (in the root-mean-square sense) by less than a prescribed tolerance.

```
## module midpoint
''' yStop = integrate (F,x,y,xStop,tol=1.0e-6)
    Modified midpoint method for solving the
    initial value problem y' = F(x,y}.
    x,y    = initial conditions
    xStop  = terminal value of x
    yStop  = y(xStop)
    F      = user-supplied function that returns the
             array F(x,y) = {y'[0],y'[1],...,y'[n-1]}.
'''
from numarray import zeros,Float64,sum
from math import sqrt

def integrate(F,x,y,xStop,tol):

    def midpoint(F,x,y,xStop,nSteps):
  # Midpoint formulas
        h = (xStop - x)/nSteps
        y0 = y
        y1 = y0 + h*F(x,y0)
        for i in range(nSteps-1):
            x = x + h
            y2 = y0 + 2.0*h*F(x,y1)
            y0 = y1
            y1 = y2
        return 0.5*(y1 + y0 + h*F(x,y2))

    def richardson(r,k):
  # Richardson's extrapolation
        for j in range(k-1,0,-1):
            const = 4.0**(k-j)
```

```
            r[j] = (const*r[j+1] - r[j])/(const - 1.0)
        return

    kMax = 51
    n = len(y)
    r = zeros((kMax,n),type=Float64)
  # Start with two integration steps
    nSteps = 2
    r[1] = midpoint(F,x,y,xStop,nSteps)
    r_old = r[1].copy()
  # Double the number of integration points
  # and refine result by Richardson extrapolation
    for k in range(2,kMax):
        nSteps = nSteps*2
        r[k] = midpoint(F,x,y,xStop,nSteps)
        richardson(r,k)
      # Compute RMS change in solution
        e = sqrt(sum((r[1] - r_old)**2)/n)
      # Check for convergence
        if e < tol: return r[1]
        r_old = r[1].copy()
    print ''Midpoint method did not converge''
```

Bulirsch–Stoer Algorithm

When used on its own, the module midpoint has a major shortcoming: the solution at points between the initial and final values of x cannot be refined by Richardson extrapolation, so that **y** is usable only at the last point. This deficiency is rectified in the Bulirsch–Stoer method. The fundamental idea behind the method is simple: apply the midpoint method in a piecewise fashion. That is, advance the solution in stages of length H, using the midpoint method with Richardson extrapolation to perform the integration in each stage. The value of H can be quite large, since the precision of the result is determined by the step length h in the midpoint method, not by H.

The original Bulirsch and Stoer technique[22] is a complex procedure that incorporates many refinements missing in our algorithm. However, the function bulStoer given below retains the essential ideas of Bulirsch and Stoer.

What are the relative merits of adaptive Runge–Kutta and Bulirsch–Stoer methods? The Runge–Kutta method is more robust, having higher tolerance for nonsmooth

---

[22] Stoer, J., and Bulirsch, R., *Introduction to Numerical Analysis*, Springer, 1980.

7.6 Bulirsch–Stoer Method

functions and stiff problems. In most applications where high precision is not required, it also tends to be more efficient. However, this is not the case in the computation of high-accuracy solutions involving smooth functions, where the Bulirsch–Stoer algorithm shines.

■ bulStoer

This function contains a simplified algorithm for the Bulirsch–Stoer method.

```
## module bulStoer
''' X,Y = bulStoer(F,x,y,xStop,H,tol=1.0e-6).
    Simplified Bulirsch-Stoer method for solving the
    initial value problem {y}' = {F(x,{y})}, where
    {y} = {y[0],y[1],...y[n-1]}.
    x,y   = initial conditions
    xStop = terminal value of x
    H     = increment of x at which results are stored
    F     = user-supplied function that returns the
            array F(x,y) = {y'[0],y'[1],...,y'[n-1]}.
'''
from numarray import array
from midpoint import *

def bulStoer(F,x,y,xStop,H,tol=1.0e-6):
    X = []
    Y = []
    X.append(x)
    Y.append(y)
    while x < xStop:
        H = min(H,xStop - x)
        y = integrate(F,x,y,x + H,tol) # Midpoint method
        x = x + H
        X.append(x)
        Y.append(y)
    return array(X),array(Y)
```

EXAMPLE 7.10

Compute the solution of the initial value problem

$$y' = \sin y \qquad y(0) = 1$$

at $x = 0.5$ with the midpoint formulas using $n = 2$ and $n = 4$, followed by Richardson extrapolation (this problem was solved with the second-order Runge–Kutta method in Example 7.3).

Solution With $n = 2$ the step length is $h = 0.25$. The midpoint formulas, Eqs. (7.26) and (7.27), yield

$$y_1 = y_0 + hf_0 = 1 + 0.25 \sin 1.0 = 1.210\,368$$

$$y_2 = y_0 + 2hf_1 = 1 + 2(0.25) \sin 1.210\,368 = 1.467\,873$$

$$y_h(0.5) = \frac{1}{2}(y_1 + y_0 + hf_2)$$

$$= \frac{1}{2}(1.210\,368 + 1.467\,873 + 0.25 \sin 1.467\,873)$$

$$= 1.463\,459$$

Using $n = 4$ we have $h = 0.125$ and the midpoint formulas become

$$y_1 = y_0 + hf_0 = 1 + 0.125 \sin 1.0 = 1.105\,184$$

$$y_2 = y_0 + 2hf_1 = 1 + 2(0.125) \sin 1.105\,184 = 1.223\,387$$

$$y_3 = y_1 + 2hf_2 = 1.105\,184 + 2(0.125) \sin 1.223\,387 = 1.340\,248$$

$$y_4 = y_2 + 2hf_3 = 1.223\,387 + 2(0.125) \sin 1.340\,248 = 1.466\,772$$

$$y_{h/2}(0.5) = \frac{1}{2}(y_4 + y_3 + hf_4)$$

$$= \frac{1}{2}(1.466\,772 + 1.340\,248 + 0.125 \sin 1.466\,772)$$

$$= 1.465\,672$$

Richardson extrapolation results in

$$y(0.5) = \frac{4y_{h/2}(0.5) - y_h(0.5)}{3} = \frac{4(1.465\,672) - 1.463\,459}{3} = 1.466\,410$$

which compares favorably with the "true" solution $y(0.5) = 1.466\,404$.

EXAMPLE 7.11

7.6 Bulirsch–Stoer Method

The differential equations governing the loop current i and the charge q on the capacitor of the electric circuit shown are

$$L\frac{di}{dt} + Ri + \frac{q}{C} = E(t) \qquad \frac{dq}{dt} = i$$

If the applied voltage E is suddenly increased from zero to 9 V, plot the resulting loop current during the first ten seconds. Use $R = 1.0\,\Omega$, $L = 2$ H and $C = 0.45$ F.

Solution Letting

$$\mathbf{y} = \begin{bmatrix} y_0 \\ y_1 \end{bmatrix} = \begin{bmatrix} q \\ i \end{bmatrix}$$

and substituting the given data, the differential equations become

$$\dot{\mathbf{y}} = \begin{bmatrix} \dot{y}_0 \\ \dot{y}_1 \end{bmatrix} = \begin{bmatrix} y_1 \\ (-Ry_1 - y_0/C + E)/L \end{bmatrix}$$

The initial conditions are

$$\mathbf{y}(0) = \begin{bmatrix} 0 \\ 0 \end{bmatrix}$$

We solved the problem with the function `bulStoer` with the increment $H = 0.5$ s:

```
## example7_11
from bulStoer import *
from numarray import array,zeros,Float64
from printSoln import *

def F(x,y):
    F = zeros((2),type=Float64)
    F[0] = y[1]
    F[1] = (-y[1] - y[0]/0.45 + 9.0)/2.0
    return F

H = 0.5
xStop = 10.0
x = 0.0
y = array([0.0, 0.0])
X,Y = bulStoer(F,x,y,xStop,H)
printSoln(X,Y,1)
raw_input(''\nPress return to exit'')
```

Skipping the numerical output, the plot of the current is

Recall that in each interval H (the spacing of open circles) the integration was performed by the modified midpoint method and refined by Richardson's extrapolation.

PROBLEM SET 7.2

1. Derive the analytical solution of the problem

$$y'' + y' - 380y = 0 \qquad y(0) = 1 \qquad y'(0) = -20$$

   Would you expect difficulties in solving this problem numerically?

2. Consider the problem

$$y' = x - 10y \qquad y(0) = 10$$

   (a) Verify that the analytical solution is $y(x) = 0.1x - 0.01 + 10.01e^{-10x}$.
   (b) Determine the step size h that you would use in numerical solution with the (nonadaptive) Runge–Kutta method.

3. ■ Integrate the initial value problem in Prob. 2 from $x = 0$ to 5 with the Runge–Kutta method using (a) $h = 0.1$, (b) $h = 0.25$ and (c) $h = 0.5$. Comment on the results.

4. ■ Integrate the initial value problem in Prob. 2 from $x = 0$ to 10 with the adaptive Runge–Kutta method.

5. ■

7.6 Bulirsch–Stoer Method

The differential equation describing the motion of the mass–spring–dashpot system is

$$\ddot{y} + \frac{c}{m}\dot{y} + \frac{k}{m}y = 0$$

where $m = 2$ kg, $c = 460$ N·s/m and $k = 450$ N/m. The initial conditions are $y(0) = 0.01$ m and $\dot{y}(0) = 0$. (a) Show that this is a stiff problem and determine a value of h that you would use in numerical integration with the nonadaptive Runge–Kutta method. (b) Carry out the integration from $t = 0$ to 0.2 s with the chosen h and plot $\dot{y}$ vs. t.

6. ■ Integrate the initial value problem specified in Prob. 5 with the adaptive Runge–Kutta method from $t = 0$ to 0.2 s and plot $\dot{y}$ vs. t.

7. ■ Compute the numerical solution of the differential equation

$$y'' = 16.81y$$

from $x = 0$ to 2 with the adaptive Runge–Kutta method. Use the initial conditions (a) $y(0) = 1.0$, $y'(0) = -4.1$; and (b) $y(0) = 1.0$, $y'(0) = -4.11$. Explain the large difference in the two solutions. *Hint*: derive the analytical solutions.

8. ■ Integrate

$$y'' + y' - y^2 = 0 \qquad y(0) = 1 \qquad y'(0) = 0$$

from $x = 0$ to 3.5. Is the sudden increase in y near the upper limit is real or an artifact caused by instability?

9. ■ Solve the stiff problem—see Eq. (7.16)

$$y'' + 1001y' + 1000y = 0 \qquad y(0) = 1 \qquad y'(0) = 0$$

from $x = 0$ to 0.2 with the adaptive Runge–Kutta method and plot y' vs. x.

10. ■ Solve

$$y'' + 2y' + 3y = 0 \qquad y(0) = 0 \qquad y'(0) = \sqrt{2}$$

with the adaptive Runge–Kutta method from $x = 0$ to 5 (the analytical solution is $y = e^{-x} \sin \sqrt{2}x$).

11. ■ Solve the differential equation

$$y'' = 2yy'$$

from $x = 0$ to 10 with the initial conditions $y(0) = 1$, $y'(0) = -1$. Plot y vs. x.

12. ■ Repeat Prob. 11 with the initial conditions $y(0) = 0$, $y'(0) = 1$ and the integration range $x = 0$ to 1.5.

13. ■ Use the adaptive Runge–Kutta method to integrate

$$y' = \left(\frac{9}{y} - y\right)x \qquad y(0) = 5$$

from $x = 0$ to 4 and plot y vs. x.

14. ■ Solve Prob. 13 with the Bulirsch–Stoer method using $H = 0.5$.

15. ■ Integrate

$$x^2 y'' + xy' + y = 0 \qquad y(1) = 0 \qquad y'(1) = -2$$

from $x = 1$ to 20, and plot y and y' vs. x. Use the Bulirsch–Stoer method.

16. ■

The magnetized iron block of mass m is attached to a spring of stiffness k and free length L. The block is at rest at $x = L$ when the electromagnet is turned on, exerting the repulsive force $F = c/x^2$ on the block. The differential equation of the resulting motion is

$$m\ddot{x} = \frac{c}{x^2} - k(x - L)$$

Determine the amplitude and the period of the motion by numerical integration with the adaptive Runge–Kutta method. Use $c = 5$ N·m², $k = 120$ N/m, $L = 0.2$ m and $m = 1.0$ kg.

17. ■

The bar ABC is attached to the vertical rod with a horizontal pin. The assembly is free to rotate about the axis of the rod. In the absence of friction, the equations of motion of the system are

$$\ddot{\theta} = \dot{\phi}^2 \sin\theta \cos\theta \qquad \ddot{\phi} = -2\dot{\theta}\dot{\phi} \cot\theta$$

The system is set into motion with the initial conditions $\theta(0) = \pi/12$ rad, $\dot{\theta}(0) = 0$, $\phi(0) = 0$ and $\dot{\phi}(0) = 20$ rad/s. Obtain a numerical solution with the adaptive Runge–Kutta method from $t = 0$ to 1.5 s and plot $\dot{\phi}$ vs. t.

18. ■ Solve the circuit problem in Example 7.11 if $R = 0$ and
$$E(t) = \begin{cases} 0 \text{ when } t < 0 \\ 9 \sin \pi t \text{ when } t \geq 0 \end{cases}$$

19. ■ Solve Prob. 21 in Problem Set 1 if $E = 240$ V (constant).

20. ■

Kirchoff's equations for the circuit in the figure are

$$L\frac{di_1}{dt} + R_1 i_1 + R_2(i_1 - i_2) = E(t)$$

$$L\frac{di_2}{dt} + R_2(i_2 - i_1) + \frac{q_2}{C} = 0$$

where

$$\frac{dq_2}{dt} = i_2$$

Using the data $R_1 = 4\,\Omega$, $R_2 = 10\,\Omega$, $L = 0.032$ H, $C = 0.53$ F and

$$E(t) = \begin{cases} 20 \text{ V if } 0 < t < 0.005 \text{ s} \\ 0 \text{ otherwise} \end{cases}$$

plot the transient loop currents i_1 and i_2 from $t = 0$ to 0.05 s.

21. ■ Consider a closed biological system populated by M number of prey and N number of predators. Volterra postulated that the two populations are related by the differential equations

$$\dot{M} = aM - bMN$$

$$\dot{N} = -cN + dMN$$

where a, b, c and d are constants. The steady-state solution is $M_0 = c/d$, $N_0 = a/b$; if numbers other than these are introduced into the system, the populations undergo periodic fluctuations. Introducing the notation

$$y_0 = M/M_0 \qquad y_1 = N/N_0$$

allows us to write the differential equations as

$$\dot{y}_0 = a(y_0 - y_0 y_1)$$
$$\dot{y}_1 = b(-y_1 + y_0 y_1)$$

Using $a = 1.0$/year, $b = 0.2$/year, $y_0(0) = 0.1$ and $y_1(0) = 1.0$, plot the two populations from $t = 0$ to 50 years.

22. ■ The equations

$$\dot{u} = -au + av$$
$$\dot{v} = cu - v - uw$$
$$\dot{w} = -bw + uv$$

known as the Lorenz equations, are encountered in theory of fluid dynamics. Letting $a = 5.0$, $b = 0.9$ and $c = 8.2$, solve these equations from $t = 0$ to 10 with the initial conditions $u(0) = 0$, $v(0) = 1.0$, $w(0) = 2.0$ and plot $u(t)$. Repeat the solution with $c = 8.3$. What conclusions can you draw from the results?

7.7 Other Methods

The methods described so far belong to a group known as *single-step methods*. The name stems from the fact that the information at a single point on the solution curve is sufficient to compute the next point. There are also *multistep methods* that utilize several points on the curve to extrapolate the solution at the next step. Well-known members of this group are the methods of Adams, Milne, Hamming and Gere. These methods were popular once, but have lost some of their luster in the last few years. Multistep methods have two shortcomings that complicate their implementation:

- The methods are not self-starting, but must be provided with the solution at the first few points by a single-step method.
- The integration formulas assume equally spaced steps, which makes it makes it difficult to change the step size.

Both of these hurdles can be overcome, but the price is complexity of the algorithm that increases with the sophistication of the method. The benefits of multistep methods are minimal—the best of them can outperform their single-step counterparts in certain problems, but these occasions are rare.

8 Two-Point Boundary Value Problems

> Solve $y'' = f(x, y, y')$, $y(a) = \alpha$, $y(b) = \beta$

8.1 Introduction

In two-point boundary value problems the auxiliary conditions associated with the differential equation, called the *boundary conditions*, are specified at two different values of x. This seemingly small departure from initial value problems has a major repercussion—it makes boundary value problems considerably more difficult to solve. In an initial value problem we were able to start at the point where the initial values were given and march the solution forward as far as needed. This technique does not work for boundary value problems, because there are not enough starting conditions available at either end point to produce a unique solution.

One way to overcome the lack of starting conditions is to guess the missing values. The resulting solution is very unlikely to satisfy boundary conditions at the other end, but by inspecting the discrepancy we can estimate what changes to make to the initial conditions before integrating again. This iterative procedure is known as the *shooting method*. The name is derived from analogy with target shooting—take a shot and observe where it hits the target, then correct the aim and shoot again.

Another means of solving two-point boundary value problems is the *finite difference method*, where the differential equations are approximated by finite differences at evenly spaced mesh points. As a consequence, a differential equation is transformed into set of simultaneous algebraic equations.

The two methods have a common problem: they give rise to nonlinear sets of equations if the differential equations are not linear. As we noted in Chapter 2, all methods of solving nonlinear equations are iterative procedures that can consume a lot of computational resources. Thus solution of nonlinear boundary value problems is not cheap. Another complication is that iterative methods need reasonably good starting values in order to converge. Since there is no set formula for determining these, an algorithm for solving nonlinear boundary value problems requires informed input; it cannot be treated as a "black box."

8.2 Shooting Method

Second-Order Differential Equation

The simplest two-point boundary value problem is a second-order differential equation with one condition specified at $x = a$ and another one at $x = b$. Here is an example of such a problem:

$$y'' = f(x, y, y'), \quad y(a) = \alpha, \quad y(b) = \beta \tag{8.1}$$

Let us now attempt to turn Eqs. (8.1) into the initial value problem

$$y'' = f(x, y, y'), \quad y(a) = \alpha, \quad y'(a) = u \tag{8.2}$$

The key to success is finding the correct value of u. This could be done by trial and error: guess u and solve the initial value problem by marching from $x = a$ to b. If the solution agrees with the prescribed boundary condition $y(b) = \beta$, we are done; otherwise we have to adjust u and try again. Clearly, this procedure is very tedious.

More systematic methods become available to us if we realize that the determination of u is a root-finding problem. Because the solution of the initial value problem depends on u, the computed value of $y(b)$ is a function of u; that is

$$y(b) = \theta(u)$$

Hence u is a root of

$$r(u) = \theta(u) - \beta = 0 \tag{8.3}$$

where $r(u)$ is the *boundary residual* (difference between the computed and specified boundary value at $x = b$). Equation (8.3) can be solved by one of the root-finding methods discussed in Chapter 4. We reject the method of bisection because it involves too many evaluations of $\theta(u)$. In the Newton–Raphson method we run into the problem of having to compute $d\theta/du$, which can be done, but not easily. That leaves Brent's algorithm as our method of choice.

Here is the procedure we use in solving nonlinear boundary value problems:

1. Specify the starting values u_1 and u_2 which *must bracket the root* u of Eq. (8.3).
2. Apply Brent's method to solve Eq. (8.3) for u. Note that each iteration requires evaluation of $\theta(u)$ by solving the differential equation as an initial value problem.

3. Having determined the value of *u*, solve the differential equations once more and record the results.

If the differential equation is linear, any root-finding method will need only one interpolation to determine *u*. But since Brent's method uses quadratic interpolation, it needs three points: u_1, u_2 and u_3, the latter being provided by a bisection step. This is wasteful, since linear interpolation with u_1 and u_2 would also result in the correct value of *u*. Therefore, we replace Brent's method with linear interpolation whenever the differential equation is linear.

■ linInterp

Here is the algorithm we use for linear interpolation:

```
## module linInterp
''' root = linInterp(f,x1,x2).
    Finds the zero of the linear function f(x) by straight
    line interpolation based on x = x1 and x2.
'''
def linInterp(f,x1,x2):
    f1 = f(x1)
    f2 = f(x2)
    return = x2 - f2*(x2 - x1)/(f2 - f1)
```

EXAMPLE 8.1
Solve the boundary value problem

$$y'' + 3yy' = 0 \quad y(0) = 0 \quad y(2) = 1$$

Solution The equivalent first-order equations are

$$\mathbf{y}' = \begin{bmatrix} y_0' \\ y_1' \end{bmatrix} = \begin{bmatrix} y_1 \\ -3y_0 y_1 \end{bmatrix}$$

with the boundary conditions

$$y_0(0) = 0 \quad y_0(2) = 1$$

Now comes the daunting task of determining the trial values of $y'(0)$. We could always pick two numbers at random and hope for the best. However, it is possible to reduce the element of chance with a little detective work. We start by making the reasonable assumption that *y* is smooth (does not wiggle) in the interval $0 \leq x \leq 2$. Next we note that *y* has to increase from 0 to 1, which requires $y' > 0$. Since both *y* and

y' are positive, we conclude that y'' must be negative in order to satisfy the differential equation. Now we are in a position to make a rough sketch of y:

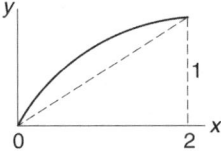

Looking at the sketch it is clear that $y'(0) > 0.5$, so that $y'(0) = 1$ and 2 appear to be reasonable values for the brackets of $y'(0)$; if they are not, Brent's method will display an error message.

In the program listed below we chose the fourth-order Runge–Kutta method for integration. It can be replaced by the adaptive version by substituting run_kut5 for run_kut4 in the import statement. Note that three user-supplied functions are needed to describe the problem at hand. Apart from the function F(x,y) that defines the differential equations, we also need the functions initCond(u) to specify the initial conditions for integration, and r(u) to provide Brent's method with the boundary condition residual. By changing a few statements in these functions, the program can be applied to any second-order boundary value problem. It also works for third-order equations if integration is started at the end where two of the three boundary conditions are specified.

```
#!/usr/bin/python
## example8_1
from numarray import zeros,Float64,array
from run_kut4 import *
from brent import *
from printSoln import *

def initCond(u):  # Init. values of [y, y']; use 'u' if unknown
    return array([0.0, u])

def r(u):         # Boundary condition residual--see Eq. (8.3)
    X,Y = integrate(F,xStart,initCond(u),xStop,h)
    y = Y[len(Y) - 1]
    r = y[0] - 1.0
    return r

def F(x,y):       # First-order differential equations
    F = zeros((2),type=Float64)
    F[0] = y[1]
```

8.2 Shooting Method

```
        F[1] = -3.0*y[0]*y[1]
        return F

xStart = 0.0            # Start of integration
xStop = 2.0             # End of integration
u1 = 1.0                # 1st trial value of unknown init. cond.
u2 = 2.0                # 2nd trial value of unknown init. cond.
h = 0.1                 # Step size
freq = 2                # Printout frequency
u = brent(r,u1,u2)      # Compute the correct initial condition
X,Y = integrate(F,xStart,initCond(u),xStop,h)
printSoln(X,Y,freq)
raw_input(''\nPress return to exit'')
```

Here is the solution :

```
       x              y[ 0 ]           y[ 1 ]
  0.0000e+000      0.0000e+000      1.5145e+000
  2.0000e-001      2.9404e-001      1.3848e+000
  4.0000e-001      5.4170e-001      1.0743e+000
  6.0000e-001      7.2187e-001      7.3287e-001
  8.0000e-001      8.3944e-001      4.5752e-001
  1.0000e+000      9.1082e-001      2.7013e-001
  1.2000e+000      9.5227e-001      1.5429e-001
  1.4000e+000      9.7572e-001      8.6471e-002
  1.6000e+000      9.8880e-001      4.7948e-002
  1.8000e+000      9.9602e-001      2.6430e-002
  2.0000e+000      1.0000e+000      1.4522e-002
```

Note that $y'(0) = 1.5145$, so that our starting values of 1.0 and 2.0 were on the mark.

EXAMPLE 8.2
Numerical integration of the initial value problem

$$y'' + 4y = 4x \qquad y(0) = 0 \qquad y'(0) = 0$$

yielded $y'(2) = 1.653\,64$. Use this information to determine the value of $y'(0)$ that would result in $y'(2) = 0$.

Solution We use linear interpolation

$$u = u_2 - \theta(u_2) \frac{u_2 - u_1}{\theta(u_2) - \theta(u_1)}$$

Two-Point Boundary Value Problems

where in our case $u = y'(0)$ and $\theta(u) = y'(2)$. So far we are given $u_1 = 0$ and $\theta(u_1) = 1.65364$. To obtain the second point, we need another solution of the initial value problem. An obvious solution is $y = x$, which gives us $y(0) = 0$ and $y'(0) = y'(2) = 1$. Thus the second point is $u_2 = 1$ and $\theta(u_2) = 1$. Linear interpolation now yields

$$y'(0) = u = 1 - (1)\frac{1-0}{1-1.65364} = 2.52989$$

EXAMPLE 8.3
Solve the third-order boundary value problem

$$y''' = 2y'' + 6xy \qquad y(0) = 2 \qquad y(5) = y'(5) = 0$$

and plot y vs. x.

Solution The first-order equations and the boundary conditions are

$$\mathbf{y}' = \begin{bmatrix} y'_0 \\ y'_1 \\ y'_2 \end{bmatrix} = \begin{bmatrix} y_1 \\ y_2 \\ 2y_2 + 6xy_0 \end{bmatrix}$$

$$y_0(0) = 2 \qquad y_0(5) = y_1(5) = 0$$

The program listed below is based on `example8_1`. Because two of the three boundary conditions are specified at the right end, we start the integration at $x = 5$ and proceed with negative h toward $x = 0$. Two of the three initial conditions are prescribed: $y_0(5) = y_1(5) = 0$, whereas the third condition $y_2(5)$ is unknown. Because the differential equation is linear, we replaced `brent` with `linInterp`. In linear interpolation the two guesses for $y_2(5)$ (u_1 and u_2) are not important, so we left them as they were in Example 8.1. The adaptive Runge-Kutta method (`run_kut5`) was chosen for the integration.

```
#!/usr/bin/python
## example8_3
from numarray import zeros,Float64,array
from run_kut5 import *
from linInterp import *
from printSoln import *

def initCond(u):  # Initial values of [y,y',y''];
                  # use 'u' if unknown
    return array([0.0, 0.0, u])

def r(u):  # Boundary condition residual--see Eq. (8.3)
    X,Y = integrate(F,xStart,initCond(u),xStop,h)
```

8.2 Shooting Method

```
        y = Y[len(Y) - 1]
        r = y[0] - 2.0
        return r

    def F(x,y):   # First-order differential equations
        F = zeros((3),type=Float64)
        F[0] = y[1]
        F[1] = y[2]
        F[2] = 2.0*y[2] + 6.0*x*y[0]
        return F

xStart = 5.0          # Start of integration
xStop = 0.0           # End of integration
u1 = 1.0              # 1st trial value of unknown init. cond.
u2 = 2.0              # 2nd trial value of unknown init. cond.
h = -0.1              # initial step size
freq = 2              # printout frequency
u = linInterp(r,u1,u2)
X,Y = integrate(F,xStart,initCond(u),xStop,h)
printSoln(X,Y,freq)
raw_input(''\nPress return to exit'')
```

We forgo the rather long printout of the solution and show just the plot:

Higher-Order Equations

Let us consider the fourth-order differential equation

$$y^{(4)} = f(x, y, y', y'', y''') \qquad (8.4a)$$

Two-Point Boundary Value Problems

with the boundary conditions

$$y(a) = \alpha_1 \quad y''(a) = \alpha_2 \quad y(b) = \beta_1 \quad y''(b) = \beta_2 \quad (8.4b)$$

To solve Eq. (8.4a) with the shooting method, we need four initial conditions at $x = a$, only two of which are specified. Denoting the unknown initial values by u_1 and u_2, we have the set of initial conditions

$$y(a) = \alpha_1 \quad y'(a) = u_1 \quad y''(a) = \alpha_2 \quad y'''(a) = u_2 \quad (8.5)$$

If Eq. (8.4a) is solved with the shooting method using the initial conditions in Eq. (8.5), the computed boundary values at $x = b$ depend on the choice of u_1 and u_2. We denote this dependence as

$$y(b) = \theta_1(u_1, u_2) \quad y''(b) = \theta_2(u_1, u_2) \quad (8.6)$$

The correct values u_1 and u_2 satisfy the given boundary conditions at $x = b$:

$$\theta_1(u_1, u_2) = \beta_1 \quad \theta_2(u_1, u_2) = \beta_2$$

or, using vector notation

$$\boldsymbol{\theta}(\mathbf{u}) = \boldsymbol{\beta} \quad (8.7)$$

These are simultaneous (generally nonlinear) equations that can be solved by the Newton–Raphson method discussed in Section 4.6. It must be pointed out again that intelligent estimates of u_1 and u_2 are needed if the differential equation is not linear.

EXAMPLE 8.4

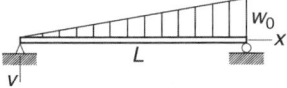

The displacement v of the simply supported beam can be obtained by solving the boundary value problem

$$\frac{d^4 v}{dx^4} = \frac{w_0}{EI}\frac{x}{L} \quad v = \frac{d^2 v}{dx^2} = 0 \text{ at } x = 0 \text{ and } x = L$$

where EI is the bending rigidity. Determine by numerical integration the slopes at the two ends and the displacement at mid-span.

Solution Introducing the dimensionless variables

$$\xi = \frac{x}{L} \quad y = \frac{EI}{w_0 L^4} v$$

8.2 Shooting Method

the problem is transformed to

$$\frac{d^4y}{d\xi^4} = \xi \qquad y = \frac{d^2y}{d\xi^2} = 0 \text{ at } \xi = 0 \text{ and } 1$$

The equivalent first-order equations and the boundary conditions are (the prime denotes $d/d\xi$)

$$\mathbf{y}' = \begin{bmatrix} y_0' \\ y_1' \\ y_2' \\ y_3' \end{bmatrix} = \begin{bmatrix} y_1 \\ y_2 \\ y_3 \\ \xi \end{bmatrix}$$

$$y_0(0) = y_2(0) = y_0(1) = y_2(1) = 0$$

The program listed below is similar to the one in Example 8.1. With appropriate changes in functions F(x,y), initCond(u) and r(u) the program can solve boundary value problems of any order greater than two. For the problem at hand we chose the Bulirsch–Stoer algorithm to do the integration because it gives us control over the printout (we need y precisely at mid-span). The nonadaptive Runge–Kutta method could also be used here, but we would have to guess a suitable step size h.

As the differential equation is linear, the solution requires only one iteration with the Newton–Raphson method. In this case the initial values $u_1 = dy/d\xi|_{x=0}$ and $u_2 = d^3y/d\xi^3|_{x=0}$ are irrelevant; convergence always occurs in one iteration.

```
#!/usr/bin/python
## example8_4
from numarray import zeros,Float64,array
from bulStoer import *
from newtonRaphson2 import *
from printSoln import *

def initCond(u):  # Initial values of [y,y',y'',y'''];
                  # use 'u' if unknown
    return array([0.0, u[0], 0.0, u[1]])

def r(u):  # Boundary condition residuals--see Eq. (8.7)
    r = zeros(len(u),type=Float64)
    X,Y = bulStoer(F,xStart,initCond(u),xStop,H)
    y = Y[len(Y) - 1]
    r[0] = y[0]
```

```
        r[1] = y[2]
        return r

def F(x,y):     # First-order differential equations
    F = zeros((4),type=Float64)
    F[0] = y[1]
    F[1] = y[2]
    F[2] = y[3]
    F[3] = x
    return F

xStart = 0.0                # Start of integration
xStop = 1.0                 # End of integration
u = array([0.0, 1.0])       # Initial guess for {u}
H = 0.5                     # Printout increment
freq = 1                    # Printout frequency
u = newtonRaphson2(r,u,1.0e-4)
X,Y = bulStoer(F,xStart,initCond(u),xStop,H)
printSoln(X,Y,freq)
raw_input(''\nPress return to exit'')
```

Here is the output:

```
    x            y[ 0 ]         y[ 1 ]         y[ 2 ]         y[ 3 ]
0.0000e+000   0.0000e+000    1.9444e-002    0.0000e+000   -1.6667e-001
5.0000e-001   6.5104e-003    1.2153e-003   -6.2500e-002   -4.1667e-002
1.0000e+000  -2.4670e-014   -2.2222e-002   -2.7190e-012    3.3333e-001
```

Noting that

$$\frac{dv}{dx} = \frac{dv}{d\xi}\frac{d\xi}{dx} = \left(\frac{w_0 L^4}{EI}\frac{dy}{d\xi}\right)\frac{1}{L} = \frac{w_0 L^3}{EI}\frac{dy}{d\xi}$$

we obtain

$$\left.\frac{dv}{dx}\right|_{x=0} = 19.444 \times 10^{-3}\frac{w_0 L^3}{EI}$$

$$\left.\frac{dv}{dx}\right|_{x=L} = -22.222 \times 10^{-3}\frac{w_0 L^3}{EI}$$

$$v|_{x=0.5L} = 6.5104 \times 10^{-3}\frac{w_0 L^4}{EI}$$

which agree with the analytical solution (easily obtained by direct integration of the differential equation).

EXAMPLE 8.5
Solve
$$y^{(4)} + \frac{4}{x}y^3 = 0$$
with the boundary conditions
$$y(0) = y'(0) = 0 \qquad y''(1) = 0 \qquad y'''(1) = 1$$
and plot y vs. x.

Solution Our first task is to handle the indeterminacy of the differential equation at the origin, where $x = y = 0$. The problem is resolved by applying L'Hospital's rule: $4y^3/x \to 12y^2 y'$ as $x \to 0$. Thus the equivalent first-order equations and the boundary conditions that we use in the solution are

$$\mathbf{y}' = \begin{bmatrix} y_0' \\ y_1' \\ y_2' \\ y_3' \end{bmatrix} = \begin{bmatrix} y_1 \\ y_2 \\ y_3 \\ \begin{cases} -12y_0^2 y_1 & \text{if } x = 0 \\ -4y_0^3/x & \text{otherwise} \end{cases} \end{bmatrix}$$

$$y_0(0) = y_1(0) = 0 \qquad y_2(1) = 0 \qquad y_3(1) = 1$$

Because the problem is nonlinear, we need reasonable estimates for $y''(0)$ and $y'''(0)$. On the basis of the boundary conditions $y''(1) = 0$ and $y'''(1) = 1$, the plot of y'' is likely to look something like this:

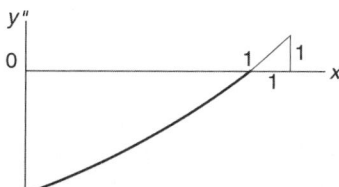

If we are right, then $y''(0) < 0$ and $y'''(0) > 0$. Based on this rather scanty information, we try $y''(0) = -1$ and $y'''(0) = 1$.

The following program uses the adaptive Runge-Kutta method (run_kut5) for integration:

```
#!/usr/bin/python
## example8_5
from numarray import zeros,Float64,array
```

Two-Point Boundary Value Problems

```
from run_kut5 import *
from newtonRaphson2 import *
from printSoln import *

def initCond(u):  # Initial values of [y,y',y'',y'''];
                  # use 'u' if unknown
    return array([0.0, 0.0, u[0], u[1]])

def r(u):  # Boundary condition residuals-- see Eq. (8.7)
    r = zeros(len(u),type=Float64)
    X,Y = integrate(F,x,initCond(u),xStop,h)
    y = Y[len(Y) - 1]
    r[0] = y[2]
    r[1] = y[3] - 1.0
    return r

def F(x,y):  # First-order differential equations
    F = zeros((4),type=Float64)
    F[0] = y[1]
    F[1] = y[2]
    F[2] = y[3]
    if x < 10.e-4: F[3] = -12.0*y[1]*y[0]**2
    else:          F[3] = -4.0*(y[0]**3)/x
    return F

x = 0.0                      # Start of integration
xStop = 1.0                  # End of integration
u = array([-1.0, 1.0])       # Initial guess for u
h = 0.1                      # Initial step size
freq = 1                     # Printout frequency
u = newtonRaphson2(r,u,1.0e-5)
X,Y = integrate(F,x,initCond(u),xStop,h)
printSoln(X,Y,freq)
raw_input(''\nPress return to exit'')
```

The results are:

```
     x           y[ 0 ]        y[ 1 ]        y[ 2 ]        y[ 3 ]
0.0000e+000   0.0000e+000   0.0000e+000  -9.7607e-001   9.7131e-001
1.0000e-001  -4.7184e-003  -9.2750e-002  -8.7893e-001   9.7131e-001
```

8.2 Shooting Method

```
3.9576e-001  -6.6403e-002  -3.1022e-001  -5.9165e-001   9.7152e-001
7.0683e-001  -1.8666e-001  -4.4722e-001  -2.8896e-001   9.7627e-001
9.8885e-001  -3.2061e-001  -4.8968e-001  -1.1144e-002   9.9848e-001
1.0000e+000  -3.2607e-001  -4.8975e-001  -6.7428e-011   1.0000e+000
```

By good fortune, our initial estimates $y''(0) = -1$ and $y'''(0) = 1$ were very close to the final values.

PROBLEM SET 8.1

1. Numerical integration of the initial value problem

$$y'' + y' - y = 0 \qquad y(0) = 0 \qquad y'(0) = 1$$

yielded $y(1) = 0.741028$. What is the value of $y'(0)$ that would result in $y(1) = 1$, assuming that $y(0)$ is unchanged?

2. The solution of the differential equation

$$y''' + y'' + 2y' = 6$$

with the initial conditions $y(0) = 2$, $y'(0) = 0$ and $y''(0) = 1$, yielded $y(1) = 3.03765$. When the solution was repeated with $y''(0) = 0$ (the other conditions being unchanged), the result was $y(1) = 2.72318$. Determine the value of $y''(0)$ so that $y(1) = 0$.

3. Roughly sketch the solution of the following boundary value problems. Use the sketch to estimate $y'(0)$ for each problem.

   (a) $y'' = -e^{-y}$ $y(0) = 1$ $y(1) = 0.5$
   (b) $y'' = 4y^2$ $y(0) = 10$ $y'(1) = 0$
   (c) $y'' = \cos(xy)$ $y(0) = 1$ $y(1) = 2$

4. Using a rough sketch of the solution estimate of $y(0)$ for the following boundary value problems.

   (a) $y'' = y^2 + xy$ $y'(0) = 0$ $y(1) = 2$
   (b) $y'' = -\dfrac{2}{x}y' - y^2$ $y'(0) = 0$ $y(1) = 2$
   (c) $y'' = -x(y')^2$ $y'(0) = 2$ $y(1) = 1$

5. Obtain a rough estimate of $y''(0)$ for the boundary value problem
$$y''' + 5y''y^2 = 0$$
$$y(0) = 0 \quad y'(0) = 1 \quad y(1) = 0$$

6. Obtain rough estimates of $y''(0)$ and $y'''(0)$ for the boundary value problem
$$y^{(4)} + 2y'' + y' \sin y = 0$$
$$y(0) = y'(0) = 0 \quad y(1) = 5 \quad y'(1) = 0$$

7. Obtain rough estimates of $\dot{x}(0)$ and $\dot{y}(0)$ for the boundary value problem
$$\ddot{x} + 2x^2 - y = 0 \quad x(0) = 1 \quad x(1) = 0$$
$$\ddot{y} + y^2 - 2x = 1 \quad y(0) = 0 \quad y(1) = 1$$

8. ■ Solve the boundary value problem
$$y'' + (1 - 0.2x)\, y^2 = 0 \quad y(0) = 0 \quad y(\pi/2) = 1$$

9. ■ Solve the boundary value problem
$$y'' + 2y' + 3y^2 = 0 \quad y(0.01) = 0 \quad y(2) = -1$$

10. ■ Solve the boundary value problem
$$y'' + \sin y + 1 = 0 \quad y(0) = 0 \quad y(\pi) = 0$$

11. ■ Solve the boundary value problem
$$y'' + \dfrac{1}{x}y' + y = 0 \quad y(0.01) = 1 \quad y'(2) = 0$$
and plot y vs. x. *Warning*: y changes very rapidly near $x = 0$.

12. ■ Solve the boundary value problem
$$y'' - (1 - e^{-x})\, y = 0 \quad y(0) = 1 \quad y(\infty) = 0$$
and plot y vs. x. *Hint*: Replace the infinity by a finite value β. Check your choice of β by repeating the solution with 1.5β. If the results change, you must increase β.

8.2 Shooting Method

13. ■ Solve the boundary value problem
$$y''' = -\frac{1}{x}y'' + \frac{1}{x^2}y' + 0.1(y')^3$$
$$y(1) = 0 \qquad y''(1) = 0 \qquad y(2) = 1$$

14. ■ Solve the boundary value problem
$$y''' + 4y'' + 6y' = 10$$
$$y(0) = y''(0) = 0 \qquad y(3) - y'(3) = 5$$

15. ■ Solve the boundary value problem
$$y''' + 2y'' + \sin y = 0$$
$$y(-1) = 0 \qquad y'(-1) = -1 \qquad y'(1) = 1$$

16. ■ Solve the differential equation in Prob. 15 with the boundary conditions
$$y(-1) = 0 \qquad y(0) = 0 \qquad y(1) = 1$$

(this is a three-point boundary value problem).

17. ■ Solve the boundary value problem
$$y^{(4)} = -xy^2$$
$$y(0) = 5 \qquad y''(0) = 0 \qquad y'(1) = 0 \qquad y'''(1) = 2$$

18. ■ Solve the boundary value problem
$$y^{(4)} = -2yy''$$
$$y(0) = y'(0) = 0 \qquad y(4) = 0 \qquad y'(4) = 1$$

19. ■

A projectile of mass m in free flight experiences the aerodynamic drag force $F_D = cv^2$, where v is the velocity. The resulting equations of motion are
$$\ddot{x} = -\frac{c}{m}v\dot{x} \qquad \ddot{y} = -\frac{c}{m}v\dot{y} - g$$
$$v = \sqrt{\dot{x}^2 + \dot{y}^2}$$

If the projectile hits a target 8 km away after a 10-s flight, determine the launch velocity v_0 and its angle of inclination θ. Use $m = 20$ kg, $c = 3.2 \times 10^{-4}$ kg/m and $g = 9.80665$ m/s^2.

20. ■

The simply supported beam carries a uniform load of intensity w_0 and the tensile force N. The differential equation for the vertical displacement v can be shown to be

$$\frac{d^4v}{dx^4} - \frac{N}{EI}\frac{d^2v}{dx^2} = \frac{w_0}{EI}$$

where EI is the bending rigidity. The boundary conditions are $v = d^2v/dx^2 = 0$ at $x = 0$ and L. Changing the variables to $\xi = x/L$ and $y = (EI/w_0L^4)v$ transforms the problem to the dimensionless form

$$\frac{d^4y}{d\xi^4} - \beta\frac{d^2y}{d\xi^2} = 1 \qquad \beta = \frac{NL^2}{EI}$$

$$y|_{\xi=0} = \frac{d^2y}{d\xi^2}\bigg|_{\xi=0} = y|_{\xi=1} = \frac{d^2y}{d\xi^2}\bigg|_{\xi=1} = 0$$

Determine the maximum displacement if (a) $\beta = 1.65929$; and (b) $\beta = -1.65929$ (N is compressive).

21. ■ Solve the boundary value problem

$$y''' + yy'' = 0 \qquad y(0) = y'(0) = 0,\ y'(\infty) = 2$$

and plot $y(x)$ and $y'(x)$. This problem arises in determining the velocity profile of the boundary layer in incompressible flow (Blasius solution).

8.3 Finite Difference Method

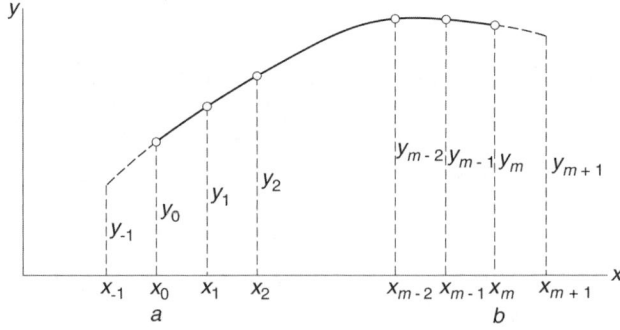

Figure 8.1. Finite difference mesh.

8.3 Finite Difference Method

In the finite difference method we divide the range of integration (a, b) into m equal subintervals of length h each, as shown in Fig. 8.1. The values of the numerical solution at the mesh points are denoted by y_i, $i = 0, 1, \ldots, m$; the purpose of the two points outside (a, b) will be explained shortly. We now make two approximations:

1. The derivatives of y in the differential equation are replaced by the finite difference expressions. It is common practice to use the first central difference approximations (see Chapter 5):

$$y'_i = \frac{y_{i+1} - y_{i-1}}{2h} \qquad y''_i = \frac{y_{i-1} - 2y_i + y_{i+1}}{h^2} \qquad \text{etc.} \qquad (8.8)$$

2. The differential equation is enforced only at the mesh points.

As a result, the differential equations are replaced by $m+1$ simultaneous algebraic equations, the unknowns being y_i, $i = 0, 1, \ldots .m$. If the differential equation is nonlinear, the algebraic equations will also be nonlinear and must be solved by the Newton–Raphson method.

Since the truncation error in a first central difference approximation is $\mathcal{O}(h^2)$, the finite difference method is not nearly as accurate as the shooting method—recall that the Runge–Kutta method has a truncation error of $O(h^5)$. Therefore, the convergence criterion specified in the Newton–Raphson method should not be too severe.

Second-Order Differential Equation

Consider the second-order differential equation

$$y'' = f(x, y, y')$$

with the boundary conditions

$$y(a) = \alpha \quad \text{or} \quad y'(a) = \alpha$$
$$y(b) = \beta \quad \text{or} \quad y'(b) = \beta$$

Approximating the derivatives at the mesh points by finite differences, the problem becomes

$$\frac{y_{i-1} - 2y_i + y_{i+1}}{h^2} = f\left(x_i, y_i, \frac{y_{i+1} - y_{i-1}}{2h}\right), \quad i = 0, 1, \ldots, m \qquad (8.9)$$

$$y_0 = \alpha \quad \text{or} \quad \frac{y_1 - y_{-1}}{2h} = \alpha \qquad (8.10\text{a})$$

$$y_m = \beta \quad \text{or} \quad \frac{y_{m+1} - y_{m-1}}{2h} = \beta \qquad (8.10\text{b})$$

Note the presence of y_{-1} and y_{m+1}, which are associated with points outside the solution domain (a, b). This "spillover" can be eliminated by using the boundary conditions. But before we do that, let us rewrite Eqs. (8.9) as

$$y_{-1} - 2y_0 + y_1 - h^2 f\left(x_0, y_0, \frac{y_1 - y_{-1}}{2h}\right) = 0 \qquad (a)$$

$$y_{i-1} - 2y_i + y_{i+1} - h^2 f\left(x_i, y_i, \frac{y_{i+1} - y_{i-1}}{2h}\right) = 0, \quad i = 1, 2, \ldots, m-1 \qquad (b)$$

$$y_{m-1} - 2y_m + y_{m+1} - h^2 f\left(x_m, y_m, \frac{y_{m+1} - y_{m-1}}{2h}\right) = 0 \qquad (c)$$

The boundary conditions on y are easily dealt with: Eq. (a) is simply replaced by $y_0 - \alpha = 0$ and Eq. (c) is replaced by $y_m - \beta = 0$. If y' are prescribed, we obtain from Eqs. (8.10) $y_{-1} = y_1 - 2h\alpha$ and $y_{m+1} = y_{m-1} + 2h\beta$, which are then substituted into Eqs. (a) and (c), respectively. Hence we finish up with $m+1$ equations in the unknowns $y_0, y_1, \ldots, y_m$:

$$\left.\begin{array}{ll} y_0 - \alpha = 0 & \text{if } y(a) = \alpha \\ -2y_0 + 2y_1 - h^2 f(x_0, y_0, \alpha) - 2h\alpha = 0 & \text{if } y'(a) = \alpha \end{array}\right\} \qquad (8.11a)$$

$$y_{i-1} - 2y_i + y_{i+1} - h^2 f\left(x_i, y_i, \frac{y_{i+1} - y_{i-1}}{2h}\right) = 0 \quad i = 1, 2, \ldots, m-1 \qquad (8.11b)$$

$$\left.\begin{array}{ll} y_m - \beta = 0 & \text{if } y(b) = \beta \\ 2y_{m-1} - 2y_m - h^2 f(x_m, y_m, \beta) + 2h\beta = 0 & \text{if } y'(b) = \beta \end{array}\right\} \qquad (8.11c)$$

EXAMPLE 8.6
Write out Eqs. (8.11) for the following linear boundary value problem using $m = 10$:

$$y'' = -4y + 4x \qquad y(0) = 0 \qquad y'(\pi/2) = 0$$

Solve these equations with a computer program.

Solution In this case $\alpha = y(0) = 0$, $\beta = y'(\pi/2) = 0$ and $f(x, y, y') = -4y + 4x$. Hence Eqs. (8.11) are

$$y_0 = 0$$
$$y_{i-1} - 2y_i + y_{i+1} - h^2(-4y_i + 4x_i) = 0, \quad i = 1, 2, \ldots, 9$$
$$2y_9 - 2y_{10} - h^2(-4y_{10} + 4x_{10}) = 0$$

8.3 Finite Difference Method

or, using matrix notation

$$\begin{bmatrix} 1 & 0 & & & & \\ 1 & -2+4h^2 & 1 & & & \\ & \ddots & \ddots & \ddots & & \\ & & 1 & -2+4h^2 & 1 & \\ & & & 2 & -2+4h^2 \end{bmatrix} \begin{bmatrix} y_0 \\ y_1 \\ \vdots \\ y_9 \\ y_{10} \end{bmatrix} = \begin{bmatrix} 0 \\ 4h^2 x_1 \\ \vdots \\ 4h^2 x_9 \\ 4h^2 x_{10} \end{bmatrix}$$

Note that the coefficient matrix is tridiagonal, so that the equations can be solved efficiently by the decomposition and back substitution routines in module LUdecomp3, described in Section 2.4. Recalling that in LUdecomp3 the diagonals of the coefficient matrix are stored in vectors **c**, **d** and **e**, we arrive at the following program:

```
#!/usr/bin/python
## example8_6
from numarray import zeros,ones,Float64,array,arange
from LUdecomp3 import *
from math import pi

def equations(x,h,m):  # Set up finite difference eqs.
    h2 = h*h
    d = ones((m + 1))*(-2.0 + 4.0*h2)
    c = ones((m),type = Float64)
    e = ones((m),type = Float64)
    b = ones((m+1))*4.0*h2*x
    d[0] = 1.0
    e[0] = 0.0
    b[0] = 0.0
    c[m-1] = 2.0
    return c,d,e,b

xStart = 0.0            # x at left end
xStop = pi/2.0          # x at right end
m = 10                  # Number of mesh spaces
h = (xStop - xStart)/m
x = arange(xStart,xStop + h,h)
c,d,e,b = equations(x,h,m)
c,d,e = LUdecomp3(c,d,e)
y = LUsolve3(c,d,e,b)
```

```
print ''\n          x                    y''
for i in range(m + 1):
    print ''%14.5e %14.5e'' %(x[i],y[i])
raw_input(''\nPress return to exit'')
```

The solution is

```
       x               y
0.00000e+000    0.00000e+000
1.57080e-001    3.14173e-001
3.14159e-001    6.12841e-001
4.71239e-001    8.82030e-001
6.28319e-001    1.11068e+000
7.85398e-001    1.29172e+000
9.42478e-001    1.42278e+000
1.09956e+000    1.50645e+000
1.25664e+000    1.54995e+000
1.41372e+000    1.56451e+000
1.57080e+000    1.56418e+000
```

The exact solution of the problem is

$$y = x - \sin 2x$$

which yields $y(\pi/2) = \pi/2 = 1.57080$. Thus the error in the numerical solution is about 0.4%. More accurate results can be achieved by increasing m. For example, with $m = 100$, we would get $y(\pi/2) = 1.57073$, which is in error by only 0.0002%.

EXAMPLE 8.7
Solve the boundary value problem

$$y'' = -3yy' \qquad y(0) = 0 \qquad y(2) = 1$$

with the finite difference method. Use $m = 10$ and compare the output with the results of the shooting method in Example 8.1.

Solution As the problem is nonlinear, Eqs. (8.11) must be solved by the Newton–Raphson method. The program listed below can be used as a model for other second-order boundary value problems. The function `residual(y)` returns the residuals of the finite difference equations, which are the left-hand sides of Eqs. (8.11). The differential equation $y'' = f(x, y, y')$ is defined in the function `F(x,y,yPrime)`. In

8.3 Finite Difference Method

this problem we chose for the initial solution $y_i = 0.5x_i$, which corresponds to the dashed straight line shown in the rough plot of y in Example 8.1. The starting values of $y_0, y_1, \ldots, y_m$ are specified by function startSoln(x). Note that we relaxed the convergence criterion in the Newton–Raphson method to 1.0×10^{-5}, which is more in line with the truncation error in the finite difference method.

```
#!/usr/bin/python
## example8_7
from numarray import zeros,Float64,array,arange
from newtonRaphson2 import *

def residual(y):   # Residuals of finite diff. Eqs. (8.11)
    r = zeros((m + 1),type=Float64)
    r[0] = y[0]
    r[m] = y[m] - 1.0
    for i in range(1,m):
        r[i] = y[i-1] - 2.0*y[i] + y[i+1]            \
             - h*h*F(x[i],y[i],(y[i+1] - y[i-1])/(2.0*h))
    return r

def F(x,y,yPrime):  # Differential eqn. y'' = F(x,y,y')
    F = -3.0*y*yPrime
    return F

def startSoln(x): # Starting solution y(x)
    y = zeros((m + 1),type=Float64)
    for i in range(m + 1): y[i] = 0.5*x[i]
    return y

xStart = 0.0            # x at left end
xStop = 2.0             # x at right end
m = 10                  # Number of mesh intervals
h = (xStop - xStart)/m
x = arange(xStart,xStop + h,h)
y = newtonRaphson2(residual,startSoln(x),1.0e-5)
print ''\n        x              y''
for i in range(m + 1):
    print ''%14.5e %14.5e'' %(x[i],y[i])
raw_input(''\nPress return to exit'')
```

Here is the output from our program together with the solution obtained in Example 8.1.

```
       x                y           y from Ex. 8.1
0.00000e+000    0.00000e+000    0.00000e+000
2.00000e-001    3.02404e-001    2.94050e-001
4.00000e-001    5.54503e-001    5.41710e-001
6.00000e-001    7.34691e-001    7.21875e-001
8.00000e-001    8.49794e-001    8.39446e-001
1.00000e+000    9.18132e-001    9.10824e-001
1.20000e+000    9.56953e-001    9.52274e-001
1.40000e+000    9.78457e-001    9.75724e-001
1.60000e+000    9.90201e-001    9.88796e-001
1.80000e+000    9.96566e-001    9.96023e-001
2.00000e+000    1.00000e+000    1.00000e+000
```

The maximum discrepancy between the solutions is 1.8% occurring at $x = 0.6$. As the shooting method used in Example 8.1 is considerably more accurate than the finite difference method, the discrepancy can be attributed to truncation errors in the finite difference solution. This error would be acceptable in many engineering problems. Again, accuracy can be increased by using a finer mesh. With $m = 100$ we can reduce the error to 0.07%, but we must question whether the tenfold increase in computation time is really worth the extra precision.

Fourth-Order Differential Equation

For the sake of brevity we limit our discussion to the special case where y' and y''' do not appear explicitly in the differential equation; that is, we consider

$$y^{(4)} = f(x, y, y'')$$

We assume that two boundary conditions are prescribed at each end of the solution domain (a, b). Problems of this form are commonly encountered in beam theory.

Again we divide the solution domain into m intervals of length h each. Replacing the derivatives of y by finite differences at the mesh points, we get the finite difference equations

$$\frac{y_{i-2} - 4y_{i-1} + 6y_i - 4y_{i+1} + y_{i+2}}{h^4} = f\left(x_i, y_i, \frac{y_{i-1} - 2y_i + y_{i+1}}{h^2}\right) \quad (8.12)$$

8.3 Finite Difference Method

where $i = 0, 1, \ldots, m$. It is more revealing to write these equations as

$$y_{-2} - 4y_{-1} + 6y_0 - 4y_1 + y_2 - h^4 f\left(x_0, y_0, \frac{y_{-1} - 2y_0 + y_1}{h^2}\right) = 0 \qquad (8.13a)$$

$$y_{-1} - 4y_0 + 6y_1 - 4y_2 + y_3 - h^4 f\left(x_1, y_1, \frac{y_0 - 2y_1 + y_2}{h^2}\right) = 0 \qquad (8.13b)$$

$$y_0 - 4y_1 + 6y_2 - 4y_3 + y_4 - h^4 f\left(x_2, y_2, \frac{y_1 - 2y_2 + y_3}{h^2}\right) = 0 \qquad (8.13c)$$

$$\vdots$$

$$y_{m-3} - 4y_{m-2} + 6y_{m-1} - 4y_m + y_{m+1} - h^4 f\left(x_{m-1}, y_{m-1}, \frac{y_{m-2} - 2y_{m-1} + y_m}{h^2}\right) = 0 \qquad (8.13d)$$

$$y_{m-2} - 4y_{m-1} + 6y_m - 4y_{m+1} + y_{m+2} - h^4 f\left(x_m, y_m, \frac{y_{m-1} - 2y_m + y_{m+1}}{h^2}\right) = 0 \qquad (8.13e)$$

We now see that there are four unknowns y_{-2}, y_{-1}, y_{m+1} and y_{m+2} that lie outside the solution domain that must be eliminated by applying the boundary conditions, a task that is facilitated by Table 8.1.

Bound. cond.	Equivalent finite difference expression
$y(a) = \alpha$	$y_0 = \alpha$
$y'(a) = \alpha$	$y_{-1} = y_1 - 2h\alpha$
$y''(a) = \alpha$	$y_{-1} = 2y_0 - y_1 + h^2\alpha$
$y'''(a) = \alpha$	$y_{-2} = 2y_{-1} - 2y_1 + y_2 - 2h^3\alpha$
$y(b) = \beta$	$y_m = \beta$
$y'(b) = \beta$	$y_{m+1} = y_{m-1} + 2h\beta$
$y''(b) = \beta$	$y_{m+1} = 2y_m - y_{m-1} + h^2\beta$
$y'''(b) = \beta$	$y_{m+2} = 2y_{m+1} - 2y_{m-1} + y_{m-2} + 2h^3\beta$

Table 8.1

The astute observer may notice that some combinations of boundary conditions will not work in eliminating the "spillover." One such combination is clearly $y(a) = \alpha_1$ and $y'''(a) = \alpha_2$. The other one is $y'(a) = \alpha_1$ and $y''(a) = \alpha_2$. In the context of beam theory, this makes sense: we can impose either a displacement y or a shear force EIy''' at a point, but it is impossible to enforce both of them simultaneously. Similarly, it makes no physical sense to prescribe both the slope y' and the bending moment EIy'' at the same point.

Two-Point Boundary Value Problems

EXAMPLE 8.8

The uniform beam of length L and bending rigidity EI is attached to rigid supports at both ends. The beam carries a concentrated load P at its mid-span. If we utilize symmetry and model only the left half of the beam, the displacement v can be obtained by solving the boundary value problem

$$EI\frac{d^4v}{dx^4} = 0$$

$$v|_{x=0} = 0 \qquad \left.\frac{dv}{dx}\right|_{x=0} = 0 \qquad \left.\frac{dv}{dx}\right|_{x=L/2} = 0 \qquad \left.EI\frac{d^3v}{dx^3}\right|_{x=L/2} = -P/2$$

Use the finite difference method to determine the displacement and the bending moment $M = -EI\,(d^2v/dx^2)$ at the mid-span (the exact values are $v = PL^3/(192EI)$ and $M = PL/8$).

Solution By introducing the dimensionless variables

$$\xi = \frac{x}{L} \qquad y = \frac{EI}{PL^3}v$$

the problem becomes

$$\frac{d^4y}{d\xi^4} = 0$$

$$y|_{\xi=0} = 0 \qquad \left.\frac{dy}{d\xi}\right|_{\xi=0} = 0 \qquad \left.\frac{dy}{d\xi}\right|_{\xi=1/2} = 0 \qquad \left.\frac{d^3y}{d\xi^3}\right|_{\xi=1/2} = -\frac{1}{2}$$

We now proceed to writing Eqs. (8.13) taking into account the boundary conditions. Referring to Table 8.1, we obtain the finite difference expressions of the boundary conditions at the left end as $y_0 = 0$ and $y_{-1} = y_1$. Hence Eqs. (8.13a) and (8.13b) become

$$y_0 = 0 \qquad \text{(a)}$$

$$-4\cancel{y_0}^{\,0} + 7y_1 - 4y_2 + y_3 = 0 \qquad \text{(b)}$$

Equation (8.13c) is

$$\cancel{y_0}^{\,0} - 4y_1 + 6y_2 - 4y_3 + y_4 = 0 \qquad \text{(c)}$$

At the midspan the boundary conditions are equivalent to $y_{m+1} = y_{m-1}$ and

$$y_{m+2} = 2y_{m+1} + y_{m-2} - 2y_{m-1} + 2h^3(-1/2) = y_{m-2} - h^3$$

8.3 Finite Difference Method

Substitution into Eqs. (8.13d) and (8.13e) yields

$$y_{m-3} - 4y_{m-2} + 7y_{m-1} - 4y_m = 0 \tag{d}$$

$$2y_{m-2} - 8y_{m-1} + 6y_m = h^3 \tag{e}$$

The coefficient matrix of Eqs. (a)–(e) can be made symmetric by dividing Eq. (e) by 2. The result is

$$\begin{bmatrix} 1 & 0 & 0 & & & & & & \\ 0 & 7 & -4 & 1 & & & & & \\ 0 & -4 & 6 & -4 & 1 & & & & \\ & \ddots & \ddots & \ddots & \ddots & \ddots & & & \\ & & 1 & -4 & 6 & -4 & 1 \\ & & & 1 & -4 & 7 & -4 \\ & & & & 1 & -4 & 3 \end{bmatrix} \begin{bmatrix} y_0 \\ y_1 \\ y_2 \\ \vdots \\ y_{m-2} \\ y_{m-1} \\ y_m \end{bmatrix} = \begin{bmatrix} 0 \\ 0 \\ 0 \\ \vdots \\ 0 \\ 0 \\ 0.5h^3 \end{bmatrix}$$

The above system of equations can be solved with the decomposition and back substitution routines in module `LUdecomp5`—see Section 2.4. Recall that `LUdecomp5` works with the vectors **d**, **e** and **f** that form the diagonals of the upper half of the matrix. The constant vector is denoted by **b**. The program that sets up and solves the equations is

```
#!/usr/bin/python
## example8_8
from numarray import zeros,ones,Float64,array,arange
from LUdecomp5 import *

def equations(x,h,m): # Set up finite difference eqs.
    h4 = h**4
    d = ones((m + 1),type = Float64)*6.0
    e = ones((m),type = Float64)*(-4.0)
    f = ones((m-1),type = Float64)
    b = zeros((m+1),type=Float64)
    d[0] = 1.0
    d[1] = 7.0
    e[0] = 0.0
    f[0] = 0.0
    d[m-1] = 7.0
    d[m] = 3.0
    b[m] = 0.5*h**3
    return d,e,f,b

xStart = 0.0           # x at left end
```

```
xStop = 0.5            # x at right end
m = 20                 # Number of mesh spaces
h = (xStop - xStart)/m
x = arange(xStart,xStop + h,h)
d,e,f,b = equations(x,h,m)
d,e,f = LUdecomp5(d,e,f)
y = LUsolve5(d,e,f,b)
print ''\n            x                    y''
for i in range(m + 1):
    print ''%14.5e %14.5e'' %(x[i],y[i])
raw_input(''\nPress return to exit'')
```

When we ran the program with $m = 20$, the last two lines of the output were

```
        x                 y
4.75000e-001      5.19531e-003
5.00000e-001      5.23438e-003
```

Thus at the mid-span we have

$$v|_{x=0.5L} = \frac{PL^3}{EI} y|_{\xi=0.5} = 5.234\,38 \times 10^{-3} \frac{PL^3}{EI}$$

$$\left.\frac{d^2v}{dx^2}\right|_{x=0.5L} = \frac{PL^3}{EI}\left(\frac{1}{L^2}\left.\frac{d^2y}{d\xi^2}\right|_{\xi=0.5}\right) \approx \frac{PL}{EI} \frac{y_{m-1} - 2y_m + y_{m+1}}{h^2}$$

$$= \frac{PL}{EI} \frac{(5.19531 - 2(5.23438) + 5.19531) \times 10^{-3}}{0.025^2}$$

$$= -0.125\,024 \frac{PL}{EI}$$

$$M|_{x=0.5L} = -EI \left.\frac{d^2v}{dx^2}\right|_{\xi=0.5} = 0.125\,024\,PL$$

In comparison, the exact solution yields

$$v|_{x=0.5L} = 5.208\,33 \times 10^{-3} \frac{PL^3}{EI}$$

$$M|_{x=0.5L} = = 0.125\,000\,PL$$

PROBLEM SET 8.2

Problems 1–5 Use first central difference approximations to transform the boundary value problem shown into simultaneous equations $\mathbf{Ay} = \mathbf{b}$.

1. $y'' = (2 + x)y$, $y(0) = 0$, $y'(1) = 5$.
2. $y'' = y + x^2$, $y(0) = 0$, $y(1) = 1$.

8.3 Finite Difference Method

3. $y'' = e^{-x}y'$, $y(0) = 1$, $y(1) = 0$.
4. $y^{(4)} = y'' - y$, $y(0) = 0$, $y'(0) = 1$, $y(1) = 0$, $y'(1) = -1$.
5. $y^{(4)} = -9y + x$, $y(0) = y''(0) = 0$, $y'(1) = y'''(1) = 0$.

Problems 6–10 Solve the given boundary value problem with the finite difference method using $m = 20$.

6. ■ $y'' = xy$, $y(1) = 1.5$ $y(2) = 3$.
7. ■ $y'' + 2y' + y = 0$, $y(0) = 0$, $y(1) = 1$. Exact solution is $y = xe^{1-x}$.
8. ■ $x^2 y'' + xy' + y = 0$, $y(1) = 0$, $y(2) = 0.638961$. Exact solution is $y = \sin(\ln x)$.
9. ■ $y'' = y^2 \sin y$, $y'(0) = 0$, $y(\pi) = 1$.
10. ■ $y'' + 2y(2xy' + y) = 0$, $y(0) = 1/2$, $y'(1) = -2/9$. Exact solution is $y = (2 + x^2)^{-1}$.

11. ■

The simply supported beam consists of three segments with the moments of inertia I_0 and I_1 as shown. A uniformly distributed load of intensity w_0 acts over the middle segment. Modeling only the *left half* of the beam, we can show that the differential equation

$$\frac{d^2 v}{dx^2} = -\frac{M}{EI}$$

for the displacement v is

$$\frac{d^2 v}{dx^2} = -\frac{w_0 L^2}{4EI_0} \times \begin{cases} \dfrac{x}{L} & \text{in } 0 < x < \dfrac{L}{4} \\ \dfrac{I_0}{I_1}\left[\dfrac{x}{L} - 2\left(\dfrac{x}{L} - \dfrac{1}{4}\right)^2\right] & \text{in } \dfrac{L}{4} < x < \dfrac{L}{2} \end{cases}$$

Introducing the dimensionless variables

$$\xi = \frac{x}{L} \qquad y = \frac{EI_0}{w_0 L^4} v \qquad \gamma = \frac{I_1}{I_0}$$

the differential equation changes to

$$\frac{d^2 y}{d\xi^2} = \begin{cases} -\dfrac{1}{4}\xi & \text{in } 0 < \xi < \dfrac{1}{4} \\ -\dfrac{1}{4\gamma}\left[\xi - 2\left(\xi - \dfrac{1}{4}\right)^2\right] & \text{in } \dfrac{1}{4} < \xi < \dfrac{1}{2} \end{cases}$$

with the boundary conditions

$$y|_{\xi=0} = \left.\frac{dy}{d\xi}\right|_{\xi=1/2} = 0$$

Use the finite difference method to determine the maximum displacement of the beam using $m = 20$ and $\gamma = 1.5$ and compare it with the exact solution

$$v_{max} = \frac{61}{9216}\frac{w_0 L^4}{E I_0}$$

12. ■

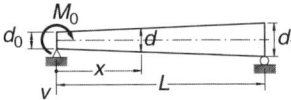

The simply supported, tapered beam has a circular cross section. A couple of magnitude M_0 is applied to the left end of the beam. The differential equation for the displacement v is

$$\frac{d^2 v}{dx^2} = -\frac{M}{EI} = -\frac{M_0(1 - x/L)}{EI_0(d/d_0)^4}$$

where

$$d = d_0\left[1 + \left(\frac{d_1}{d_0} - 1\right)\frac{x}{L}\right] \qquad I_0 = \frac{\pi d_0^4}{64}$$

Substituting

$$\xi = \frac{x}{L} \qquad y = \frac{EI_0}{M_0 L^2}v \qquad \delta = \frac{d_1}{d_0}$$

the differential equation changes to

$$\frac{d^2 y}{d\xi^2} = -\frac{1-\xi}{[1+(\delta-1)\xi]^4}$$

with the boundary conditions

$$y|_{\xi=0} = y|_{\xi=1} = 0$$

Solve the problem with the finite difference method using $\delta = 1.5$ and $m = 20$; plot y vs. ξ. The exact solution is

$$y = -\frac{(3 + 2\delta\xi - 3\xi)\xi^2}{6(1 + \delta\xi - \xi)^2} + \frac{\xi}{3\delta}$$

13. ■ Solve Example 8.4 by the finite difference method with $m = 20$. *Hint*: Compute the end slopes from the second noncentral differences in Tables 5.3.

14. ■ Solve Prob. 20 in Problem Set 8.1 with the finite difference method. Use $m = 20$.

8.3 Finite Difference Method

15. ■

The simply supported beam of length L is resting on an elastic foundation of stiffness $k\,\text{N/m}^2$. The displacement v of the beam due to the uniformly distributed load of intensity w_0 N/m is given by the solution of the boundary value problem

$$EI\frac{d^4v}{dx^4} + kv = w_0, \quad v|_{x=0} = \frac{d^2y}{dx^2}\bigg|_{x=0} = v|_{x=L} = \frac{d^2v}{dx^2}\bigg|_{x=L} = 0$$

The nondimensional form of the problem is

$$\frac{d^4y}{d\xi^4} + \gamma y = 1, \quad y|_{\xi=0} = \frac{d^2y}{dx^2}\bigg|_{\xi=0} = y|_{\xi=1} = \frac{d^2y}{dx^2}\bigg|_{\xi=1} = 0$$

where

$$\xi = \frac{x}{L} \qquad y = \frac{EI}{w_0 L^4}v \qquad \gamma = \frac{kL^4}{EI}$$

Solve this problem by the finite difference method with $\gamma = 10^5$ and plot y vs. ξ.

16. ■ Solve Prob. 15 if the ends of the beam are free and the load is confined to the middle half of the beam. Consider only the left half of the beam, in which case the nondimensional form of the problem is

$$\frac{d^4y}{d\xi^4} + \gamma y = \begin{cases} 0 & \text{in } 0 < \xi < 1/4 \\ 1 & \text{in } 1/4 < \xi < 1/2 \end{cases}$$

$$\frac{d^2y}{d\xi^2}\bigg|_{\xi=0} = \frac{d^3y}{d\xi^3}\bigg|_{\xi=0} = \frac{dy}{d\xi}\bigg|_{\xi=1/2} = \frac{d^3y}{d\xi^3}\bigg|_{\xi=1/2} = 0$$

17. ■ The general form of a linear, second-order boundary value problem is

$$y'' = r(x) + s(x)y + t(x)y'$$

$$y(a) = \alpha \text{ or } y'(a) = \alpha$$
$$y(b) = \beta \text{ or } y'(b) = \beta$$

Write a program that solves this problem with the finite difference method for any user-specified $r(x)$, $s(x)$ and $t(x)$. Test the program by solving Prob. 8.

9 Symmetric Matrix Eigenvalue Problems

> Find λ for which nontrivial solutions of $\mathbf{Ax} = \lambda\mathbf{x}$ exist.

9.1 Introduction

The *standard form* of the matrix eigenvalue problem is

$$\mathbf{Ax} = \lambda\mathbf{x} \tag{9.1}$$

where $\mathbf{A}$ is a given $n \times n$ matrix. The problem is to find the scalar λ and the vector $\mathbf{x}$. Rewriting Eq. (9.1) in the form

$$(\mathbf{A} - \lambda\mathbf{I})\mathbf{x} = \mathbf{0} \tag{9.2}$$

it becomes apparent that we are dealing with a system of n homogeneous equations. An obvious solution is the trivial one $\mathbf{x} = \mathbf{0}$. A nontrivial solution can exist only if the determinant of the coefficient matrix vanishes; that is, if

$$|\mathbf{A} - \lambda\mathbf{I}| = 0 \tag{9.3}$$

Expansion of the determinant leads to the polynomial equation, also known as the *characteristic equation*

$$a_0 + a_1\lambda + a_2\lambda^2 + \cdots + a_n\lambda^n = 0$$

which has the roots λ_i, $i = 1, 2, \ldots, n$, called the *eigenvalues* of the matrix $\mathbf{A}$. The solutions $\mathbf{x}_i$ of $(\mathbf{A} - \lambda_i\mathbf{I})\mathbf{x} = \mathbf{0}$ are known as the *eigenvectors*.

As an example, consider the matrix

$$\mathbf{A} = \begin{bmatrix} 1 & -1 & 0 \\ -1 & 2 & -1 \\ 0 & -1 & 1 \end{bmatrix} \tag{a}$$

9.1 Introduction

The characteristic equation is

$$|\mathbf{A} - \lambda\mathbf{I}| = \begin{vmatrix} 1-\lambda & -1 & 0 \\ -1 & 2-\lambda & -1 \\ 0 & -1 & 1-\lambda \end{vmatrix} = -3\lambda + 4\lambda^2 - \lambda^3 = 0 \qquad (b)$$

The roots of this equation are $\lambda_1 = 0$, $\lambda_2 = 1$, $\lambda_3 = 3$. To compute the eigenvector corresponding the λ_3, we substitute $\lambda = \lambda_3$ into Eq. (9.2), obtaining

$$\begin{bmatrix} -2 & -1 & 0 \\ -1 & -1 & -1 \\ 0 & -1 & -2 \end{bmatrix} \begin{bmatrix} x_1 \\ x_2 \\ x_3 \end{bmatrix} = \begin{bmatrix} 0 \\ 0 \\ 0 \end{bmatrix} \qquad (c)$$

We know that the determinant of the coefficient matrix is zero, so that the equations are not linearly independent. Therefore, we can assign an arbitrary value to any one component of **x** and use two of the equations to compute the other two components. Choosing $x_1 = 1$, the first equation of Eq. (c) yields $x_2 = -2$ and from the third equation we get $x_3 = 1$. Thus the eigenvector associated with λ_3 is

$$\mathbf{x}_3 = \begin{bmatrix} 1 \\ -2 \\ 1 \end{bmatrix}$$

The other two eigenvectors

$$\mathbf{x}_2 = \begin{bmatrix} 1 \\ 0 \\ -1 \end{bmatrix} \qquad \mathbf{x}_1 = \begin{bmatrix} 1 \\ 1 \\ 1 \end{bmatrix}$$

can be obtained in the same manner.

It is sometimes convenient to display the eigenvectors as columns of a matrix **X**. For the problem at hand, this matrix is

$$\mathbf{X} = \begin{bmatrix} \mathbf{x}_1 & \mathbf{x}_2 & \mathbf{x}_3 \end{bmatrix} = \begin{bmatrix} 1 & 1 & 1 \\ 1 & 0 & -2 \\ 1 & -1 & 1 \end{bmatrix}$$

It is clear from the above example that the magnitude of an eigenvector is indeterminate; only its direction can be computed from Eq. (9.2). It is customary to *normalize* the eigenvectors by assigning a unit magnitude to each vector. Thus the normalized eigenvectors in our example are

$$\mathbf{X} = \begin{bmatrix} 1/\sqrt{3} & 1/\sqrt{2} & 1/\sqrt{6} \\ 1/\sqrt{3} & 0 & -2/\sqrt{6} \\ 1/\sqrt{3} & -1/\sqrt{2} & 1/\sqrt{6} \end{bmatrix}$$

Throughout this chapter we assume that the eigenvectors are normalized.

Here are some useful properties of eigenvalues and eigenvectors, given without proof:

- All eigenvalues of a symmetric matrix are real.
- All eigenvalues of a symmetric, positive-definite matrix are real and positive.
- The eigenvectors of a symmetric matrix are orthonormal; that is, $\mathbf{X}^T\mathbf{X} = \mathbf{I}$.
- If the eigenvalues of $\mathbf{A}$ are λ_i, then the eigenvalues of $\mathbf{A}^{-1}$ are λ_i^{-1}.

Eigenvalue problems that originate from physical problems often end up with a symmetric $\mathbf{A}$. This is fortunate, because symmetric eigenvalue problems are easier to solve than their nonsymmetric counterparts (which may have complex eigenvalues). In this chapter we largely restrict our discussion to eigenvalues and eigenvectors of symmetric matrices.

Common sources of eigenvalue problems are the analysis of vibrations and stability. These problems often have the following characteristics:

- The matrices are large and sparse (e.g., have a banded structure).
- We need to know only the eigenvalues; if eigenvectors are required, only a few of them are of interest.

A useful eigenvalue solver must be able to utilize these characteristics to minimize the computations. In particular, it should be flexible enough to compute only what we need and no more.

9.2 Jacobi Method

Jacobi method is a relatively simple iterative procedure that extracts *all* the eigenvalues and eigenvectors of a symmetric matrix. Its utility is limited to small matrices (less than 20×20), because the computational effort increases very rapidly with the size of the matrix. The main strength of the method is its robustness—it seldom fails to deliver.

Similarity Transformation and Diagonalization

Consider the standard matrix eigenvalue problem

$$\mathbf{A}\mathbf{x} = \lambda \mathbf{x} \qquad (9.4)$$

where $\mathbf{A}$ is symmetric. Let us now apply the transformation

$$\mathbf{x} = \mathbf{P}\mathbf{x}^* \qquad (9.5)$$

9.2 Jacobi Method

where **P** is a nonsingular matrix. Substituting Eq. (9.5) into Eq. (9.4) and premultiplying each side by $\mathbf{P}^{-1}$, we get

$$\mathbf{P}^{-1}\mathbf{APx}^* = \lambda \mathbf{P}^{-1}\mathbf{Px}^*$$

or

$$\mathbf{A}^*\mathbf{x}^* = \lambda \mathbf{x}^* \tag{9.6}$$

where $\mathbf{A}^* = \mathbf{P}^{-1}\mathbf{AP}$. Because λ was untouched by the transformation, the eigenvalues of **A** are also the eigenvalues of $\mathbf{A}^*$. Matrices that have the same eigenvalues are deemed to be *similar*, and the transformation between them is called a *similarity transformation*.

Similarity transformations are frequently used to change an eigenvalue problem to a form that is easier to solve. Suppose that we managed by some means to find a **P** that diagonalizes $\mathbf{A}^*$. Equations (9.6) then are

$$\begin{bmatrix} A_{11}^* - \lambda & 0 & \cdots & 0 \\ 0 & A_{22}^* - \lambda & \cdots & 0 \\ \vdots & \vdots & \ddots & \vdots \\ 0 & 0 & \cdots & A_{nn}^* - \lambda \end{bmatrix} \begin{bmatrix} x_1^* \\ x_2^* \\ \vdots \\ x_n^* \end{bmatrix} = \begin{bmatrix} 0 \\ 0 \\ \vdots \\ 0 \end{bmatrix}$$

which have the solutions

$$\lambda_1 = A_{11}^* \quad \lambda_2 = A_{22}^* \quad \cdots \quad \lambda_n = A_{nn}^* \tag{9.7}$$

$$\mathbf{x}_1^* = \begin{bmatrix} 1 \\ 0 \\ \vdots \\ 0 \end{bmatrix} \quad \mathbf{x}_2^* = \begin{bmatrix} 0 \\ 1 \\ \vdots \\ 0 \end{bmatrix} \quad \cdots \quad \mathbf{x}_n^* = \begin{bmatrix} 0 \\ 0 \\ \vdots \\ 1 \end{bmatrix}$$

or

$$\mathbf{X}^* = \begin{bmatrix} \mathbf{x}_1^* & \mathbf{x}_2^* & \cdots & \mathbf{x}_n^* \end{bmatrix} = \mathbf{I}$$

According to Eq. (9.5) the eigenvectors of **A** are

$$\mathbf{X} = \mathbf{PX}^* = \mathbf{PI} = \mathbf{P} \tag{9.8}$$

Hence the transformation matrix **P** contains the eigenvectors of **A**, and the eigenvalues of **A** are the diagonal terms of $\mathbf{A}^*$.

Jacobi Rotation

A special similarity transformation is the plane rotation

$$\mathbf{x} = \mathbf{Rx}^* \tag{9.9}$$

where

$$\mathbf{R} = \begin{bmatrix} 1 & 0 & 0 & 0 & 0 & 0 & 0 & 0 \\ 0 & 1 & 0 & 0 & 0 & 0 & 0 & 0 \\ 0 & 0 & c & 0 & 0 & s & 0 & 0 \\ 0 & 0 & 0 & 1 & 0 & 0 & 0 & 0 \\ 0 & 0 & 0 & 0 & 1 & 0 & 0 & 0 \\ 0 & 0 & -s & 0 & 0 & c & 0 & 0 \\ 0 & 0 & 0 & 0 & 0 & 0 & 1 & 0 \\ 0 & 0 & 0 & 0 & 0 & 0 & 0 & 1 \end{bmatrix} \begin{matrix} \\ \\ k \\ \\ \\ \ell \\ \\ \\ \end{matrix} \qquad (9.10)$$

is called the *Jacobi rotation matrix*. Note that $\mathbf{R}$ is an identity matrix modified by the terms $c = \cos\theta$ and $s = \sin\theta$ appearing at the intersections of columns/rows k and ℓ, where θ is the rotation angle. The rotation matrix has the useful property of being *orthogonal*, meaning that

$$\mathbf{R}^{-1} = \mathbf{R}^T \qquad (9.11)$$

One consequence of orthogonality is that the transformation in Eq. (9.9) has the essential characteristic of a rotation: it preserves the magnitude of the vector; that is, $|\mathbf{x}| = |\mathbf{x}^*|$.

The similarity transformation corresponding to the plane rotation in Eq. (9.9) is

$$\mathbf{A}^* = \mathbf{R}^{-1}\mathbf{A}\mathbf{R} = \mathbf{R}^T\mathbf{A}\mathbf{R} \qquad (9.12)$$

The matrix $\mathbf{A}^*$ not only has the same eigenvalues as the original matrix $\mathbf{A}$, but thanks to orthogonality of $\mathbf{R}$, it is also symmetric. The transformation in Eq. (9.12) changes only the rows/columns k and ℓ of $\mathbf{A}$. The formulas for these changes are

$$\begin{aligned} A^*_{kk} &= c^2 A_{kk} + s^2 A_{\ell\ell} - 2cs\, A_{k\ell} \\ A^*_{\ell\ell} &= c^2 A_{\ell\ell} + s^2 A_{kk} + 2cs\, A_{k\ell} \\ A^*_{k\ell} &= A^*_{\ell k} = (c^2 - s^2) A_{k\ell} + cs(A_{kk} - A_{\ell\ell}) \\ A^*_{ki} &= A^*_{ik} = cA_{ki} - sA_{\ell i}, \quad i \neq k, \quad i \neq \ell \\ A^*_{\ell i} &= A^*_{i\ell} = cA_{\ell i} + sA_{ki}, \quad i \neq k, \quad i \neq \ell \end{aligned} \qquad (9.13)$$

Jacobi Diagonalization

The angle θ in the Jacobi rotation matrix can be chosen so that $A^*_{k\ell} = A^*_{\ell k} = 0$. This suggests the following idea: why not diagonalize $\mathbf{A}$ by looping through all the off-diagonal terms and zero them one by one? This is exactly what Jacobi diagonalization does. However, there is a major snag—the transformation that annihilates an

9.2 Jacobi Method

off-diagonal term also undoes some of the previously created zeroes. Fortunately, it turns out that the off-diagonal terms that reappear will be smaller than before. Thus Jacobi method is an iterative procedure that repeatedly applies Jacobi rotations until the off-diagonal terms have virtually vanished. The final transformation matrix $\mathbf{P}$ is the accumulation of individual rotations $\mathbf{R}_i$:

$$\mathbf{P} = \mathbf{R}_1 \cdot \mathbf{R}_2 \cdot \mathbf{R}_3 \ldots \tag{9.14}$$

The columns of $\mathbf{P}$ finish up being the eigenvectors of $\mathbf{A}$ and the diagonal elements of $\mathbf{A}^* = \mathbf{P}^T \mathbf{A} \mathbf{P}$ become the eigenvectors.

Let us now look at the details of a Jacobi rotation. From Eq. (9.13) we see that $A^*_{k\ell} = 0$ if

$$(c^2 - s^2) A_{k\ell} + cs(A_{kk} - A_{\ell\ell}) = 0 \tag{a}$$

Using the trigonometric identities $c^2 - s^2 = \cos^2\theta - \sin^2\theta = \cos 2\theta$ and $cs = \cos\theta \sin\theta = (1/2)\sin 2\theta$, we obtain from Eq. (a)

$$\tan 2\theta = -\frac{2 A_{k\ell}}{A_{kk} - A_{\ell\ell}} \tag{b}$$

which could be solved for θ, followed by computation of $c = \cos\theta$ and $s = \sin\theta$. However, the procedure described below leads to better algorithm.[23]

Introducing the notation

$$\phi = \cot 2\theta = -\frac{A_{kk} - A_{\ell\ell}}{2 A_{k\ell}} \tag{9.15}$$

and utilizing the trigonometric identity

$$\tan 2\theta = \frac{2t}{(1 - t^2)}$$

where $t = \tan\theta$, we can write Eq. (b) as

$$t^2 + 2\phi t - 1 = 0$$

which has the roots

$$t = -\phi \pm \sqrt{\phi^2 + 1}$$

It has been found that the root $|t| \leq 1$, which corresponds to $|\theta| \leq 45°$, leads to the more stable transformation. Therefore, we choose the plus sign if $\phi > 0$ and the minus sign if $\phi \leq 0$, which is equivalent to using

$$t = \text{sgn}(\phi)\left(-|\phi| + \sqrt{\phi^2 + 1}\right)$$

---

[23] The procedure is adapted from Press, W. H., et al., *Numerical Recipes in Fortran*, 2nd ed, Cambridge University Press, 1992.

Symmetric Matrix Eigenvalue Problems

To forestall excessive roundoff error if ϕ is large, we multiply both sides of the equation by $|\phi| + \sqrt{\phi^2 + 1}$ and solve for t, which yields

$$t = \frac{\text{sgn}(\phi)}{|\phi| + \sqrt{\phi^2 + 1}} \tag{9.16a}$$

In the case of very large ϕ, we should replace Eq. (9.16a) by the approximation

$$t = \frac{1}{2\phi} \tag{9.16b}$$

to prevent overflow in the computation of ϕ^2. Having computed t, we can use the trigonometric relationship $\tan\theta = \sin\theta/\cos\theta = \sqrt{1 - \cos^2\theta}/\cos\theta$ to obtain

$$c = \frac{1}{\sqrt{1 + t^2}} \qquad s = tc \tag{9.17}$$

We now improve the transformation formulas in Eqs. (9.13). Solving Eq. (a) for $A_{\ell\ell}$, we obtain

$$A_{\ell\ell} = A_{kk} + A_{k\ell}\frac{c^2 - s^2}{cs} \tag{c}$$

Replacing all occurrences of $A_{\ell\ell}$ by Eq. (c) and simplifying, we can write the transformation formulas in Eqs.(9.13) as

$$A_{kk}^* = A_{kk} - tA_{k\ell}$$
$$A_{\ell\ell}^* = A_{\ell\ell} + tA_{k\ell}$$
$$A_{k\ell}^* = A_{\ell k}^* = 0 \tag{9.18}$$
$$A_{ki}^* = A_{ik}^* = A_{ki} - s(A_{\ell i} + \tau A_{ki}), \quad i \neq k, \quad i \neq \ell$$
$$A_{\ell i}^* = A_{i\ell}^* = A_{\ell i} + s(A_{ki} - \tau A_{\ell i}), \quad i \neq k, \quad i \neq \ell$$

where

$$\tau = \frac{s}{1 + c} \tag{9.19}$$

The introduction of τ allowed us to express each formula in the form (original value) + (change), which is helpful in reducing the roundoff error.

At the start of Jacobi's diagonalization process the transformation matrix **P** is initialized to the identity matrix. Each Jacobi's rotation changes this matrix from **P** to $\mathbf{P}^* = \mathbf{PR}$. The corresponding changes in the elements of **P** can be shown to be (only the columns k and ℓ are affected)

$$P_{ik}^* = P_{ik} - s(P_{i\ell} + \tau P_{ik}) \tag{9.20}$$
$$P_{i\ell}^* = P_{i\ell} + s(P_{ik} - \tau P_{i\ell})$$

9.2 Jacobi Method

We still have to decide the order in which the off-diagonal elements of **A** are to be eliminated. Jacobi's original idea was to attack the largest element since this results in fewest number of rotations. The problem here is that **A** has to be searched for the largest element after every rotation, which is a time-consuming process. If the matrix is large, it is faster to sweep through it by rows or columns and annihilate every element above some threshold value. In the next sweep the threshold is lowered and the process repeated. We adopt Jacobi's original scheme because of its simpler implementation.

In summary, the Jacobi diagonalization procedure, which uses only the upper half of the matrix, is:

1. Find the largest (absolute value) off-diagonal element $A_{k\ell}$ in the upper half of **A**.
2. Compute ϕ, t, c and s from Eqs. (9.15)–(9.17).
3. Compute τ from Eq. (9.19).
4. Modify the elements in the upper half of **A** according to Eqs. (9.18).
5. Update the transformation matrix **P** using Eqs. (9.20).
6. Repeat steps 1–5 until the $A_{k\ell} < \varepsilon$, where ε is the error tolerance.

■ jacobi

This function computes all eigenvalues λ_i and eigenvectors $\mathbf{x}_i$ of a symmetric, $n \times n$ matrix **A** by the Jacobi method. The algorithm works exclusively with the upper triangular part of **A**, which is destroyed in the process. The principal diagonal of **A** is replaced by the eigenvalues, and the columns of the transformation matrix **P** become the normalized eigenvectors.

```
## module jacobi
''' lam,x = jacobi(a,tol = 1.0e-9).
    Solution of std. eigenvalue problem [a]{x} = lambda{x}
    by Jacobi's method. Returns eigenvalues in vector {lam}
    and the eigenvectors as columns of matrix [x].
'''
from numarray import array,identity,diagonal
from math import sqrt

def jacobi(a,tol = 1.0e-9):

    def maxElem(a): # Find largest off-diag. element a[k,l]
        n = len(a)
        aMax = 0.0
```

Symmetric Matrix Eigenvalue Problems

```
        for i in range(n-1):
            for j in range(i+1,n):
                if abs(a[i,j]) >= aMax:
                    aMax = abs(a[i,j])
                    k = i; l = j
        return aMax,k,l

    def rotate(a,p,k,l): # Rotate to make a[k,l] = 0
        n = len(a)
        aDiff = a[l,l] - a[k,k]
        if abs(a[k,l]) < abs(aDiff)*1.0e-36: t = a[k,l]/aDiff
        else:
            phi = aDiff/(2.0*a[k,l])
            t = 1.0/(abs(phi) + sqrt(phi**2 + 1.0))
            if phi < 0.0: t = -t
        c = 1.0/sqrt(t**2 + 1.0); s = t*c
        tau = s/(1.0 + c)
        temp = a[k,l]
        a[k,l] = 0.0
        a[k,k] = a[k,k] - t*temp
        a[l,l] = a[l,l] + t*temp
        for i in range(k):      # Case of i < k
            temp = a[i,k]
            a[i,k] = temp - s*(a[i,l] + tau*temp)
            a[i,l] = a[i,l] + s*(temp - tau*a[i,l])
        for i in range(k+1,l):  # Case of k < i < l
            temp = a[k,i]
            a[k,i] = temp - s*(a[i,l] + tau*a[k,i])
            a[i,l] = a[i,l] + s*(temp - tau*a[i,l])
        for i in range(l+1,n):  # Case of i > l
            temp = a[k,i]
            a[k,i] = temp - s*(a[l,i] + tau*temp)
            a[l,i] = a[l,i] + s*(temp - tau*a[l,i])
        for i in range(n):      # Update transformation matrix
            temp = p[i,k]
            p[i,k] = temp - s*(p[i,l] + tau*p[i,k])
            p[i,l] = p[i,l] + s*(temp - tau*p[i,l])

    n = len(a)
    maxRot = 5*(n**2)           # Set limit on number of rotations
```

9.2 Jacobi Method

```
    p = identity(n)*1.0     # Initialize transformation matrix
    for i in range(maxRot): # Jacobi rotation loop
        aMax,k,l = maxElem(a)
        if aMax < tol: return diagonal(a),p
        rotate(a,p,k,l)
    print 'Jacobi method did not converge'
```

■ sortJacobi

The eigenvalues/eigenvectors returned by jacobi are not ordered. The function listed below can be used to sort the eigenvalues and eigenvectors into ascending order of eigenvalues.

```
## module sortJacobi
''' sortJacobi(lam,x).
    Sorts the eigenvalues {lam} and eigenvectors [x]
    in order of ascending eigenvalues.
'''
import swap

def sortJacobi(lam,x):
    n = len(lam)
    for i in range(n-1):
        index = i
        val = lam[i]
        for j in range(i+1,n):
            if lam[j] < val:
                index = j
                val = lam[j]
        if index != i:
            swap.swapRows(lam,i,index)
            swap.swapCols(x,i,index)
```

Transformation to Standard Form

Physical problems often give rise to eigenvalue problems of the form

$$\mathbf{A}\mathbf{x} = \lambda \mathbf{B}\mathbf{x} \qquad (9.21)$$

where $\mathbf{A}$ and $\mathbf{B}$ are symmetric $n \times n$ matrices. We assume that $\mathbf{B}$ is also positive definite. Such problems must be transformed into the standard form before they can be solved by Jacobi diagonalization.

As **B** is symmetric and positive definite, we can apply Choleski decomposition $\mathbf{B} = \mathbf{LL}^T$, where **L** is a lower-triangular matrix (see Section 2.3). Then we introduce the transformation

$$\mathbf{x} = (\mathbf{L}^{-1})^T \mathbf{z} \tag{9.22}$$

Substituting into Eq. (9.21), we get

$$\mathbf{A}(\mathbf{L}^{-1})^T \mathbf{z} = \lambda \mathbf{LL}^T (\mathbf{L}^{-1})^T \mathbf{z}$$

Premultiplying both sides by $\mathbf{L}^{-1}$ results in

$$\mathbf{L}^{-1}\mathbf{A}(\mathbf{L}^{-1})^T \mathbf{z} = \lambda \mathbf{L}^{-1}\mathbf{LL}^T (\mathbf{L}^{-1})^T \mathbf{z}$$

Because $\mathbf{L}^{-1}\mathbf{L} = \mathbf{L}^T (\mathbf{L}^{-1})^T = \mathbf{I}$, the last equation reduces to the standard form

$$\mathbf{Hz} = \lambda \mathbf{z} \tag{9.23}$$

where

$$\mathbf{H} = \mathbf{L}^{-1}\mathbf{A}(\mathbf{L}^{-1})^T \tag{9.24}$$

An important property of this transformation is that it does not destroy the symmetry of the matrix; i.e., a symmetric **A** results in a symmetric **H**.

Here is the general procedure for solving eigenvalue problems of the form $\mathbf{Ax} = \lambda \mathbf{Bx}$:

1. Use Choleski decomposition $\mathbf{B} = \mathbf{LL}^T$ to compute **L**.
2. Compute $\mathbf{L}^{-1}$ (a triangular matrix can be inverted with relatively small computational effort).
3. Compute **H** from Eq. (9.24).
4. Solve the standard eigenvalue problem $\mathbf{Hz} = \lambda \mathbf{z}$ (e.g., using the Jacobi method).
5. Recover the eigenvectors of the original problem from Eq. (9.22): $\mathbf{x} = (\mathbf{L}^{-1})^T \mathbf{z}$. Note that the eigenvalues were untouched by the transformation.

An important special case is where **B** is a diagonal matrix:

$$\mathbf{B} = \begin{bmatrix} \beta_1 & 0 & \cdots & 0 \\ 0 & \beta_2 & \cdots & 0 \\ \vdots & \vdots & \ddots & \vdots \\ 0 & 0 & \cdots & \beta_n \end{bmatrix} \tag{9.25}$$

9.2 Jacobi Method

Here

$$\mathbf{L} = \begin{bmatrix} \beta_1^{1/2} & 0 & \cdots & 0 \\ 0 & \beta_2^{1/2} & \cdots & 0 \\ \vdots & \vdots & \ddots & \vdots \\ 0 & 0 & \cdots & \beta_n^{1/2} \end{bmatrix} \qquad \mathbf{L}^{-1} = \begin{bmatrix} \beta_1^{-1/2} & 0 & \cdots & 0 \\ 0 & \beta_2^{-1/2} & \cdots & 0 \\ \vdots & \vdots & \ddots & \vdots \\ 0 & 0 & \cdots & \beta_n^{-1/2} \end{bmatrix} \qquad (9.26a)$$

and

$$H_{ij} = \frac{A_{ij}}{\sqrt{\beta_i \beta_j}} \qquad (9.26b)$$

■ stdForm

Given the matrices **A** and **B**, the function stdForm returns **H** and the transformation matrix $\mathbf{T} = (\mathbf{L}^{-1})^T$. The inversion of **L** is carried out by invert (the triangular shape of **L** allows this to be done by back substitution). Note that original **A**, **B** and **L** are destroyed.

```
## module stdForm
''' h,t = stdForm(a,b).
    Transforms the eigenvalue problem [a]{x} = lambda[b]{x}
    to the standard form [h]{z} = lambda{z}. The eigenvectors
    are related by {x} = [t]{z}.
'''
from numarray import dot,matrixmultiply,transpose
from choleski import *

def stdForm(a,b):

    def invert(L): # Inverts lower triangular matrix L
        n = len(L)
        for j in range(n-1):
            L[j,j] = 1.0/L[j,j]
            for i in range(j+1,n):
                L[i,j] = -dot(L[i,j:i],L[j:i,j])/L[i,i]
        L[n-1,n-1] = 1.0/L[n-1,n-1]

    n = len(a)
    L = choleski(b)
    invert(L)
    h = matrixmultiply(b,matrixmultiply(a,transpose(L)))
    return h,transpose(L)
```

Symmetric Matrix Eigenvalue Problems

EXAMPLE 9.1

The stress matrix (tensor) corresponding to the state of stress shown is

$$\mathbf{S} = \begin{bmatrix} 80 & 30 & 0 \\ 30 & 40 & 0 \\ 0 & 0 & 60 \end{bmatrix} \text{MPa}$$

(each row of the matrix consists of the three stress components acting on a coordinate plane). It can be shown that the eigenvalues of **S** are the *principal stresses* and the eigenvectors are normal to the *principal planes*. (1) Determine the principal stresses by diagonalizing **S** with one Jacobi rotation and (2) compute the eigenvectors.

Solution of Part(1) To eliminate S_{12} we must apply a rotation in the 1–2 plane. With $k = 1$ and $\ell = 2$ Eq. (9.15) is

$$\phi = -\frac{S_{11} - S_{22}}{2S_{12}} = -\frac{80 - 40}{2(30)} = -\frac{2}{3}$$

Equation (9.16a) then yields

$$t = \frac{\text{sgn}(\phi)}{|\phi| + \sqrt{\phi^2 + 1}} = \frac{-1}{2/3 + \sqrt{(2/3)^2 + 1}} = -0.535\,18$$

According to Eqs. (9.18), the changes in **S** due to the rotation are

$$S_{11}^* = S_{11} - tS_{12} = 80 - (-0.535\,18)(30) = 96.055 \text{ MPa}$$

$$S_{22}^* = S_{22} + tS_{12} = 40 + (-0.535\,18)(30) = 23.945 \text{ MPa}$$

$$S_{12}^* = S_{21}^* = 0$$

Hence the diagonalized stress matrix is

$$\mathbf{S}^* = \begin{bmatrix} 96.055 & 0 & 0 \\ 0 & 23.945 & 0 \\ 0 & 0 & 60 \end{bmatrix}$$

where the diagonal terms are the principal stresses.

9.2 Jacobi Method

Solution of Part (2) To compute the eigenvectors, we start with Eqs. (9.17) and (9.19), which yield

$$c = \frac{1}{\sqrt{1+t^2}} = \frac{1}{\sqrt{1+(-0.535\,18)^2}} = 0.881\,68$$

$$s = tc = (-0.535\,18)(0.881\,68) = -0.471\,86$$

$$\tau = \frac{s}{1+c} = \frac{-0.471\,86}{1+0.881\,68} = -0.250\,77$$

We obtain the changes in the transformation matrix $\mathbf{P}$ from Eqs. (9.20). Because $\mathbf{P}$ is initialized to the identity matrix, the first equation gives us

$$P_{11}^* = P_{11} - s(P_{12} + \tau P_{11})$$
$$= 1 - (-0.471\,86)[0 + (-0.250\,77)(1)] = 0.881\,67$$
$$P_{21}^* = P_{21} - s(P_{22} + \tau P_{21})$$
$$= 0 - (-0.471\,86)[1 + (-0.250\,77)(0)] = 0.471\,86$$

Similarly, the second equation of Eqs. (9.20) yields

$$P_{12}^* = -0.471\,86 \qquad P_{22}^* = 0.881\,67$$

The third row and column of $\mathbf{P}$ are not affected by the transformation. Thus

$$\mathbf{P}^* = \begin{bmatrix} 0.88167 & -0.47186 & 0 \\ 0.47186 & 0.88167 & 0 \\ 0 & 0 & 1 \end{bmatrix}$$

The columns of $\mathbf{P}^*$ are the eigenvectors of $\mathbf{S}$.

EXAMPLE 9.2

(1) Show that the analysis of the electric circuit shown leads to a matrix eigenvalue problem. (2) Determine the circular frequencies and the relative amplitudes of the currents.

Solution of Part (1) Kirchoff's equations for the three loops are

$$L\frac{di_1}{dt} + \frac{q_1 - q_2}{3C} = 0$$

$$L\frac{di_2}{dt} + \frac{q_2 - q_1}{3C} + \frac{q_2 - q_3}{C} = 0$$

$$2L\frac{di_3}{dt} + \frac{q_3 - q_2}{C} + \frac{q_3}{C} = 0$$

Symmetric Matrix Eigenvalue Problems

Differentiating and substituting $dq_k/dt = i_k$, we get

$$\frac{1}{3}i_1 - \frac{1}{3}i_2 = -LC\frac{d^2 i_1}{dt^2}$$

$$-\frac{1}{3}i_1 + \frac{4}{3}i_2 - i_3 = -LC\frac{d^2 i_2}{dt^2}$$

$$-i_2 + 2i_3 = -2LC\frac{d^2 i_3}{dt^2}$$

These equations admit the solution

$$i_k(t) = u_k \sin \omega t$$

where ω is the circular frequency of oscillation (measured in rad/s) and u_k are the relative amplitudes of the currents. Substitution into Kirchoff's equations yields $\mathbf{Au} = \lambda \mathbf{Bu}$ ($\sin \omega t$ cancels out), where

$$\mathbf{A} = \begin{bmatrix} 1/3 & -1/3 & 0 \\ -1/3 & 4/3 & -1 \\ 0 & -1 & 2 \end{bmatrix} \quad \mathbf{B} = \begin{bmatrix} 1 & 0 & 0 \\ 0 & 1 & 0 \\ 0 & 0 & 2 \end{bmatrix} \quad \lambda = LC\omega^2$$

which represents an eigenvalue problem of the nonstandard form.

Solution of Part (2) Since $\mathbf{B}$ is a diagonal matrix, we can readily transform the problem into the standard form $\mathbf{Hz} = \lambda \mathbf{z}$. From Eq. (9.26a) we get

$$\mathbf{L}^{-1} = \begin{bmatrix} 1 & 0 & 0 \\ 0 & 1 & 0 \\ 0 & 0 & 1/\sqrt{2} \end{bmatrix}$$

and Eq. (9.26b) yields

$$\mathbf{H} = \begin{bmatrix} 1/3 & -1/3 & 0 \\ -1/3 & 4/3 & -1/\sqrt{2} \\ 0 & -1/\sqrt{2} & 1 \end{bmatrix}$$

The eigenvalues and eigenvectors of $\mathbf{H}$ can now be obtained with the Jacobi method. Skipping the details, we obtain the following results:

$$\lambda_1 = 0.147\,79 \qquad \lambda_2 = 0.582\,35 \qquad \lambda_3 = 1.936\,53$$

$$\mathbf{z}_1 = \begin{bmatrix} 0.810\,27 \\ 0.451\,02 \\ 0.374\,23 \end{bmatrix} \quad \mathbf{z}_2 = \begin{bmatrix} 0.562\,74 \\ -0.420\,40 \\ -0.711\,76 \end{bmatrix} \quad \mathbf{z}_3 = \begin{bmatrix} 0.163\,70 \\ -0.787\,30 \\ 0.594\,44 \end{bmatrix}$$

The eigenvectors of the original problem are recovered from Eq. (9.22): $\mathbf{y}_i = (\mathbf{L}^{-1})^T \mathbf{z}_i$, which yields

$$\mathbf{u}_1 = \begin{bmatrix} 0.810\,27 \\ 0.451\,02 \\ 0.264\,62 \end{bmatrix} \quad \mathbf{u}_2 = \begin{bmatrix} 0.562\,74 \\ -0.420\,40 \\ -0.503\,29 \end{bmatrix} \quad \mathbf{u}_3 = \begin{bmatrix} 0.163\,70 \\ -0.787\,30 \\ 0.420\,33 \end{bmatrix}$$

These vectors should now be normalized (each $\mathbf{z}_i$ was normalized, but the transformation to $\mathbf{u}_i$ does not preserve the magnitudes of vectors). The circular frequencies are $\omega_i = \sqrt{\lambda_i/(LC)}$, so that

$$\omega_1 = \frac{0.3844}{\sqrt{LC}} \qquad \omega_2 = \frac{0.7631}{\sqrt{LC}} \qquad \omega_3 = \frac{1.3916}{\sqrt{LC}}$$

EXAMPLE 9.3

The propped cantilever beam carries a compressive axial load P. The lateral displacement $u(x)$ of the beam can be shown to satisfy the differential equation

$$u^{(4)} + \frac{P}{EI} u'' = 0 \tag{a}$$

where EI is the bending rigidity. The boundary conditions are

$$u(0) = u'(0) = 0 \qquad u(L) = u'(L) = 0 \tag{b}$$

(1) Show that displacement analysis of the beam results in a matrix eigenvalue problem if the derivatives are approximated by finite differences. (2) Use the Jacobi method to compute the lowest three buckling loads and the corresponding eigenvectors.

Solution of Part (1) We divide the beam into $n+1$ segments of length $L/(n+1)$ each as shown. Replacing the derivatives of u in Eq. (a) by central finite differences of $\mathcal{O}(h^2)$ at the interior nodes (nodes 1 to n), we obtain

$$\frac{u_{i-2} - 4u_{i-1} + 6u_i - 4u_{i+1} + u_{i+2}}{h^4}$$

$$= \frac{P}{EI} \frac{-u_{i-1} + 2u_i - u_{i+1}}{h^2}, \, i = 1, 2, \ldots, n$$

After multiplication by h^4, the equations become

$$u_{-1} - 4u_0 + 6u_1 - 4u_2 + u_3 = \lambda(-u_0 + 2u_1 - u_2)$$

$$u_0 - 4u_1 + 6u_2 - 4u_3 + u_4 = \lambda(-u_1 + 2u_2 - u_3)$$

$$\vdots \tag{c}$$

Symmetric Matrix Eigenvalue Problems

$$u_{n-3} - 4u_{n-2} + 6u_{n-1} - 4u_n + u_{n+1} = \lambda(-u_{n-2} + 2u_{n-1} - u_n)$$
$$u_{n-2} - 4u_{n-1} + 6u_n - 4u_{n+1} + u_{n+2} = \lambda(-u_{n-1} + 2u_n - u_{n+1})$$

where

$$\lambda = \frac{Ph^2}{EI} = \frac{PL^2}{(n+1)^2 EI}$$

The displacements u_{-1}, u_0, u_{n+1} and u_{n+2} can be eliminated by using the prescribed boundary conditions. Referring to Table 8.1, we obtain the finite difference approximations to the boundary conditions:

$$u_0 = 0 \qquad u_{-1} = -u_1 \qquad u_{n+1} = 0 \qquad u_{n+2} = u_n$$

Substitution into Eqs. (c) yields the matrix eigenvalue problem $\mathbf{A}\mathbf{x} = \lambda \mathbf{B}\mathbf{x}$, where

$$\mathbf{A} = \begin{bmatrix} 5 & -4 & 1 & 0 & 0 & \cdots & 0 \\ -4 & 6 & -4 & 1 & 0 & \cdots & 0 \\ 1 & -4 & 6 & -4 & 1 & \cdots & 0 \\ \vdots & \ddots & \ddots & \ddots & \ddots & \ddots & \vdots \\ 0 & \cdots & 1 & -4 & 6 & -4 & 1 \\ 0 & \cdots & 0 & 1 & -4 & 6 & -4 \\ 0 & \cdots & 0 & 0 & 1 & -4 & 7 \end{bmatrix}$$

$$\mathbf{B} = \begin{bmatrix} 2 & -1 & 0 & 0 & 0 & \cdots & 0 \\ -1 & 2 & -1 & 0 & 0 & \cdots & 0 \\ 0 & -1 & 2 & -1 & 0 & \cdots & 0 \\ \vdots & \ddots & \vdots & \ddots & \ddots & \ddots & \vdots \\ 0 & \cdots & 0 & -1 & 2 & -1 & 0 \\ 0 & \cdots & 0 & 0 & -1 & 2 & -1 \\ 0 & \cdots & 0 & 0 & 0 & -1 & 2 \end{bmatrix}$$

Solution of Part (2) The problem with the Jacobi method is that it insists on finding *all* the eigenvalues and eigenvectors. It is also incapable of exploiting banded structures of matrices. Thus the program listed below does much more work than necessary for the problem at hand. More efficient methods of solution will be introduced later in this chapter.

```
#!/usr/bin/python
## example9_3
from numarray import array,zeros,Float64
from stdForm import *
from jacobi import *
from sortJacobi import *
```

9.2 Jacobi Method

```
n = 10
a = zeros((n,n),type=Float64)
b = zeros((n,n),type=Float64)
for i in range(n):
    a[i,i] = 6.0
    b[i,i] = 2.0
a[0,0] = 5.0
a[n-1,n-1] = 7.0
for i in range(n-1):
    a[i,i+1] = -4.0
    a[i+1,i] = -4.0
    b[i,i+1] = -1.0
    b[i+1,i] = -1.0
for i in range(n-2):
    a[i,i+2] = 1.0
    a[i+2,i] = 1.0

h,t = stdForm(a,b)          # Convert to std. form
lam,z = jacobi(h)            # Solve by Jacobi mthd.
x = matrixmultiply(t,z)      # Eigenvectors of orig. prob.
for i in range(n):           # Normalize eigenvectors
    xMag = sqrt(dot(x[:,i],x[:,i]))
    x[:,i] = x[:,i]/xMag
sortJacobi(lam,x)            # Arrange in ascending order
print ''Eigenvalues:\n'',lam[0:3]
print ''\nEigenvectors:\n'',x[:,0:3]
raw_input(''\n Press return to exit'')
```

Running the program with $n = 10$ resulted in the following output:

```
Eigenvalues:
[ 0.16410379   0.47195675   0.90220118]


Eigenvectors:
[[ 0.16410119  -0.18476623   0.30699491]
 [ 0.30618978  -0.26819121   0.36404289]
 [ 0.40786549  -0.19676237   0.14669942]
 [ 0.45735999   0.00994855  -0.12192373]
 [ 0.45146805   0.26852252  -0.1724502 ]
 [ 0.39607358   0.4710634    0.06772929]
 [ 0.30518404   0.53612023   0.40894875]
```

```
    [ 0.19863178   0.44712859   0.57038382]
    [ 0.09881943   0.26022826   0.43341183]
    [ 0.0270436    0.07776771   0.1486333  ]]
```

The first three mode shapes, which represent the relative displacements of the buckled beam, are plotted below (we appended the zero end displacements to the eigenvectors before plotting the points).

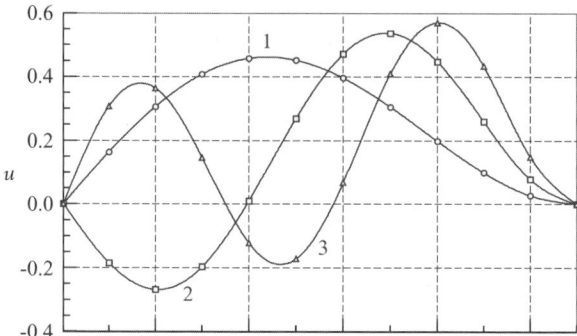

The buckling loads are given by $P_i = (n+1)^2 \lambda_i EI/L^2$. Thus

$$P_1 = \frac{(11)^2 \, (0.164\,103\,7) \, EI}{L^2} = 19.857 \frac{EI}{L^2}$$

$$P_2 = \frac{(11)^2 \, (0.471\,956\,75) \, EI}{L^2} = 57.107 \frac{EI}{L^2}$$

$$P_3 = \frac{(11)^2 \, (0.902\,201\,18) \, EI}{L^2} = 109.17 \frac{EI}{L^2}$$

The analytical values are $P_1 = 20.19 EI/L^2$, $P_2 = 59.68 EI/L^2$ and $P_3 = 118.9 EI/L^2$. It can be seen that the error introduced by the finite difference approximation increases with the mode number (the error in P_{i+1} is larger than in P_i). Of course, the accuracy of the finite difference model can be improved by using larger n, but beyond $n = 20$ the cost of computation with the Jacobi method becomes rather high.

9.3 Inverse Power and Power Methods

Inverse Power Method

The inverse power method is a simple and efficient algorithm that finds the smallest eigenvalue λ_1 and the corresponding eigenvector $\mathbf{x}_1$ of

$$\mathbf{A}\mathbf{x} = \lambda \mathbf{x} \tag{9.27}$$

9.3 Inverse Power and Power Methods

The method works like this:

1. Let **v** be an approximation to $\mathbf{x}_1$ (a random vector of unit magnitude will do).
2. Solve
$$\mathbf{Az} = \mathbf{v} \tag{9.28}$$
for the vector **z**.
3. Compute $|\mathbf{z}|$.
4. Let $\mathbf{v} = \mathbf{z}/|\mathbf{z}|$ and repeat steps 2–4 until the change in **v** is negligible.

At the conclusion of the procedure, $|\mathbf{z}| = \pm 1/\lambda_1$ and $\mathbf{v} = \mathbf{x}_1$. The sign of λ_1 is determined as follows: if **z** changes sign between successive iterations, λ_1 is negative; otherwise, λ_1 is positive.

Let us now investigate why the method works. Since the eigenvectors $\mathbf{x}_i$ of Eq. (9.27) are orthonormal (linearly independent), they can be used as the basis for any n-dimensional vector. Thus **v** and **z** admit the unique representations

$$\mathbf{v} = \sum_{i=1}^{n} v_i \mathbf{x}_i \qquad \mathbf{z} = \sum_{i=1}^{n} z_i \mathbf{x}_i \tag{a}$$

where v_i and z_i are the components of **v** and **z** with respect to the eigenvectors $\mathbf{x}_i$. Substitution into Eq. (9.28) yields

$$\mathbf{A} \sum_{i=1}^{n} z_i \mathbf{x}_i - \sum_{i=1}^{n} v_i \mathbf{x}_i = \mathbf{0}$$

But $\mathbf{Ax}_i = \lambda_i \mathbf{x}_i$, so that

$$\sum_{i=1}^{n} (z_i \lambda_i - v_i) \mathbf{x}_i = \mathbf{0}$$

Hence

$$z_i = \frac{v_i}{\lambda_i}$$

It follows from Eq. (a) that

$$\mathbf{z} = \sum_{i=1}^{n} \frac{v_i}{\lambda_i} \mathbf{x}_i = \frac{1}{\lambda_1} \sum_{i=1}^{n} v_i \frac{\lambda_1}{\lambda_i} \mathbf{x}_i$$
$$= \frac{1}{\lambda_1} \left(v_1 \mathbf{x}_1 + v_2 \frac{\lambda_1}{\lambda_2} \mathbf{x}_2 + v_3 \frac{\lambda_1}{\lambda_3} \mathbf{x}_3 + \cdots \right) \tag{9.29}$$

Since $\lambda_1/\lambda_i < 1$ ($i \neq 1$), we observe that the coefficient of $\mathbf{x}_1$ has become more prominent in **z** than it was in **v**; hence **z** is a better approximation to $\mathbf{x}_1$. This completes the first iterative cycle.

Symmetric Matrix Eigenvalue Problems

In subsequent cycles we set $\mathbf{v} = \mathbf{z}/|\mathbf{z}|$ and repeat the process. Each iteration will increase the dominance of the first term in Eq. (9.29) so that the process converges to

$$\mathbf{z} = \frac{1}{\lambda_1} v_1 \mathbf{x}_1 = \frac{1}{\lambda_1} \mathbf{x}_1$$

(at this stage $v_1 = 1$ since $\mathbf{v} = \mathbf{x}_1$, so that $v_1 = 1, v_2 = v_3 = \cdots = 0$).

The inverse power method also works with the nonstandard eigenvalue problem

$$\mathbf{A}\mathbf{x} = \lambda \mathbf{B}\mathbf{x} \tag{9.30}$$

provided that Eq. (9.28) is replaced by

$$\mathbf{A}\mathbf{z} = \mathbf{B}\mathbf{v} \tag{9.31}$$

The alternative is, of course, to transform the problem to standard form before applying the power method.

Eigenvalue Shifting

By inspection of Eq. (9.29) we see that the speed of convergence is determined by the strength of the inequality $|\lambda_1/\lambda_2| < 1$ (the second term in the equation). If $|\lambda_2|$ is well separated from $|\lambda_1|$, the inequality is strong and the convergence is rapid. On the other hand, close proximity of these two eigenvalues results in very slow convergence.

The rate of convergence can be improved by a technique called *eigenvalue shifting*. If we let

$$\lambda = \lambda^* + s \tag{9.32}$$

where s is a predetermined "shift," the eigenvalue problem in Eq. (9.27) is transformed to

$$\mathbf{A}\mathbf{x} = (\lambda^* + s)\mathbf{x}$$

or

$$\mathbf{A}^*\mathbf{x} = \lambda^*\mathbf{x} \tag{9.33}$$

where

$$\mathbf{A}^* = \mathbf{A} - s\mathbf{I} \tag{9.34}$$

Solving the transformed problem in Eq. (9.33) by the inverse power method yields λ_1^* and $\mathbf{x}_1$, where λ_1^* is the smallest eigenvalue of $\mathbf{A}^*$. The corresponding eigenvalue of the original problem, $\lambda = \lambda_1^* + s$, is thus the *eigenvalue closest to s*.

Eigenvalue shifting has two applications. An obvious one is the determination of the eigenvalue closest to a certain value s. For example, if the working speed of a shaft is s rpm, it is imperative to assure that there are no natural frequencies (which are related to the eigenvalues) close to that speed.

9.3 Inverse Power and Power Methods

Eigenvalue shifting is also be used to speed up convergence. Suppose that we are computing the smallest eigenvalue λ_1 of the matrix **A**. The idea is to introduce a shift s that makes λ_1^*/λ_2^* as small as possible. Since $\lambda_1^* = \lambda_1 - s$, we should choose $s \approx \lambda_1$ ($s = \lambda_1$ should be avoided to prevent division by zero). Of course, this method works only if we have a prior estimate of λ_1.

The inverse power method with eigenvalue shifting is a particularly powerful tool for finding eigenvectors if the eigenvalues are known. By shifting very close to an eigenvalue, the corresponding eigenvector can be computed in one or two iterations.

Power Method

The power method converges to the eigenvalue *farthest from zero* and the associated eigenvector. It is very similar to the inverse power method; the only difference between the two methods is the interchange of **v** and **z** in Eq. (9.28). The outline of the procedure is:

1. Let **v** be an approximation to x_1 (a random vector of unit magnitude will do).
2. Compute the vector

$$\mathbf{z} = \mathbf{A}\mathbf{v} \tag{9.35}$$

3. Compute $|\mathbf{z}|$.
4. Let $\mathbf{v} = \mathbf{z}/|\mathbf{z}|$ and repeat steps 2–4 until the change in **v** is negligible.

At the conclusion of the procedure, $|\mathbf{z}|' = \pm\lambda_n$ and $\mathbf{v} = \mathbf{x}_n$ (the sign of λ_n is determined in the same way as in the inverse power method).

■ inversePower

Given the matrix **A** and the shift s, the function inversePower returns the eigenvalue of **A** closest to s and the corresponding eigenvector. The matrix $\mathbf{A}^* = \mathbf{A} - s\mathbf{I}$ is decomposed as soon as it is formed, so that only the solution phase (forward and back substitution) is needed in the iterative loop. If **A** is banded, the efficiency of the program could be improved by replacing LUdecomp and LUsolve by functions that specialize in banded matrices (e.g., LUdecomp5 and LUsolve5)—see Example 9.6. The program line that forms $\mathbf{A}^*$ must also be modified to be compatible with the storage scheme used for **A**.

```
## module inversePower
''' lam,x = inversePower(a,s,tol=1.0e-6).
    Inverse power method for solving the eigenvalue problem
    [a]{x} = lam{x}. Returns 'lam' closest to 's' and the
    corresponding eigenvector {x}.
'''
```

Symmetric Matrix Eigenvalue Problems

```
from numarray import zeros,Float64,dot,identity
from LUdecomp import *
from math import sqrt
from random import random

def inversePower(a,s,tol=1.0e-6):
    n = len(a)
    aStar = a - identity(n)*s    # Form [a*] = [a] - s[I]
    aStar = LUdecomp(aStar)      # Decompose [a*]
    x = zeros((n),type=Float64)
    for i in range(n):           # Seed [x] with random numbers
        x[i] = random()
    xMag = sqrt(dot(x,x))        # Normalize [x]
    x =x/xMag
    for i in range(50):          # Begin iterations
        xOld = x.copy()          # Save current [x]
        x = LUsolve(aStar,x)     # Solve [a*][x] = [xOld]
        xMag = sqrt(dot(x,x))    # Normalize [x]
        x = x/xMag
        if dot(xOld,x) < 0.0:    # Detect change in sign of [x]
            sign = -1.0
            x = -x
        else: sign = 1.0
        if sqrt(dot(xOld - x,xOld - x)) < tol:
            return s + sign/xMag,x
    print 'Inverse power method did not converge'
```

EXAMPLE 9.4
The stress matrix describing the state of stress at a point is

$$\mathbf{S} = \begin{bmatrix} -30 & 10 & 20 \\ 10 & 40 & -50 \\ 20 & -50 & -10 \end{bmatrix} \text{MPa}$$

Determine the largest principal stress (the eigenvalue of **S** farthest from zero) by the power method.

Solution First iteration:

Let $v = \begin{bmatrix} 1 & 0 & 0 \end{bmatrix}^T$ be the initial guess for the eigenvector. Then

$$\mathbf{z} = \mathbf{Sv} = \begin{bmatrix} -30 & 10 & 20 \\ 10 & 40 & -50 \\ 20 & -50 & -10 \end{bmatrix} \begin{bmatrix} 1 \\ 0 \\ 0 \end{bmatrix} = \begin{bmatrix} -30.0 \\ 10.0 \\ 20.0 \end{bmatrix}$$

9.3 Inverse Power and Power Methods

$$|\mathbf{z}| = \sqrt{30^2 + 10^2 + 20^2} = 37.417$$

$$\mathbf{v} = \frac{\mathbf{z}}{|\mathbf{z}|} = \begin{bmatrix} -30.0 \\ 10.0 \\ 20.0 \end{bmatrix} \frac{1}{37.417} = \begin{bmatrix} -0.801\,77 \\ 0.267\,26 \\ 0.534\,52 \end{bmatrix}$$

Second iteration:

$$\mathbf{z} = \mathbf{Sv} = \begin{bmatrix} -30 & 10 & 20 \\ 10 & 40 & -50 \\ 20 & -50 & -10 \end{bmatrix} \begin{bmatrix} -0.801\,77 \\ 0.267\,26 \\ 0.534\,52 \end{bmatrix} = \begin{bmatrix} 37.416 \\ -24.053 \\ -34.744 \end{bmatrix}$$

$$|\mathbf{z}| = \sqrt{37.416^2 + 24.053^2 + 34.744^2} = 56.442$$

$$\mathbf{v} = \frac{\mathbf{z}}{|\mathbf{z}|} = \begin{bmatrix} 37.416 \\ -24.053 \\ -34.744 \end{bmatrix} \frac{1}{56.442} = \begin{bmatrix} 0.66291 \\ -0.42615 \\ -0.61557 \end{bmatrix}$$

Third iteration:

$$\mathbf{z} = \mathbf{Sv} = \begin{bmatrix} -30 & 10 & 20 \\ 10 & 40 & -50 \\ 20 & -50 & -10 \end{bmatrix} \begin{bmatrix} 0.66291 \\ -0.42615 \\ -0.61557 \end{bmatrix} = \begin{bmatrix} -36.460 \\ 20.362 \\ 40.721 \end{bmatrix}$$

$$|\mathbf{z}| = \sqrt{36.460^2 + 20.362^2 + 40.721^2} = 58.328$$

$$\mathbf{v} = \frac{\mathbf{z}}{|\mathbf{z}|} = \begin{bmatrix} -36.460 \\ 20.362 \\ 40.721 \end{bmatrix} \frac{1}{58.328} = \begin{bmatrix} -0.62509 \\ 0.34909 \\ 0.69814 \end{bmatrix}$$

At this point the approximation of the eigenvalue we seek is $\lambda = -58.328$ MPa (the negative sign is determined by the sign reversal of $\mathbf{z}$ between iterations). This is actually close to the second-largest eigenvalue $\lambda_2 = -58.39$ MPa! By continuing the iterative process we would eventually end up with the largest eigenvalue $\lambda_3 = 70.94$ MPa. But since $|\lambda_2|$ and $|\lambda_3|$ are rather close, the convergence is too slow from this point on for manual labor. Here is a program that does the calculations for us:

```
#!/usr/bin/python
## example9_4
from numarray import array,matrixmultiply,dot
from math import sqrt

s = array([[-30.0,  10.0,  20.0], \
           [ 10.0,  40.0, -50.0], \
           [ 20.0, -50.0, -10.0]])
```

Symmetric Matrix Eigenvalue Problems

```
v = array([1.0, 0.0, 0.0])
for i in range(100):
    vOld = v.copy()
    z = matrixmultiply(s,v)
    zMag = sqrt(dot(z,z))
    v = z/zMag
    if dot(vOld,v) < 0.0:
        sign = -1.0
        v = -v
    else: sign = 1.0
    if sqrt(dot(vOld - v,vOld - v)) < 1.0e-6: break
lam = sign*zMag
print ''Number of iterations ='',i
print ''Eigenvalue ='',lam
raw_input(''\nPrint press return to exit'')
```

The results are:

```
Number of iterations = 92
Eigenvalue = 70.9434833068
```

Note that it took 92 iterations to reach convergence!

EXAMPLE 9.5
Determine the smallest eigenvalue λ_1 and the corresponding eigenvector of

$$\mathbf{A} = \begin{bmatrix} 11 & 2 & 3 & 1 & 4 \\ 2 & 9 & 3 & 5 & 2 \\ 3 & 3 & 15 & 4 & 3 \\ 1 & 5 & 4 & 12 & 4 \\ 4 & 2 & 3 & 4 & 17 \end{bmatrix}$$

Use the inverse power method with eigenvalue shifting knowing that $\lambda_1 \approx 5$.

Solution
```
#!/usr/bin/python##
example9_5
from numarray import array
from inversePower import *

s = 5.0
```

9.3 Inverse Power and Power Methods

```
a = array([[ 11.0, 2.0,  3.0,  1.0,  4.0], \
           [  2.0, 9.0,  3.0,  5.0,  2.0], \
           [  3.0, 3.0, 15.0,  4.0,  3.0], \
           [  1.0, 5.0,  4.0, 12.0,  4.0], \
           [  4.0, 2.0,  3.0,  4.0, 17.0]])
lam,x = inversePower(a,s)
print ''Eigenvalue ='',lam
print ''\nEigenvector:\n'',x
raw_input(''\nPrint press return to exit'')
```

Here is the output:

```
Eigenvalue = 4.87394637865

Eigenvector:
[-0.26726603  0.74142854  0.05017271 -0.59491453  0.14970633]
```

Convergence was achieved with 4 iterations. Without the eigenvalue shift 26 iterations would be required.

EXAMPLE 9.6

Unlike Jacobi diagonalization, the inverse power method lends itself to eigenvalue problems of banded matrices. Write a program that computes the smallest buckling load of the beam described in Example 9.3, making full use of the banded forms. Run the program with 100 interior nodes ($n = 100$).

Solution The function inversePower5 listed below returns the smallest eigenvalue and the corresponding eigenvector of $\mathbf{Ax} = \lambda \mathbf{Bx}$, where $\mathbf{A}$ is a pentadiagonal matrix and $\mathbf{B}$ is a sparse matrix (in this problem it is tridiagonal). The matrix $\mathbf{A}$ is input by its diagonals $\mathbf{d}$, $\mathbf{e}$ and $\mathbf{f}$ as was done in Section 2.4 in conjunction with the LU decomposition. The algorithm for inversePower5 does not use $\mathbf{B}$ directly, but calls the function Bv(v) that supplies the product $\mathbf{Bv}$. Eigenvalue shifting is not used.

```
## module inversePower5
''' lam,x = inversePower5(Bv,d,e,f,tol=1.0e-6).
    Inverse power method for solving the eigenvalue problem
    [A]{x} = lam[B]{x}, where [A] = [f\e\d\e\f] is
    pentadiagonal and [B] is sparse.. User must supply the
    function Bv(v) that returns the vector [B]{v}.
'''
```

Symmetric Matrix Eigenvalue Problems

```
from numarray import zeros,Float64,dot
from LUdecomp5 import *
from math import sqrt
from random import random

def inversePower5(Bv,d,e,f,tol=1.0e-6):
    n = len(d)
    d,e,f = LUdecomp5(d,e,f)
    x = zeros((n),type=Float64)
    for i in range(n):              # Seed {v} with random numbers
        x[i] = random()
    xMag = sqrt(dot(x,x))           # Normalize {v}
    x = x/xMag
    for i in range(30):             # Begin iterations
        xOld = x.copy()             # Save current {v}
        x = Bv(xOld)                # Compute [B]{v}
        x = LUsolve5(d,e,f,x)       # Solve [A]{z} = [B]{v}
        xMag = sqrt(dot(x,x))       # Normalize {z}
        x = x/xMag
        if dot(xOld,x) < 0.0:       # Detect change in sign of {x}
            sign = -1.0
            x = -x
        else: sign = 1.0
        if sqrt(dot(xOld - x,xOld - x)) < tol:
            return sign/xMag,x
    print 'Inverse power method did not converge'
```

The program that utilizes `inversePower5` is

```
#!/usr/bin/python
## example9_6
from numarray import ones
from inversePower5 import *

def Bv(v):              # Computes {z} = [B]{v}
    n = len(v)
    z = zeros((n),type=Float64)
    z[0] = 2.0*v[0] - v[1]
    for i in range(1,n-1):
        z[i] = -v[i-1] + 2.0*v[i] - v[i+1]
```

9.3 Inverse Power and Power Methods

```
        z[n-1] = -v[n-2] + 2.0*v[n-1]
    return z

n = 100                      # Number of interior nodes
d = ones((n))*6.0            # Specify diagonals of [A] = [f\e\d\e\f]
d[0] = 5.0
d[n-1] = 7.0
e = ones((n-1))*(-4.0)
f = ones((n-2))*1.0
lam,x = inversePower5(Bv,d,e,f)
print ''PL^2/EI ='',lam*(n+1)**2
raw_input(''\nPress return to exit'')
```

The output, shown below, is in excellent agreement with the analytical value.

PL^2/EI = 20.1867355603

PROBLEM SET 9.1

1. Given

$$A = \begin{bmatrix} 7 & 3 & 1 \\ 3 & 9 & 6 \\ 1 & 6 & 8 \end{bmatrix} \quad B = \begin{bmatrix} 4 & 0 & 0 \\ 0 & 9 & 0 \\ 0 & 0 & 4 \end{bmatrix}$$

convert the eigenvalue problem $Ax = \lambda Bx$ to the standard form $Hz = \lambda z$. What is the relationship between x and z?

2. Convert the eigenvalue problem $Ax = \lambda Bx$, where

$$A = \begin{bmatrix} 4 & -1 & 0 \\ -1 & 4 & -1 \\ 0 & -1 & 4 \end{bmatrix} \quad B = \begin{bmatrix} 2 & -1 & 0 \\ -1 & 2 & -1 \\ 0 & -1 & 1 \end{bmatrix}$$

to the standard form.

3. An eigenvalue of the problem in Prob. 2 is roughly 2.5. Use the inverse power method with eigenvalue shifting to compute this eigenvalue to four decimal places. Start with $x = \begin{bmatrix} 1 & 0 & 0 \end{bmatrix}^T$. *Hint*: two iterations should be sufficient.

4. The stress matrix at a point is

$$S = \begin{bmatrix} 150 & -60 & 0 \\ -60 & 120 & 0 \\ 0 & 0 & 80 \end{bmatrix} \text{MPa}$$

Compute the principal stresses (eigenvalues of S).

5.

The two pendulums are connected by a spring which is undeformed when the pendulums are vertical. The equations of motion of the system can be shown to be

$$kL(\theta_2 - \theta_1) - mg\theta_1 = mL\ddot{\theta}_1$$
$$-kL(\theta_2 - \theta_1) - 2mg\theta_2 = 2mL\ddot{\theta}_2$$

where θ_1 and θ_2 are the angular displacements and k is the spring stiffness. Determine the circular frequencies of vibration and the relative amplitudes of the angular displacements. Use $m = 0.25$ kg, $k = 20$ N/m, $L = 0.75$ m and $g = 9.80665$ m/s^2.

6.

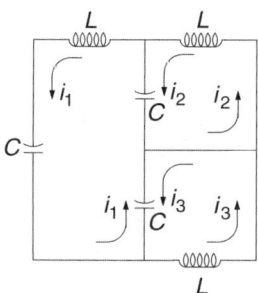

Kirchoff's laws for the electric circuit are

$$3i_1 - i_2 - i_3 = -LC\frac{d^2 i_1}{dt^2}$$
$$-i_1 + i_2 = -LC\frac{d^2 i_2}{dt^2}$$
$$-i_1 + i_3 = -LC\frac{d^2 i_3}{dt^2}$$

Compute the circular frequencies of the circuit and the relative amplitudes of the loop currents.

9.3 Inverse Power and Power Methods

7. Compute the matrix $\mathbf{A}^*$ that results from annihilation of A_{14} and A_{41} in the matrix

$$\mathbf{A} = \begin{bmatrix} 4 & -1 & 0 & 1 \\ -1 & 6 & -2 & 0 \\ 0 & -2 & 3 & 2 \\ 1 & 0 & 2 & 4 \end{bmatrix}$$

by a Jacobi rotation.

8. ■ Use the Jacobi method to determine the eigenvalues and eigenvectors of

$$\mathbf{A} = \begin{bmatrix} 4 & -1 & -2 \\ -1 & 3 & 3 \\ -2 & 3 & 1 \end{bmatrix}$$

9. ■ Find the eigenvalues and eigenvectors of

$$\mathbf{A} = \begin{bmatrix} 4 & -2 & 1 & -1 \\ -2 & 4 & -2 & 1 \\ 1 & -2 & 4 & -2 \\ -1 & 1 & -2 & 4 \end{bmatrix}$$

with the Jacobi method.

10. ■ Use the power method to compute the largest eigenvalue and the corresponding eigenvector of the matrix $\mathbf{A}$ given in Prob. 9.

11. ■ Find the smallest eigenvalue and the corresponding eigenvector of the matrix $\mathbf{A}$ in Prob. 9. Use the inverse power method.

12. ■ Let

$$\mathbf{A} = \begin{bmatrix} 1.4 & 0.8 & 0.4 \\ 0.8 & 6.6 & 0.8 \\ 0.4 & 0.8 & 5.0 \end{bmatrix} \quad \mathbf{B} = \begin{bmatrix} 0.4 & -0.1 & 0.0 \\ -0.1 & 0.4 & -0.1 \\ 0.0 & -0.1 & 0.4 \end{bmatrix}$$

Find the eigenvalues and eigenvectors of $\mathbf{Ax} = \lambda \mathbf{Bx}$ by the Jacobi method.

13. ■ Use the inverse power method to compute the smallest eigenvalue in Prob. 12.

14. ■ Use the Jacobi method to compute the eigenvalues and eigenvectors of the matrix

$$\mathbf{A} = \begin{bmatrix} 11 & 2 & 3 & 1 & 4 & 2 \\ 2 & 9 & 3 & 5 & 2 & 1 \\ 3 & 3 & 15 & 4 & 3 & 2 \\ 1 & 5 & 4 & 12 & 4 & 3 \\ 4 & 2 & 3 & 4 & 17 & 5 \\ 2 & 1 & 2 & 3 & 5 & 8 \end{bmatrix}$$

15. ■ Find the eigenvalues of $\mathbf{Ax} = \lambda \mathbf{Bx}$ by the Jacobi method, where

$$\mathbf{A} = \begin{bmatrix} 6 & -4 & 1 & 0 \\ -4 & 6 & -4 & 1 \\ 1 & -4 & 6 & -4 \\ 0 & 1 & -4 & 7 \end{bmatrix} \qquad \mathbf{B} = \begin{bmatrix} 1 & -2 & 3 & -1 \\ -2 & 6 & -2 & 3 \\ 3 & -2 & 6 & -2 \\ -1 & 3 & -2 & 9 \end{bmatrix}$$

Warning: **B** is not positive definite.

16. ■

The figure shows a cantilever beam with a superimposed finite difference mesh. If $u(x, t)$ is the lateral displacement of the beam, the differential equation of motion governing bending vibrations is

$$u^{(4)} = -\frac{\gamma}{EI}\ddot{u}$$

where γ is the mass per unit length and EI is the bending rigidity. The boundary conditions are $u(0, t) = u'(0, t) = u''(L, t) = u'''(L, t) = 0$. With $u(x, t) = y(x)\sin\omega t$ the problem becomes

$$y^{(4)} = \frac{\omega^2 \gamma}{EI} y \qquad y(0) = y'(0) = y''(L) = y'''(L) = 0$$

The corresponding finite difference equations are

$$\begin{bmatrix} 7 & -4 & 1 & 0 & 0 & \cdots & 0 \\ -4 & 6 & -4 & 1 & 0 & \cdots & 0 \\ 1 & -4 & 6 & -4 & 1 & \cdots & 0 \\ \vdots & \ddots & \ddots & \ddots & \ddots & \ddots & \vdots \\ 0 & \cdots & 1 & -4 & 6 & -4 & 1 \\ 0 & \cdots & 0 & 1 & -4 & 5 & -2 \\ 0 & \cdots & 0 & 0 & 1 & -2 & 1 \end{bmatrix} \begin{bmatrix} y_1 \\ y_2 \\ y_3 \\ \vdots \\ y_{n-2} \\ y_{n-1} \\ y_n \end{bmatrix} = \lambda \begin{bmatrix} y_1 \\ y_2 \\ y_3 \\ \vdots \\ y_{n-2} \\ y_{n-1} \\ y_n/2 \end{bmatrix}$$

where

$$\lambda = \frac{\omega^2 \gamma}{EI}\left(\frac{L}{n}\right)^4$$

(a) Write down the matrix **H** of the standard form $\mathbf{Hz} = \lambda \mathbf{z}$ and the transformation matrix **P** as in $\mathbf{y} = \mathbf{Pz}$. (b) Write a program that computes the lowest two circular frequencies of the beam and the corresponding mode shapes (eigenvectors) using the Jacobi method. Run the program with $n = 10$. *Note*: the analytical solution for the lowest circular frequency is $\omega_1 = (3.515/L^2)\sqrt{EI/\gamma}$.

17. ■

The simply supported column in Fig. (a) consists of three segments with the bending rigidities shown. If only the first buckling mode is of interest, it is sufficient to model half of the beam as shown in Fig. (b). The differential equation for the lateral displacement $u(x)$ is

$$u'' = -\frac{P}{EI}u$$

with the boundary conditions $u(0) = u'(L/2) = 0$. The corresponding finite difference equations are

$$\begin{bmatrix} 2 & -1 & 0 & 0 & 0 & 0 & 0 & \cdots & 0 \\ -1 & 2 & -1 & 0 & 0 & 0 & 0 & \cdots & 0 \\ 0 & -1 & 2 & -1 & 0 & 0 & 0 & \cdots & 0 \\ 0 & 0 & -1 & 2 & -1 & 0 & 0 & \cdots & 0 \\ 0 & 0 & 0 & -1 & 2 & -1 & 0 & \cdots & 0 \\ 0 & 0 & 0 & 0 & -1 & 2 & -1 & \cdots & 0 \\ \vdots & \vdots & \vdots & \vdots & \vdots & \ddots & \ddots & \ddots & \vdots \\ 0 & \cdots & 0 & 0 & 0 & 0 & -1 & 2 & -1 \\ 0 & \cdots & 0 & 0 & 0 & 0 & 0 & -1 & 1 \end{bmatrix} \begin{bmatrix} u_1 \\ u_2 \\ u_3 \\ u_4 \\ u_5 \\ u_6 \\ \vdots \\ u_9 \\ u_{10} \end{bmatrix} = \lambda \begin{bmatrix} u_1 \\ u_2 \\ u_3 \\ u_4 \\ u_5/1.5 \\ u_6/2 \\ \vdots \\ u_9/2 \\ u_{10}/4 \end{bmatrix}$$

where

$$\lambda = \frac{P}{EI_0}\left(\frac{L}{20}\right)^2$$

Write a program that computes the lowest buckling load P of the column with the inverse power method. Utilize the banded forms of the matrices.

18. ■

The springs supporting the three-bar linkage are undeformed when the linkage is horizontal. The equilibrium equations of the linkage in the presence of the

horizontal force P can be shown to be

$$\begin{bmatrix} 6 & 5 & 3 \\ 3 & 3 & 2 \\ 1 & 1 & 1 \end{bmatrix} \begin{bmatrix} \theta_1 \\ \theta_2 \\ \theta_3 \end{bmatrix} = \frac{P}{kL} \begin{bmatrix} 1 & 1 & 1 \\ 0 & 1 & 1 \\ 0 & 0 & 1 \end{bmatrix} \begin{bmatrix} \theta_1 \\ \theta_2 \\ \theta_3 \end{bmatrix}$$

where k is the spring stiffness. Determine the smallest buckling load P and the corresponding mode shape. *Hint*: the equations can easily rewritten in the standard form $\mathbf{A}\theta = \lambda\theta$, where $\mathbf{A}$ is symmetric.

19. ■

The differential equations of motion for the mass–spring system are

$$k(-2u_1 + u_2) = m\ddot{u}_1$$
$$k(u_1 - 2u_2 + u_3) = 3m\ddot{u}_2$$
$$k(u_2 - 2u_3) = 2m\ddot{u}_3$$

where $u_i(t)$ is the displacement of mass i from its equilibrium position and k is the spring stiffness. Determine the circular frequencies of vibration and the corresponding mode shapes.

20. ■

Kirchoff's equations for the circuit are

$$L\frac{d^2 i_1}{dt^2} + \frac{1}{C}i_1 + \frac{2}{C}(i_1 - i_2) = 0$$

$$L\frac{d^2 i_2}{dt^2} + \frac{2}{C}(i_2 - i_1) + \frac{3}{C}(i_2 - i_3) = 0$$

$$L\frac{d^2 i_3}{dt^2} + \frac{3}{C}(i_3 - i_2) + \frac{4}{C}(i_3 - i_4) = 0$$

$$L\frac{d^2 i_4}{dt^2} + \frac{4}{C}(i_4 - i_3) + \frac{5}{C}i_4 = 0$$

Find the circular frequencies of the currents.

9.4 Householder Reduction to Tridiagonal Form

21. ■

Determine the circular frequencies of oscillation for the circuit shown, given the Kirchoff equations

$$L\frac{d^2 i_1}{dt^2} + L\left(\frac{d^2 i_1}{dt^2} - \frac{d^2 i_2}{dt^2}\right) + \frac{1}{C} i_1 = 0$$

$$L\left(\frac{d^2 i_2}{dt^2} - \frac{d^2 i_1}{dt^2}\right) + L\left(\frac{d^2 i_2}{dt^2} - \frac{d^2 i_3}{dt^2}\right) + \frac{2}{C} i_2 = 0$$

$$L\left(\frac{d^2 i_3}{dt^2} - \frac{d^2 i_2}{dt^2}\right) + L\left(\frac{d^2 i_3}{dt^2} - \frac{d^2 i_4}{dt^2}\right) + \frac{3}{C} i_3 = 0$$

$$L\left(\frac{d^2 i_4}{dt^2} - \frac{d^2 i_3}{dt^2}\right) + L\frac{d^2 i_4}{dt^2} + \frac{4}{C} i_4 = 0$$

22. ■ Several iterative methods exist for finding the eigenvalues of a matrix **A**. One of these is the *LR method*, which requires the matrix to be symmetric and positive definite. Its algorithm very simple:

    Let $\mathbf{A}_0 = \mathbf{A}$
    do with $i = 0, 1, 2, \ldots$
        Use Choleski's decomposition $\mathbf{A}_i = \mathbf{L}_i \mathbf{L}_i^T$ to compute $\mathbf{L}_i$
        Form $\mathbf{A}_{i+1} = \mathbf{L}_i^T \mathbf{L}_i$
    end do

It can be shown that the diagonal elements of $\mathbf{A}_{i+1}$ converge to the eigenvalues of **A**. Write a program that implements the LR method and test it with

$$\mathbf{A} = \begin{bmatrix} 4 & 3 & 1 \\ 3 & 4 & 2 \\ 1 & 2 & 3 \end{bmatrix}$$

9.4 Householder Reduction to Tridiagonal Form

It was mentioned before that similarity transformations can be used to transform an eigenvalue problem to a form that is easier to solve. The most desirable of the "easy" forms is, of course, the diagonal form that results from the Jacobi method. However, the Jacobi method requires about $10n^3$ to $20n^3$ multiplications, so that the amount of computation increases very rapidly with n. We are generally better off by reducing the matrix to the tridiagonal form, which can be done in precisely $n - 2$ transformations by the Householder method. Once the tridiagonal form is achieved, we still have to extract the eigenvalues and the eigenvectors, but there are effective means of dealing with that, as we see the next article.

Householder Matrix

Each Householder transformation utilizes the *Householder matrix*

$$\mathbf{Q} = \mathbf{I} - \frac{\mathbf{u}\mathbf{u}^T}{H} \tag{9.36}$$

where $\mathbf{u}$ is a vector and

$$H = \frac{1}{2}\mathbf{u}^T\mathbf{u} = \frac{1}{2}|\mathbf{u}|^2 \tag{9.37}$$

Note that $\mathbf{u}\mathbf{u}^T$ in Eq. (9.36) is the outer product; that is, a matrix with the elements $\left(\mathbf{u}\mathbf{u}^T\right)_{ij} = u_i u_j$. Since $\mathbf{Q}$ is obviously symmetric ($\mathbf{Q}^T = \mathbf{Q}$), we can write

$$\mathbf{Q}^T\mathbf{Q} = \mathbf{Q}\mathbf{Q} = \left(\mathbf{I} - \frac{\mathbf{u}\mathbf{u}^T}{H}\right)\left(\mathbf{I} - \frac{\mathbf{u}\mathbf{u}^T}{H}\right) = \mathbf{I} - 2\frac{\mathbf{u}\mathbf{u}^T}{H} + \frac{\mathbf{u}\left(\mathbf{u}^T\mathbf{u}\right)\mathbf{u}^T}{H^2}$$

$$= \mathbf{I} - 2\frac{\mathbf{u}\mathbf{u}^T}{H} + \frac{\mathbf{u}(2H)\mathbf{u}^T}{H^2} = \mathbf{I}$$

which shows that $\mathbf{Q}$ is also orthogonal.

Now let $\mathbf{x}$ be an arbitrary vector and consider the transformation $\mathbf{Q}\mathbf{x}$. Choosing

$$\mathbf{u} = \mathbf{x} + k\mathbf{e}_1 \tag{9.38}$$

where

$$k = \pm|\mathbf{x}| \qquad \mathbf{e}_1 = \begin{bmatrix} 1 & 0 & 0 & \cdots & 0 \end{bmatrix}^T$$

we get

$$\mathbf{Q}\mathbf{x} = \left(\mathbf{I} - \frac{\mathbf{u}\mathbf{u}^T}{H}\right)\mathbf{x} = \left[\mathbf{I} - \frac{\mathbf{u}(\mathbf{x} + k\mathbf{e}_1)^T}{H}\right]\mathbf{x}$$

$$= \mathbf{x} - \frac{\mathbf{u}\left(\mathbf{x}^T\mathbf{x} + k\mathbf{e}_1^T\mathbf{x}\right)}{H} = \mathbf{x} - \frac{\mathbf{u}\left(k^2 + kx_1\right)}{H}$$

But

$$2H = (\mathbf{x} + k\mathbf{e}_1)^T(\mathbf{x} + k\mathbf{e}_1) = |\mathbf{x}|^2 + k\left(\mathbf{x}^T\mathbf{e}_1 + \mathbf{e}_1^T\mathbf{x}\right) + k^2\mathbf{e}_1^T\mathbf{e}_1$$

$$= k^2 + 2kx_1 + k^2 = 2\left(k^2 + kx_1\right)$$

9.4 Householder Reduction to Tridiagonal Form

so that

$$\mathbf{Q}\mathbf{x} = \mathbf{x} - \mathbf{u} = -k\mathbf{e}_1 = \begin{bmatrix} -k & 0 & 0 & \cdots & 0 \end{bmatrix}^T \quad (9.39)$$

Hence the transformation eliminates all elements of $\mathbf{x}$ except the first one.

Householder Reduction of a Symmetric Matrix

Let us now apply the following transformation to a symmetric $n \times n$ matrix $\mathbf{A}$:

$$\mathbf{P}_1 \mathbf{A} = \begin{bmatrix} 1 & \mathbf{0}^T \\ \mathbf{0} & \mathbf{Q} \end{bmatrix} \begin{bmatrix} A_{11} & \mathbf{x}^T \\ \mathbf{x} & \mathbf{A}' \end{bmatrix} = \begin{bmatrix} A_{11} & \mathbf{x}^T \\ \mathbf{Q}\mathbf{x} & \mathbf{Q}\mathbf{A}' \end{bmatrix} \quad (9.40)$$

Here $\mathbf{x}$ represents the first column of $\mathbf{A}$ with the first element omitted, and $\mathbf{A}'$ is simply $\mathbf{A}$ with its first row and column removed. The matrix $\mathbf{Q}$ of dimensions $(n-1) \times (n-1)$ is constructed using Eqs. (9.36)–(9.38). Referring to Eq. (9.39), we see that the transformation reduces the first column of $\mathbf{A}$ to

$$\begin{bmatrix} A_{11} \\ \mathbf{Q}\mathbf{x} \end{bmatrix} = \begin{bmatrix} A_{11} \\ -k \\ 0 \\ \vdots \\ 0 \end{bmatrix}$$

The transformation

$$\mathbf{A} \leftarrow \mathbf{P}_1 \mathbf{A} \mathbf{P}_1 = \begin{bmatrix} A_{11} & (\mathbf{Q}\mathbf{x})^T \\ \mathbf{Q}\mathbf{x} & \mathbf{Q}\mathbf{A}'\mathbf{Q} \end{bmatrix} \quad (9.41)$$

thus tridiagonalizes the first row as well as the first column of $\mathbf{A}$. Here is a diagram of the transformation for a 4×4 matrix:

$$\begin{bmatrix} 1 & 0 & 0 & 0 \\ 0 & & & \\ 0 & & \mathbf{Q} & \\ 0 & & & \end{bmatrix} \cdot \begin{bmatrix} A_{11} & A_{12} & A_{13} & A_{14} \\ A_{21} & & & \\ A_{31} & & \mathbf{A}' & \\ A_{41} & & & \end{bmatrix} \cdot \begin{bmatrix} 1 & 0 & 0 & 0 \\ 0 & & & \\ 0 & & \mathbf{Q} & \\ 0 & & & \end{bmatrix}$$

$$= \begin{bmatrix} A_{11} & -k & 0 & 0 \\ -k & & & \\ 0 & & \mathbf{Q}\mathbf{A}'\mathbf{Q} & \\ 0 & & & \end{bmatrix}$$

The second row and column of $\mathbf{A}$ are reduced next by applying the transformation to the 3×3 lower right portion of the matrix. This transformation can be expressed as $\mathbf{A} \leftarrow \mathbf{P}_2 \mathbf{A} \mathbf{P}_2$, where now

$$\mathbf{P}_2 = \begin{bmatrix} \mathbf{I}_2 & \mathbf{0}^T \\ \mathbf{0} & \mathbf{Q} \end{bmatrix} \tag{9.42}$$

In Eq. (9.42) $\mathbf{I}_2$ is a 2×2 identity matrix and $\mathbf{Q}$ is a $(n-2) \times (n-2)$ matrix constructed by choosing for $\mathbf{x}$ the bottom $n-2$ elements of the second column of $\mathbf{A}$. It takes a total of $n-2$ transformation with

$$\mathbf{P}_i = \begin{bmatrix} \mathbf{I}_i & \mathbf{0}^T \\ \mathbf{0} & \mathbf{Q} \end{bmatrix}, \quad i = 1, 2, \ldots, n-2$$

to attain the tridiagonal form.

It is wasteful to form $\mathbf{P}_i$ and then carry out the matrix multiplication $\mathbf{P}_i\mathbf{A}\mathbf{P}_i$. We note that

$$\mathbf{A'Q} = \mathbf{A'}\left(\mathbf{I} - \frac{\mathbf{u}\mathbf{u}^T}{H}\right) = \mathbf{A'} - \frac{\mathbf{A'u}}{H}\mathbf{u}^T = \mathbf{A'} - \mathbf{v}\mathbf{u}^T$$

where

$$\mathbf{v} = \frac{\mathbf{A'u}}{H} \tag{9.43}$$

Therefore,

$$\mathbf{Q}\mathbf{A'Q} = \left(\mathbf{I} - \frac{\mathbf{u}\mathbf{u}^T}{H}\right)(\mathbf{A'} - \mathbf{v}\mathbf{u}^T) = \mathbf{A'} - \mathbf{v}\mathbf{u}^T - \frac{\mathbf{u}\mathbf{u}^T}{H}(\mathbf{A'} - \mathbf{v}\mathbf{u}^T)$$

$$= \mathbf{A'} - \mathbf{v}\mathbf{u}^T - \frac{\mathbf{u}(\mathbf{u}^T\mathbf{A'})}{H} + \frac{\mathbf{u}(\mathbf{u}^T\mathbf{v})\mathbf{u}^T}{H}$$

$$= \mathbf{A'} - \mathbf{v}\mathbf{u}^T - \mathbf{u}\mathbf{v}^T + 2g\mathbf{u}\mathbf{u}^T$$

where

$$g = \frac{\mathbf{u}^T\mathbf{v}}{2H} \tag{9.44}$$

Letting

$$\mathbf{w} = \mathbf{v} - g\mathbf{u} \tag{9.45}$$

it can be easily verified that the transformation can be written as

$$\mathbf{Q}\mathbf{A'Q} = \mathbf{A'} - \mathbf{w}\mathbf{u}^T - \mathbf{u}\mathbf{w}^T \tag{9.46}$$

which gives us the following computational procedure which is to be carried out with $i = 1, 2, \ldots, n-2$:

1. Let $\mathbf{A'}$ be the $(n-i) \times (n-i)$ lower right-hand portion of $\mathbf{A}$.
2. Let $\mathbf{x} = \begin{bmatrix} A_{i+1,i} & A_{i+2,i} & \cdots & A_{n,i} \end{bmatrix}^T$ (the column of length $n-i$ just to the left of $\mathbf{A'}$).
3. Compute $|\mathbf{x}|$. Let $k = |\mathbf{x}|$ if $x_1 > 0$ and $k = -|\mathbf{x}|$ if $x_1 < 0$ (this choice of sign minimizes the roundoff error).

9.4 Householder Reduction to Tridiagonal Form

4. Let $\mathbf{u} = \begin{bmatrix} k+x_1 & x_2 & x_3 & \cdots & x_{n-i} \end{bmatrix}^T$.
5. Compute $H = |\mathbf{u}|^2$.
6. Compute $\mathbf{v} = \mathbf{A}'\mathbf{u}/H$.
7. Compute $g = \mathbf{u}^T\mathbf{v}/(2H)$.
8. Compute $\mathbf{w} = \mathbf{v} - g\mathbf{u}$.
9. Compute the transformation $\mathbf{A}' \leftarrow \mathbf{A}' - \mathbf{w}^T\mathbf{u} - \mathbf{u}^T\mathbf{w}$.
10. Set $A_{i,i+1} = A_{i+1,i} = -k$.

Accumulated Transformation Matrix

Since we used similarity transformations, the eigenvalues of the tridiagonal matrix are the same as those of the original matrix. However, to determine the eigenvectors **X** of original **A** we must use the transformation

$$\mathbf{X} = \mathbf{P}\mathbf{X}_{\text{tridiag}}$$

where **P** is the accumulation of the individual transformations:

$$\mathbf{P} = \mathbf{P}_1 \mathbf{P}_2 \cdots \mathbf{P}_{n-2}$$

We build up the accumulated transformation matrix by initializing **P** to a $n \times n$ identity matrix and then applying the transformation

$$\mathbf{P} \leftarrow \mathbf{P}\mathbf{P}_i = \begin{bmatrix} \mathbf{P}_{11} & \mathbf{P}_{12} \\ \mathbf{P}_{21} & \mathbf{P}_{22} \end{bmatrix} \begin{bmatrix} \mathbf{I}_i & \mathbf{0}^T \\ \mathbf{0} & \mathbf{Q} \end{bmatrix} = \begin{bmatrix} \mathbf{P}_{11} & \mathbf{P}_{21}\mathbf{Q} \\ \mathbf{P}_{12} & \mathbf{P}_{22}\mathbf{Q} \end{bmatrix} \quad \text{(b)}$$

with $i = 1, 2, \ldots, n-2$. It can be seen that each multiplication affects only the rightmost $n - i$ columns of **P** (since the first row of $\mathbf{P}_{12}$ contains only zeroes, it can also be omitted in the multiplication). Using the notation

$$\mathbf{P}' = \begin{bmatrix} \mathbf{P}_{12} \\ \mathbf{P}_{22} \end{bmatrix}$$

we have

$$\begin{bmatrix} \mathbf{P}_{12}\mathbf{Q} \\ \mathbf{P}_{22}\mathbf{Q} \end{bmatrix} = \mathbf{P}'\mathbf{Q} = \mathbf{P}'\left(\mathbf{I} - \frac{\mathbf{u}\mathbf{u}^T}{H}\right) = \mathbf{P}' - \frac{\mathbf{P}'\mathbf{u}}{H}\mathbf{u}^T = \mathbf{P}' - \mathbf{y}\mathbf{u}^T \quad (9.47)$$

where

$$\mathbf{y} = \frac{\mathbf{P}'\mathbf{u}}{H} \quad (9.48)$$

The procedure for carrying out the matrix multiplication in Eq. (b) is:

- Retrieve **u** (in our triangularization procedure the **u**'s are stored in the columns of the lower triangular portion of **A**).

Symmetric Matrix Eigenvalue Problems

- Compute $H = |\mathbf{u}|^2$
- Compute $\mathbf{y} = \mathbf{P}'\mathbf{u}/H$.
- Compute the transformation $\mathbf{P}' \leftarrow \mathbf{P}' - \mathbf{y}\mathbf{u}^T$.

■ householder

The function householder in this module does the triangulization. It returns (**d**, **c**), where **d** and **c** are vectors that contain the elements of the principal diagonal and the subdiagonal, respectively. Only the upper triangular portion is reduced to the triangular form. The part below the principal diagonal is used to store the vectors **u**. This is done automatically by the statement u = a[k+1:n,k] which does not create a new object u, but simply sets up a reference to a[k+1:n,k] (makes a deep copy). Thus any changes made to u are reflected in a[k+1:n,k].

The function computeP returns the accumulated transformation matrix **P**. There is no need to call it if only the eigenvalues are to be computed.

```
## module householder
''' d,c = householder(a).
    Householder similarity transformation of matrix [a] to
    the tridiagonal form [c\d\c].

    p = computeP(a).
    Computes the accumulated transformation matrix [p]
    after calling householder(a).
'''
from numarray import dot,matrixmultiply,diagonal, \
     outerproduct,identity
from math import sqrt

def householder(a):
    n = len(a)
    for k in range(n-2):
        u = a[k+1:n,k]
        uMag = sqrt(dot(u,u))
        if u[0] < 0.0: uMag = -uMag
        u[0] = u[0] + uMag
        h = dot(u,u)/2.0
        v = matrixmultiply(a[k+1:n,k+1:n],u)/h
        g = dot(u,v)/(2.0*h)
        v = v - g*u
        a[k+1:n,k+1:n] = a[k+1:n,k+1:n] - outerproduct(v,u) \
                       - outerproduct(u,v)
```

9.4 Householder Reduction to Tridiagonal Form

```
        a[k,k+1] = -uMag
    return diagonal(a),diagonal(a,1)

def computeP(a):
    n = len(a)
    p = identity(n)*1.0
    for k in range(n-2):
        u = a[k+1:n,k]
        h = dot(u,u)/2.0
        v = matrixmultiply(p[1:n,k+1:n],u)/h
        p[1:n,k+1:n] = p[1:n,k+1:n] - outerproduct(v,u)
    return p
```

EXAMPLE 9.7
Transform the matrix

$$\mathbf{A} = \begin{bmatrix} 7 & 2 & 3 & -1 \\ 2 & 8 & 5 & 1 \\ 3 & 5 & 12 & 9 \\ -1 & 1 & 9 & 7 \end{bmatrix}$$

into tridiagonal form using Householder reduction.

Solution Reduce the first row and column:

$$\mathbf{A}' = \begin{bmatrix} 8 & 5 & 1 \\ 5 & 12 & 9 \\ 1 & 9 & 7 \end{bmatrix} \qquad \mathbf{x} = \begin{bmatrix} 2 \\ 3 \\ -1 \end{bmatrix} \qquad k = |\mathbf{x}| = 3.7417$$

$$\mathbf{u} = \begin{bmatrix} k + x_1 \\ x_2 \\ x_3 \end{bmatrix} = \begin{bmatrix} 5.7417 \\ 3 \\ -1 \end{bmatrix} \qquad H = \frac{1}{2}|\mathbf{u}|^2 = 21.484$$

$$\mathbf{u}\mathbf{u}^T = \begin{bmatrix} 32.967 & 17.225 & -5.7417 \\ 17.225 & 9 & -3 \\ -5.7417 & -3 & 1 \end{bmatrix}$$

$$\mathbf{Q} = \mathbf{I} - \frac{\mathbf{u}\mathbf{u}^T}{H} = \begin{bmatrix} -0.53450 & -0.80176 & 0.26725 \\ -0.80176 & 0.58108 & 0.13964 \\ 0.26725 & 0.13964 & 0.95345 \end{bmatrix}$$

$$\mathbf{Q}\mathbf{A}'\mathbf{Q} = \begin{bmatrix} 10.642 & -0.1388 & -9.1294 \\ -0.1388 & 5.9087 & 4.8429 \\ -9.1294 & 4.8429 & 10.4480 \end{bmatrix}$$

Symmetric Matrix Eigenvalue Problems

$$A \leftarrow \begin{bmatrix} A_{11} & (\mathbf{Qx})^T \\ \mathbf{Qx} & \mathbf{QA'Q} \end{bmatrix} = \begin{bmatrix} 7 & -3.7417 & 0 & 0 \\ -3.7417 & 10.642 & -0.1388 & -9.1294 \\ 0 & -0.1388 & 5.9087 & 4.8429 \\ 0 & -9.1294 & 4.8429 & 10.4480 \end{bmatrix}$$

In the last step we used the formula $\mathbf{Qx} = \begin{bmatrix} -k & 0 & \cdots & 0 \end{bmatrix}^T$.

Reduce the second row and column:

$$\mathbf{A'} = \begin{bmatrix} 5.9087 & 4.8429 \\ 4.8429 & 10.4480 \end{bmatrix} \qquad \mathbf{x} = \begin{bmatrix} -0.1388 \\ -9.1294 \end{bmatrix} \qquad k = -|\mathbf{x}| = -9.1305$$

where the negative sign on k was determined by the sign of x_1.

$$\mathbf{u} = \begin{bmatrix} k + x_1 \\ x_2 \end{bmatrix} = \begin{bmatrix} -9.2693 \\ -9.1294 \end{bmatrix} \qquad H = \frac{1}{2}|\mathbf{u}|^2 = 84.633$$

$$\mathbf{uu}^T = \begin{bmatrix} 85.920 & 84.623 \\ 84.623 & 83.346 \end{bmatrix}$$

$$\mathbf{Q} = \mathbf{I} - \frac{\mathbf{uu}^T}{H} = \begin{bmatrix} 0.01521 & -0.99988 \\ -0.99988 & 0.01521 \end{bmatrix}$$

$$\mathbf{QA'Q} = \begin{bmatrix} 10.594 & 4.772 \\ 4.772 & 5.762 \end{bmatrix}$$

$$\mathbf{A} \leftarrow \begin{bmatrix} A_{11} & A_{12} & \mathbf{0}^T \\ A_{21} & A_{22} & (\mathbf{Qx})^T \\ \mathbf{0} & \mathbf{Qx} & \mathbf{QA'Q} \end{bmatrix} = \begin{bmatrix} 7 & -3.742 & 0 & 0 \\ -3.742 & 10.642 & 9.131 & 0 \\ 0 & 9.131 & 10.594 & 4.772 \\ 0 & -0 & 4.772 & 5.762 \end{bmatrix}$$

EXAMPLE 9.8
Use the function householder to tridiagonalize the matrix in Example 9.7; also determine the transformation matrix **P**.

Solution
```
#!/usr/bin/python
## example9_8
from numarray import array,matrixmultiply
from householder3 import *

a = array([[ 7.0, 2.0,  3.0, -1.0], \
           [ 2.0, 8.0,  5.0,  1.0], \
```

```
                [ 3.0,  5.0, 12.0,  9.0], \
                [-1.0,  1.0,  9.0,  7.0]])
d,c = householder(a)
print ''Principal diagonal {d}:\n'',d
print ''\nSubdiagonal {c}:\n'',c
print ''\nTransformation matrix [P]:''
print computeP(a)
raw_input(''\nPress return to exit'')
```

The results of running the above program are:

```
Principal diagonal {d}:
[  7.          10.64285714  10.59421525   5.76292761]

Subdiagonal {c}:
[-3.74165739  9.13085149  4.77158058]

Transformation matrix [P]:
[[ 1.          0.          0.          0.        ]
 [ 0.         -0.53452248 -0.25506831  0.80574554]
 [ 0.         -0.80178373 -0.14844139 -0.57888514]
 [ 0.          0.26726124 -0.95546079 -0.12516436]]
```

9.5 Eigenvalues of Symmetric Tridiagonal Matrices

Sturm Sequence

In principle, the eigenvalues of a matrix $\mathbf{A}$ can be determined by finding the roots of the characteristic equation $|\mathbf{A} - \lambda\mathbf{I}| = 0$. This method is impractical for large matrices, since the evaluation of the determinant involves $n^3/3$ multiplications. However, if the matrix is tridiagonal (we also assume it to be symmetric), its characteristic polynomial

$$P_n(\lambda) = |\mathbf{A}-\lambda\mathbf{I}| = \begin{vmatrix} d_1-\lambda & c_1 & 0 & 0 & \cdots & 0 \\ c_1 & d_2-\lambda & c_2 & 0 & \cdots & 0 \\ 0 & c_2 & d_3-\lambda & c_3 & \cdots & 0 \\ 0 & 0 & c_3 & d_4-\lambda & \cdots & 0 \\ \vdots & \vdots & \vdots & \vdots & \ddots & \vdots \\ 0 & 0 & \cdots & 0 & c_{n-1} & d_n-\lambda \end{vmatrix}$$

can be computed with only $3(n-1)$ multiplications using the following sequence of operations:

$$P_0(\lambda) = 1$$
$$P_1(\lambda) = d_1 - \lambda \tag{9.49}$$
$$P_i(\lambda) = (d_i - \lambda)P_{i-1}(\lambda) - c_{i-1}^2 P_{i-2}(\lambda), \quad i = 2, 3, \ldots, n$$

The polynomials $P_0(\lambda), P_1(\lambda), \ldots, P_n(\lambda)$ form a *Sturm sequence* that has the following property:

- The number of sign changes in the sequence $P_0(a), P_1(a), \ldots, P_n(a)$ is equal to the number of roots of $P_n(\lambda)$ that are smaller than a. If a member $P_i(a)$ of the sequence is zero, its sign is to be taken opposite to that of $P_{i-1}(a)$.

As we see later, Sturm sequence property makes it possible to bracket the eigenvalues of a tridiagonal matrix.

■ sturmSeq

Given **d**, **c** and λ, the function sturmSeq returns the Sturm sequence

$$P_0(\lambda), P_1(\lambda), \ldots, P_n(\lambda)$$

The function numLambdas returns the number of sign changes in the sequence (as noted before, this equals the number of eigenvalues that are smaller than λ).

```
## module sturmSeq
''' p = sturmSeq(c,d,lam).
    Returns the Sturm sequence {p[0],p[1],...,p[n]}
    associated with the characteristic polynomial
    |[A] - lam[I]| = 0, where [A] = [c\d\c] is a n x n
    tridiagonal matrix.

    numLam = numLambdas(p).
    Returns the number of eigenvalues of a tridiagonal
    matrix [A] = [c\d\c] that are smaller than 'lam'.
    Uses the Sturm sequence {p} obtained from 'sturmSeq'.
'''
from numarray import ones, Float64

def sturmSeq(d,c,lam):
    n = len(d) + 1
```

9.5 Eigenvalues of Symmetric Tridiagonal Matrices

```
    p = ones((n),type=Float64)
    p[1] = d[0] - lam
    for i in range(2,n):
        p[i] = (d[i-1] - lam)*p[i-1] - (c[i-2]**2)*p[i-2]
    return p

def numLambdas(p):
    n = len(p)
    signOld = 1
    numLam = 0
    for i in range(1,n):
        if p[i] > 0.0: sign = 1
        elif p[i] < 0.0: sign = -1
        else: sign = -signOld
        if sign*signOld < 0: numLam = numLam + 1
        signOld = sign
    return numLam
```

EXAMPLE 9.9
Use the Sturm sequence property to show that the smallest eigenvalue of **A** is in the interval (0.25, 0.5), where

$$\mathbf{A} = \begin{bmatrix} 2 & -1 & 0 & 0 \\ -1 & 2 & -1 & 0 \\ 0 & -1 & 2 & -1 \\ 0 & 0 & -1 & 2 \end{bmatrix}$$

Solution Taking $\lambda = 0.5$, we have $d_i - \lambda = 1.5$ and $c_{i-1}^2 = 1$ and the Sturm sequence in Eqs. (9.49) becomes

$$P_0(0.5) = 1$$
$$P_1(0.5) = 1.5$$
$$P_2(0.5) = 1.5(1.5) - 1 = 1.25$$
$$P_3(0.5) = 1.5(1.25) - 1.5 = 0.375$$
$$P_4(0.5) = 1.5(0.375) - 1.25 = -0.6875$$

Since the sequence contains one sign change, there exists one eigenvalue smaller than 0.5.

Repeating the process with $\lambda = 0.25$, we get $d_i - \lambda = 1.75$ and $c_{i-1}^2 = 1$, which results in the Sturm sequence

$$P_0(0.25) = 1$$
$$P_1(0.25) = 1.75$$
$$P_2(0.25) = 1.75(1.75) - 1 = 2.0625$$
$$P_3(0.25) = 1.75(2.0625) - 1.75 = 1.8594$$
$$P_4(0.25) = 1.75(1.8594) - 2.0625 = 1.1915$$

There are no sign changes in the sequence, so that all the eigenvalues are greater than 0.25. We thus conclude that $0.25 < \lambda_1 < 0.5$.

Gerschgorin's Theorem

Gerschgorin's theorem is useful in determining the *global bounds* on the eigenvalues of an $n \times n$ matrix **A**. The term "global" means the bounds that enclose all the eigenvalues. We give here a simplified version for a symmetric matrix.

- If λ is an eigenvalue of **A**, then

$$a_i - r_i \leq \lambda \leq a_i + r_i, \quad i = 1, 2, \ldots, n$$

where

$$a_i = A_{ii} \qquad r_i = \sum_{\substack{j=1 \\ j \neq i}}^{n} |A_{ij}| \tag{9.50}$$

It follows that the limits on the smallest and the largest eigenvalues are given by

$$\lambda_{\min} \geq \min_i(a_i - r_i) \qquad \lambda_{\max} \leq \max_i(a_i + r_i) \tag{9.51}$$

■ gerschgorin

The function gerschgorin returns the lower and upper global bounds on the eigenvalues of a symmetric tridiagonal matrix $\mathbf{A} = [\mathbf{c}\backslash\mathbf{d}\backslash\mathbf{c}]$.

```
## module gerschgorin
''' lamMin,lamMax = gerschgorin(d,c).
    Applies Gerschgorin's theorem to find the global bounds on
    the eigenvalues of a tridiagomal matrix [A] = [c\d\c].
'''
def gerschgorin(d,c):
```

9.5 Eigenvalues of Symmetric Tridiagonal Matrices

```
n = len(d)
lamMin = d[0] - abs(c[0])
lamMax = d[0] + abs(c[0])
for i in range(1,n-1):
    lam = d[i] - abs(c[i]) - abs(c[i-1])
    if lam < lamMin: lamMin = lam
    lam = d[i] + abs(c[i]) + abs(c[i-1])
    if lam > lamMax: lamMax = lam
lam = d[n-1] - abs(c[n-2])
if lam < lamMin: lamMin = lam
lam = d[n-1] + abs(c[n-2])
if lam > lamMax: lamMax = lam
return lamMin,lamMax
```

EXAMPLE 9.10
Use Gerschgorin's theorem to determine the bounds on the eigenvalues of the matrix

$$\mathbf{A} = \begin{bmatrix} 4 & -2 & 0 \\ -2 & 4 & -2 \\ 0 & -2 & 5 \end{bmatrix}$$

Solution Referring to Eqs. (9.50), we get

$$a_1 = 4 \quad a_2 = 4 \quad a_3 = 5$$
$$r_1 = 2 \quad r_2 = 4 \quad r_3 = 2$$

Hence

$$\lambda_{\min} \geq \min(a_i - r_i) = 4 - 4 = 0$$
$$\lambda_{\max} \leq \max(a_i + r_i) = 4 + 4 = 8$$

Bracketing Eigenvalues

The Sturm sequence property together with Gerschgorin's theorem provides us convenient tools for bracketing each eigenvalue of a symmetric tridiagonal matrix.

■ lamRange

The function lamRange brackets the N smallest eigenvalues of a symmetric tridiagonal matrix $\mathbf{A} = [\mathbf{c}\backslash\mathbf{d}\backslash\mathbf{c}]$. It returns the sequence $r_0, r_1, \ldots, r_N$, where each interval (r_{i-1}, r_i) contains exactly one eigenvalue. The algorithm first finds the bounds on all the eigenvalues by Gerschgorin's theorem. Then the method of bisection in conjunction with the Sturm sequence property is used to determine $r_N, r_{N-1}, \ldots, r_0$ in that order.

```
## module lamRange
''' r = lamRange(d,c,N).
    Returns the sequence {r[0],r[1],...,r[N]} that
    separates the N lowest eigenvalues of the tridiagonal
    matrix [A] = [c\d\c]; that is, r[i] < lam[i] < r[i+1].
'''
from numarray import ones,Float64
from sturmSeq import *
from gerschgorin import *

def lamRange(d,c,N):
    lamMin,lamMax = gerschgorin(d,c)
    r = ones((N+1),type=Float64)
    r[0] = lamMin
  # Search for eigenvalues in descending order
    for k in range(N,0,-1):
      # First bisection of interval(lamMin,lamMax)
        lam = (lamMax + lamMin)/2.0
        h = (lamMax - lamMin)/2.0
        for i in range(1000):
          # Find number of eigenvalues less than lam
            p = sturmSeq(d,c,lam)
            numLam = numLambdas(p)
          # Bisect again & find the half containing lam
            h = h/2.0
            if numLam < k: lam = lam + h
            elif numLam > k: lam = lam - h
            else: break
      # If eigenvalue located, change the upper limit
      # of search and record it in [r]
        lamMax = lam
        r[k] = lam
    return r
```

EXAMPLE 9.11
Bracket each eigenvalue of the matrix **A** in Example 9.10.

9.5 Eigenvalues of Symmetric Tridiagonal Matrices

Solution In Example 9.10 we found that all the eigenvalues lie in (0, 8). We now bisect this interval and use the Sturm sequence to determine the number of eigenvalues in (0, 4). With $\lambda = 4$, the sequence is—see Eqs. (9.49)

$$P_0(4) = 1$$
$$P_1(4) = 4 - 4 = 0$$
$$P_2(4) = (4-4)(0) - 2^2(1) = -4$$
$$P_3(4) = (5-4)(-4) - 2^2(0) = -4$$

Since a zero value is assigned the sign opposite to that of the preceding member, the signs in this sequence are $(+, -, -, -)$. The one sign change shows the presence of one eigenvalue in (0, 4).

Next we bisect the interval (4, 8) and compute the Sturm sequence with $\lambda = 6$:

$$P_0(6) = 1$$
$$P_1(6) = 4 - 6 = -2$$
$$P_2(6) = (4-6)(-2) - 2^2(1) = 0$$
$$P_3(6) = (5-6)(0) - 2^2(-2) = 8$$

In this sequence the signs are $(+, -, +, +)$, indicating two eigenvalues in (0, 6).
Therefore

$$0 \leq \lambda_1 \leq 4 \qquad 4 \leq \lambda_2 \leq 6 \qquad 6 \leq \lambda_3 \leq 8$$

Computation of Eigenvalues

Once the desired eigenvalues are bracketed, they can be found by determining the roots of $P_n(\lambda) = 0$ with bisection or Brent's method.

■ eigenvals3

The function `eigenvals3` computes the N smallest eigenvalues of a symmetric tridiagonal matrix with the method of Brent.

```
## module eigenvals3
''' lam = eigenvals3(d,c,N).
    Returns the N smallest eigenvalues of a
    tridiagonal matrix [A] = [c\d\c].
'''
from lamRange import *
```

Symmetric Matrix Eigenvalue Problems

```
from brent import *
from sturmSeq import sturmSeq
from numarray import zeros,Float64

def eigenvals3(d,c,N):

    def f(x):               # f(x) = |[A] - x[I]|
        p = sturmSeq(d,c,x)
        return p[len(p)-1]

    lam = zeros((N),type=Float64)
    r = lamRange(d,c,N)     # Bracket eigenvalues
    for i in range(N):      # Solve by Brent's method
        lam[i] = brent(f,r[i],r[i+1])
    return lam
```

EXAMPLE 9.12

Use `eigenvals3` to determine the three smallest eigenvalues of the 100×100 matrix

$$\mathbf{A} = \begin{bmatrix} 2 & -1 & 0 & \cdots & 0 \\ -1 & 2 & -1 & \cdots & 0 \\ 0 & -1 & 2 & \cdots & 0 \\ \vdots & \vdots & \ddots & \ddots & \vdots \\ 0 & 0 & \cdots & -1 & 2 \end{bmatrix}$$

Solution

```
#!/usr/bin/python
## example9_12
from numarray import ones,Float64
from eigenvals3 import *

N = 3
n = 100
d = ones((n))*2.0
c = ones((n-1))*(-1.0)
lambdas = eigenvals3(d,c,N)
print lambdas
raw_input(''\nPress return to exit'')
```

Here are the eigenvalues:

```
[ 0.00096744  0.00386881  0.0087013 ]
```

9.5 Eigenvalues of Symmetric Tridiagonal Matrices

Computation of Eigenvectors

If the eigenvalues are known (approximate values will be good enough), the best means of computing the corresponding eigenvectors is the inverse power method with eigenvalue shifting. This method was discussed before, but the algorithm listed did not take advantage of banding. Here we present a version of the method written for symmetric tridiagonal matrices.

■ **inversePower3**

This function is very similar to inversePower listed in Art. 9.3, but it executes much faster since it exploits the tridiagonal structure of the matrix.

```
## module inversePower3
''' lam,x = inversePower3(d,c,s,tol=1.0e-6).
    Inverse power method applied to a tridiagonal matrix
    [A] = [c\d\c]. Returns the eigenvalue closest to 's'
    and the corresponding eigenvector.
'''
from numarray import dot,zeros,Float64
from LUdecomp3 import *
from math import sqrt
from random import random

def inversePower3(d,c,s,tol=1.0e-6):
    n = len(d)
    e = c.copy()
    cc = c.copy()                  # Save original [c]
    dStar = d - s                  # Form [A*] = [A] - s[I]
    LUdecomp3(cc,dStar,e)          # Decompose [A*]
    x = zeros((n),type=Float64)
    for i in range(n):             # Seed [x] with random numbers
        x[i] = random()
    xMag = sqrt(dot(x,x))          # Normalize [x]
    x =x/xMag
    flag = 0
    for i in range(30):            # Begin iterations
        xOld = x.copy()            # Save current [x]
        LUsolve3(cc,dStar,e,x)     # Solve [A*][x] = [xOld]
        xMag = sqrt(dot(x,x))      # Normalize [x]
        x = x/xMag
```

Symmetric Matrix Eigenvalue Problems

```
            if dot(xOld,x) < 0.0:    # Detect change in sign of [x]
                sign = -1.0
                x = -x
            else: sign = 1.0
            if sqrt(dot(xOld - x,xOld - x)) < tol:
                return s + sign/xMag,x
    print 'Inverse power method did not converge'
```

EXAMPLE 9.13
Compute the 10th smallest eigenvalue of the matrix **A** given in Example 9.12.

Solution The following program extracts the Nth eigenvalue of **A** by the inverse power method with eigenvalue shifting:

```
#!/usr/bin/python
## example9_13
from numarray import ones
from lamRange import *
from inversePower3 import *

N = 10
n = 100
d = ones((n))*2.0
c = ones((n-1))*(-1.0)
r = lamRange(d,c,N)              # Bracket N smallest eigenvalues
s = (r[N-1] + r[N])/2.0          # Shift to midpoint of Nth bracket
lam,x = inversePower3(d,c,s)     # Inverse power method
print ''Eigenvalue No.'',N,'' ='',lam
raw_input(''\nPress return to exit'')
```

The result is

```
Eigenvalue No. 10  = 0.0959737849345
```

EXAMPLE 9.14
Compute the three smallest eigenvalues and the corresponding eigenvectors of the matrix **A** in Example 9.5.

Solution
```
#!/usr/bin/python
## example9_14
```

9.5 Eigenvalues of Symmetric Tridiagonal Matrices

```
from householder3 import *
from eigenvals3 import *
from inversePower3 import *
from numarray import array,zeros,Float64,matrixmultiply

N = 3    # Number of eigenvalues requested
a = array([[ 11.0, 2.0,  3.0,  1.0,  4.0], \
           [  2.0, 9.0,  3.0,  5.0,  2.0], \
           [  3.0, 3.0, 15.0,  4.0,  3.0], \
           [  1.0, 5.0,  4.0, 12.0,  4.0], \
           [  4.0, 2.0,  3.0,  4.0, 17.0]])
xx = zeros((len(a),N),type=Float64)
d,c = householder(a)                 # Tridiagonalize [A]
p = computeP(a)                      # Compute transformation matrix
lambdas = eigenvals3(d,c,N)          # Compute eigenvalues
for i in range(N):
    s = lambdas[i]*1.0000001         # Shift very close to eigenvalue
    lam,x = inversePower3(d,c,s)     # Compute eigenvector [x]
    xx[:,i] = x                      # Place [x] in array [xx]
xx = matrixmultiply(p,xx)            # Recover eigenvectors of [A]
print ''Eigenvalues:\n'',lambdas
print ''\nEigenvectors:\n'',xx
raw_input(''Press return to exit'')

Eigenvalues:
[  4.87394638   8.66356791  10.93677451]

Eigenvectors:
[[ 0.26726603  0.72910002  0.50579164]
 [-0.74142854  0.41391448 -0.31882387]
 [-0.05017271 -0.4298639   0.52077788]
 [ 0.59491453  0.06955611 -0.60290543]
 [-0.14970633 -0.32782151 -0.08843985]]
```

PROBLEM SET 9.2

1. Use Gerschgorin's theorem to determine bounds on the eigenvalues of

(a) $\mathbf{A} = \begin{bmatrix} 10 & 4 & -1 \\ 4 & 2 & 3 \\ -1 & 3 & 6 \end{bmatrix}$ (b) $\mathbf{B} = \begin{bmatrix} 4 & 2 & -2 \\ 2 & 5 & 3 \\ -2 & 3 & 4 \end{bmatrix}$

2. Use the Sturm sequence to show that

$$A = \begin{bmatrix} 5 & -2 & 0 & 0 \\ -2 & 4 & -1 & 0 \\ 0 & -1 & 4 & -2 \\ 0 & 0 & -2 & 5 \end{bmatrix}$$

has one eigenvalue in the interval (2, 4).

3. Bracket each eigenvalue of

$$A = \begin{bmatrix} 4 & -1 & 0 \\ -1 & 4 & -1 \\ 0 & -1 & 4 \end{bmatrix}$$

4. Bracket each eigenvalue of

$$A = \begin{bmatrix} 6 & 1 & 0 \\ 1 & 8 & 2 \\ 0 & 2 & 9 \end{bmatrix}$$

5. Bracket every eigenvalue of

$$A = \begin{bmatrix} 2 & -1 & 0 & 0 \\ -1 & 2 & -1 & 0 \\ 0 & -1 & 2 & -1 \\ 0 & 0 & -1 & 1 \end{bmatrix}$$

6. Tridiagonalize the matrix

$$A = \begin{bmatrix} 12 & 4 & 3 \\ 4 & 9 & 3 \\ 3 & 3 & 15 \end{bmatrix}$$

with Householder's reduction.

7. Use Householder's reduction to transform the matrix

$$A = \begin{bmatrix} 4 & -2 & 1 & -1 \\ -2 & 4 & -2 & 1 \\ 1 & -2 & 4 & -2 \\ -1 & 1 & -2 & 4 \end{bmatrix}$$

to tridiagonal form.

9.5 Eigenvalues of Symmetric Tridiagonal Matrices

8. ■ Compute all the eigenvalues of

$$A = \begin{bmatrix} 6 & 2 & 0 & 0 & 0 \\ 2 & 5 & 2 & 0 & 0 \\ 0 & 2 & 7 & 4 & 0 \\ 0 & 0 & 4 & 6 & 1 \\ 0 & 0 & 0 & 1 & 3 \end{bmatrix}$$

9. ■ Find the smallest two eigenvalues of

$$A = \begin{bmatrix} 4 & -1 & 0 & 1 \\ -1 & 6 & -2 & 0 \\ 0 & -2 & 3 & 2 \\ 1 & 0 & 2 & 4 \end{bmatrix}$$

10. ■ Compute the three smallest eigenvalues of

$$A = \begin{bmatrix} 7 & -4 & 3 & -2 & 1 & 0 \\ -4 & 8 & -4 & 3 & -2 & 1 \\ 3 & -4 & 9 & -4 & 3 & -2 \\ -2 & 3 & -4 & 10 & -4 & 3 \\ 1 & -2 & 3 & -4 & 11 & -4 \\ 0 & 1 & -2 & 3 & -4 & 12 \end{bmatrix}$$

and the corresponding eigenvectors.

11. ■ Find the two smallest eigenvalues of the 6 × 6 Hilbert matrix

$$A = \begin{bmatrix} 1 & 1/2 & 1/3 & \cdots & 1/6 \\ 1/2 & 1/3 & 1/4 & \cdots & 1/7 \\ 1/3 & 1/4 & 1/5 & \cdots & 1/8 \\ \vdots & \vdots & \vdots & \ddots & \vdots \\ 1/6 & 1/7 & 1/8 & \cdots & 1/11 \end{bmatrix}$$

Recall that this matrix is ill-conditioned.

12. ■ Rewrite the function lamRange(d,c,N) so that it will bracket the *N largest* eigenvalues of a tridiagonal matrix. Use this function to bracket the two largest eigenvalues of the Hilbert matrix in Example 9.11.

13. ■

The differential equations of motion of the mass–spring system are

$$k(-2u_1 + u_2) = m\ddot{u}_1$$

$$k(u_1 - 2u_2 + u_3) = 3m\ddot{u}_2$$

$$k(u_2 - 2u_3) = 2m\ddot{u}_3$$

where $u_i(t)$ is the displacement of mass i from its equilibrium position and k is the spring stiffness. Substituting $u_i(t) = y_i \sin \omega t$, we obtain the matrix eigenvalue problem

$$\begin{bmatrix} 2 & -1 & 0 \\ -1 & 2 & -1 \\ 0 & -1 & 2 \end{bmatrix} \begin{bmatrix} y_1 \\ y_2 \\ y_3 \end{bmatrix} = \frac{m\omega^2}{k} \begin{bmatrix} 1 & 0 & 0 \\ 0 & 3 & 0 \\ 0 & 0 & 2 \end{bmatrix} \begin{bmatrix} y_1 \\ y_2 \\ y_3 \end{bmatrix}$$

Determine the circular frequencies ω and the corresponding relative amplitudes y_i of vibration.

14. ∎

The figure shows n identical masses connected by springs of different stiffnesses. The equation governing free vibration of the system is $\mathbf{Au} = m\omega^2 \mathbf{u}$, where ω is the circular frequency and

$$\mathbf{A} = \begin{bmatrix} k_1 + k_2 & -k_2 & 0 & 0 & \cdots & 0 \\ -k_2 & k_2 + k_3 & -k_3 & 0 & \cdots & 0 \\ 0 & -k_3 & k_3 + k_4 & -k_4 & \cdots & 0 \\ \vdots & \vdots & \ddots & \ddots & \ddots & \vdots \\ 0 & \cdots & 0 & -k_{n-1} & k_{n-1} + k_n & -k_n \\ 0 & \cdots & 0 & 0 & -k_n & k_n \end{bmatrix}$$

Given the spring stiffnesses $\mathbf{k} = \begin{bmatrix} k_1 & k_2 & \cdots & k_n \end{bmatrix}^T$, write a program that computes the N lowest eigenvalues $\lambda = m\omega^2$ and the corresponding eigenvectors. Run the program with $N = 4$ and

$$\mathbf{k} = \begin{bmatrix} 400 & 400 & 400 & 0.2 & 400 & 400 & 200 \end{bmatrix}^T \text{ kN/m}$$

Note that the system is weakly coupled, k_4 being small. Do the results make sense?

15. ∎

9.5 Eigenvalues of Symmetric Tridiagonal Matrices

The differential equation of motion of the axially vibrating bar is

$$u'' = \frac{\rho}{E}\ddot{u}$$

where $u(x, t)$ is the axial displacement, ρ represents the mass density and E is the modulus of elasticity. The boundary conditions are $u(0, t) = u'(L, t) = 0$. Letting $u(x, t) = y(x)\sin\omega t$, we obtain

$$y'' = -\omega^2 \frac{\rho}{E} y \qquad y(0) = y'(L) = 0$$

The corresponding finite difference equations are

$$\begin{bmatrix} 2 & -1 & 0 & 0 & \cdots & 0 \\ -1 & 2 & -1 & 0 & \cdots & 0 \\ 0 & -1 & 2 & -1 & \cdots & 0 \\ \vdots & \vdots & \ddots & \ddots & \ddots & \vdots \\ 0 & 0 & \cdots & -1 & 2 & -1 \\ 0 & 0 & \cdots & 0 & -1 & 1 \end{bmatrix} \begin{bmatrix} y_1 \\ y_2 \\ y_3 \\ \vdots \\ y_{n-1} \\ y_n \end{bmatrix} = \left(\frac{\omega L}{n}\right)^2 \frac{\rho}{E} \begin{bmatrix} y_1 \\ y_2 \\ y_3 \\ \vdots \\ y_{n-1} \\ y_n/2 \end{bmatrix}$$

(a) If the standard form of these equations is $\mathbf{Hz} = \lambda\mathbf{z}$, write down $\mathbf{H}$ and the transformation matrix $\mathbf{P}$ in $\mathbf{y} = \mathbf{Pz}$. (b) Compute the lowest circular frequency of the bar with $n = 10, 100$ and 1000 utilizing the module inversePower3. *Note*: the analytical solution is $\omega_1 = \pi\sqrt{E/\rho}/(2L)$.

16. ■

The simply supported column is resting on an elastic foundation of stiffness k (N/m per meter length). An axial force P acts on the column. The differential equation and the boundary conditions for the lateral displacement u are

$$u^{(4)} + \frac{P}{EI}u'' + \frac{k}{EI}u = 0$$

$$u(0) = u''(0) = u(L) = u''(L) = 0$$

Using the mesh shown, the finite difference approximation of these equations is

$$(5 + \alpha)u_1 - 4u_2 + u_3 = \lambda(2u_1 - u_2)$$

$$-4u_1 + (6 + \alpha)u_2 - 4u_3 + u_4 = \lambda(-u_1 + 2u_2 + u_3)$$

$$u_1 - 4u_2 + (6+\alpha)u_3 - 4u_4 + u_5 = \lambda(-u_2 + 2u_3 - u_4)$$

$$\vdots$$

$$u_{n-3} - 4u_{n-2} + (6+\alpha)u_{n-1} - 4u_n = \lambda(-u_{n-2} + 2u_{n-1} - u_n)$$

$$u_{n-2} - 4u_{n-1} + (5+\alpha)u_n = \lambda(-u_{n-1} + 2u_n)$$

where

$$\alpha = \frac{kh^4}{EI} = \frac{1}{(n+1)^4}\frac{kL^4}{EI} \qquad \lambda = \frac{Ph^2}{EI} = \frac{1}{(n+1)^2}\frac{PL^2}{EI}$$

Write a program that computes the lowest three buckling loads P and the corresponding mode shapes. Run the program with $kL^4/(EI) = 1000$ and $n = 25$.

17. ■ Find smallest five eigenvalues of the 20×20 matrix

$$\mathbf{A} = \begin{bmatrix} 2 & 1 & 0 & 0 & \cdots & 0 & 1 \\ 1 & 2 & 1 & 0 & \cdots & 0 & 0 \\ 0 & 1 & 2 & 1 & \cdots & 0 & 0 \\ \vdots & \vdots & \ddots & \ddots & \ddots & \vdots & \vdots \\ 0 & 0 & \cdots & 1 & 2 & 1 & 0 \\ 0 & 0 & \cdots & 0 & 1 & 2 & 1 \\ 1 & 0 & \cdots & 0 & 0 & 1 & 2 \end{bmatrix}$$

Note: this is a difficult matrix that has many pairs of double eigenvalues.

9.6 Other Methods

On occasions when all the eigenvalues and eigenvectors of a matrix are required, the *QR algorithm* is a worthy contender. It is based on the decomposition $\mathbf{A} = \mathbf{QR}$ where $\mathbf{Q}$ and $\mathbf{R}$ are orthogonal and upper triangular matrices, respectively. The decomposition is carried out in conjuction with Householder transformation. There is also a *QL algorithm* : $\mathbf{A} = \mathbf{QL}$ that works in the same manner, but here $\mathbf{L}$ is a lower triangular matrix.

Schur's factorization is another solid technique for determining the eigenvalues of $\mathbf{A}$. Here the decomposition is $\mathbf{A} = \mathbf{Q}^T\mathbf{UQ}$, where $\mathbf{Q}$ is orthogonal and $\mathbf{U}$ is an upper triangular matrix. The diagonal terms of $\mathbf{U}$ are the eigenvalues of $\mathbf{A}$.

The *LR algorithm* is probably the fastest means of computing the eigenvalues; it is also very simple to implement—see Prob. 22 of Problem Set 9.1. But its stability is inferior to the other methods.

10 Introduction to Optimization

> Find **x** that minimizes $F(\mathbf{x})$ subject to $g(\mathbf{x}) = 0$, $h(\mathbf{x}) \geq 0$

10.1 Introduction

Optimization is the term often used for minimizing or maximizing a function. It is sufficient to consider the problem of minimization only; maximization of $F(\mathbf{x})$ is achieved by simply minimizing $-F(\mathbf{x})$. In engineering, optimization is closely related to design. The function $F(\mathbf{x})$, called the *merit function* or *objective function*, is the quantity that we wish to keep as small as possible, such as cost or weight. The components of **x**, known as the *design variables*, are the quantities that we are free to adjust. Physical dimensions (lengths, areas, angles, etc.) are common examples of design variables.

Optimization is a large topic with many books dedicated to it. The best we can do in limited space is to introduce a few basic methods that are good enough for problems that are reasonably well behaved and don't involve too many design variables. By omitting the more sophisticated methods, we may actually not miss all that much. All optimization algorithms are unreliable to a degree—any one of them may work on one problem and fail on another. As a rule of thumb, by going up in sophistication we gain computational efficiency, but not necessarily reliability.

The algorithms for minimization are iterative procedures that require starting values of the design variables **x**. If $F(\mathbf{x})$ has several local minima, the initial choice of **x** determines which of these will be computed. There is no guaranteed way of finding the global optimal point. One suggested procedure is to make several computer runs using different starting points and pick the best result.

More often than not, the design variables are also subjected to restrictions, or *constraints*, which may have the form of equalities or inequalities. As an example, take the minimum weight design of a roof truss that has to carry a certain loading.

Assume that the layout of the members is given, so that the design variables are the cross-sectional areas of the members. Here the design is dominated by inequality constraints that consist of prescribed upper limits on the stresses and possibly the displacements.

The majority of available methods are designed for *unconstrained optimization*, where no restrictions are placed on the design variables. In these problems the minima, if they exist, are stationary points (points where gradient vector of $F(\mathbf{x})$ vanishes). In the more difficult problem of *constrained optimization* the minima are usually located where the $F(\mathbf{x})$ surface meets the constraints. There are special algorithms for constrained optimization, but they are not easily accessible due to their complexity and specialization. One way to tackle a problem with constraints is to use an unconstrained optimization algorithm, but modify the merit function so that any violation of constraints is heavily penalized.

Consider the problem of minimizing $F(\mathbf{x})$ where the design variables are subject to the constraints

$$g_i(\mathbf{x}) = 0, \quad i = 1, 2, \ldots, M \tag{10.1a}$$

$$h_j(\mathbf{x}) \leq 0, \quad j = 1, 2, \ldots, N \tag{10.1b}$$

We choose the new merit function be

$$F^*(\mathbf{x}) - F(\mathbf{x}) + \lambda P(\mathbf{x}) \tag{10.2a}$$

where

$$P(\mathbf{x}) = \sum_{i=1}^{M} [g_i(x)]^2 + \sum_{j=1}^{N} \{\max[0, h_j(\mathbf{x})]\}^2 \tag{10.2b}$$

is the *penalty function* and λ is a multiplier. The function $\max(a, b)$ returns the larger of a and b. It is evident that $P(\mathbf{x}) = 0$ if no constraints are violated. Violation of a constraint imposes a penalty proportional to the square of the violation. Hence the minimization algorithm tends to avoid the violations, the degree of avoidance being dependent on the magnitude of λ. If λ is small, optimization will proceed faster because there is more "space" in which the procedure can operate, but there may be significant violation of constraints. On the other hand, a large λ can result in a poorly conditioned procedure, but the constraints will be tightly enforced. It is advisable to run the optimization program with λ that is on the small side. If the results show unacceptable constraint violation, increase λ and run the program again, starting with the results of the previous run.

An optimization procedure may also become ill-conditioned when the constraints have widely different magnitudes. This problem can be alleviated by *scaling*

10.2 Minimization Along a Line

the offending constraints; that is, multiplying the constraint equations by suitable constants.

10.2 Minimization Along a Line

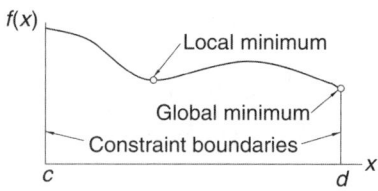

Figure 10.1. Example of local and global minima.

Consider the problem of minimizing a function $f(x)$ of a single variable x with the constraints $c \leq x \leq d$. A hypothetical plot of the function is shown in Fig. 10.1. There are two minimum points: a stationary point characterized by $f'(x) = 0$ that represents a local minimum, and a global minimum at the constraint boundary. It appears that finding the global minimum is simple. All the stationary points could be located by finding the roots of $df/dx = 0$, and each constraint boundary may be checked for a global minimum by evaluating $f(c)$ and $f(d)$. Then why do we need an optimization algorithm? We need it if $f(\mathbf{x})$ is difficult or impossible to differentiate; for example, if f represents a complex computer algorithm.

Bracketing

Before a minimization algorithm can be entered, the minimum point must be bracketed. The procedure of bracketing is simple: start with an initial value of x_0 and move *downhill* computing the function at $x_1, x_2, x_3, \ldots$ until we reach the point x_n where $f(x)$ increases for the first time. The minimum point is now bracketed in the interval (x_{n-2}, x_n). What should the step size $h_i = x_{i+1} - x_i$ be? It is not a good idea have a constant h_i since it often results in too many steps. A more efficient scheme is to increase the size with every step, the goal being to reach the minimum quickly, even if the resulting bracket is wide. We chose to increase the step size by a constant factor; that is, we use $h_{i+1} = ch_i$, $c > 1$.

Golden Section Search

The golden section search is the counterpart of bisection used in finding roots of equations. Suppose that the minimum of $f(x)$ has been bracketed in the interval (a, b) of length h. To telescope the interval, we evaluate the function at $x_1 = b - Rh$

and $x_2 = a + Rh$, as shown in Fig. 10.2(a). The constant R will be determined shortly. If $f_1 > f_2$ as indicated in the figure, the minimum lies in (x_1, b); otherwise it is located in (a, x_2).

Figure 10.2. Golden section telescoping.

Assuming that $f_1 > f_2$, we set $a \leftarrow x_1$ and $x_1 \leftarrow x_2$, which yields a new interval (a, b) of length $h' = Rh$, as illustrated in Fig. 10.2(b). To carry out the next telescoping operation we evaluate the function at $x_2 = a + Rh'$ and repeat the process.

The procedure works only if Figs. 10.1(a) and (b) are similar; i.e., if the same constant R locates x_1 and x_2 in both figures. Referring to Fig. 10.2(a), we note that $x_2 - x_1 = 2Rh - h$. The same distance in Fig. 10.2(b) is $x_1 - a = h' - Rh'$. Equating the two, we get

$$2Rh - h = h' - Rh'$$

Substituting $h' = Rh$ and cancelling h yields

$$2R - 1 = R(1 - R)$$

the solution of which is the *golden ratio*[24]:

$$R = \frac{-1 + \sqrt{5}}{2} = 0.618\,033\,989\ldots \tag{10.3}$$

Note that each telescoping decreases the interval containing the minimum by the factor R, which is not as good as the factor is 0.5 in bisection. However, the golden search method achieves this reduction with *one function evaluation*, whereas two evaluations would be needed in bisection.

---

[24] R is the ratio of the sides of a "golden rectangle," considered by ancient Greeks to have the perfect proportions.

10.2 Minimization Along a Line

The number of telescoping required to reduce h from $|b - a|$ to an error tolerance ε is given by

$$|b - a|\, R^n = \varepsilon$$

which yields

$$n = \frac{\ln(\varepsilon / |b - a|)}{\ln R} = -2.078\,087 \ln \frac{\varepsilon}{|b - a|} \tag{10.4}$$

■ goldSearch

This module contains the bracketing and the golden section search algorithms. For the factor that multiplies successive search intervals in bracket we chose $c = 1 + R$.

```
## module goldSearch
''' a,b = bracket(f,xStart,h)
    Finds the brackets (a,b) of a minimum point of the
    user-supplied scalar function f(x).
    The search starts downhill from xStart with a step
    length h.

    x,fMin = search(f,a,b,tol=1.0e-6)
    Golden section method for determining x that minimizes
    the user-supplied scalar function f(x).
    The minimum must be bracketed in (a,b).
'''
from math import log

def bracket(f,x1,h):
    c = 1.618033989
    f1 = f(x1)
    x2 = x1 + h; f2 = f(x2)
  # Determine downhill direction and change sign of h if needed
    if f2 > f1:
        h = -h
        x2 = x1 + h; f2 = f(x2)
      # Check if minimum between x1 - h and x1 + h
        if f2 > f1: return x2,x1 - h
  # Search loop
    for i in range (100):
        h = c*h
```

```
            x3 = x2 + h; f3 = f(x3)
            if f3 > f2: return x1,x3
            x1 = x2; x2 = x3
            f1 = f2; f2 = f3
        print ''Bracket did not find a mimimum''

def search(f,a,b,tol=1.0e-9):
    nIter = -2.078087*log(tol/abs(b-a)) # Eq. (10.4)
    R = 0.618033989
    C = 1.0 - R
  # First telescoping
    x1 = R*a + C*b; x2 = C*a + R*b
    f1 = f(x1); f2 = f(x2)
  # Main loop
    for i in range(nIter):
        if f1 > f2:
            a = x1
            x1 = x2; f1 = f2
            x2 = C*a + R*b; f2 = f(x2)
        else:
            b = x2
            x2 = x1; f2 = f1
            x1 = R*a + C*b; f1 = f(x1)
    if f1 < f2: return x1,f1
    else: return x2,f2
```

EXAMPLE 10.1
Use goldSearch to find x that minimizes

$$f(x) = 1.6x^3 + 3x^2 - 2x$$

subject to the constraint $x \geq 0$. Compare the result with the analytical solution.

Solution This is a constrained minimization problem. The minimum of $f(x)$ is either a stationary point in $x \geq 0$, or located at the constraint boundary $x = 0$. We handle the constraint with the penalty function method by minimizing $f(x) + \lambda \, [\min(0, x)]^2$.

Starting at $x = 1$ and choosing $h = 0.01$ for the first step size in bracket (both choices being rather arbitrary), we arrive at the following program:

```
#!/usr/bin/python
## example10_1
```

10.2 Minimization Along a Line

```
from goldSearch import *

def f(x):
    lam = 1.0        # Constraint multiplier
    c = min(0.0, x)  # Constraint function
    return 1.6*x**3 + 3.0*x**2 - 2.0*x + lam*c**2

xStart = 1.0
h = 0.01
x1,x2 = bracket(f,xStart,h)
x,fMin = search(f,x1,x2)
print ''x ='',x
print ''f(x) ='',fMin
raw_input (''\nPress return to exit'')
```

The result is

```
x = 0.27349402621
f(x) = -0.28985978555
```

Since the minimum was found to be a stationary point, the constraint was not active. Therefore, the penalty function was superfluous, but we did not know that at the beginning.

The locations of stationary points are obtained analytically by solving

$$f'(x) = 4.8x^2 + 6x - 2 = 0$$

The positive root of this equation is $x = 0.273\,494$. As this is the only positive root, there are no other stationary points in $x \geq 0$ that we must check out. The only other possible location of a minimum is the constraint boundary $x = 0$. But here $f(0) = 0$ is larger than the function at the stationary point, leading to the conclusion that the global minimum occurs at $x = 0.273\,494$.

EXAMPLE 10.2

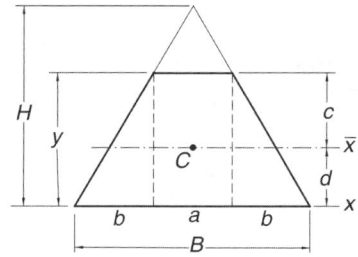

Introduction to Optimization

The trapezoid shown is the cross section of a beam. It is formed by removing the top from a triangle of base $B = 48$ mm and height $H = 60$ mm. The problem is the find the height y of the trapezoid that maximizes the section modulus

$$S = I_{\bar{x}}/c$$

where $I_{\bar{x}}$ is the second moment of the cross-sectional area about the axis that passes through the centroid C of the cross section. By optimizing the section modulus, we minimize the maximum bending stress $\sigma_{\max} = M/S$ in the beam, M being the bending moment.

Solution Considering the area of the trapezoid as a composite of a rectangle and two triangles, we find the section modulus through the following sequence of computations:

Base of rectangle	$a = B(H-y)/H$
Base of triangle	$b = (B-a)/2$
Area	$A = (B+a)y/2$
First moment of area about x-axis	$Q_x = (ay)y/2 + 2(by/2)y/3$
Location of centroid	$d = Q_x/A$
Distance involved in S	$c = y - d$
Second moment of area about x-axis	$I_x = ay^3/3 + 2(by^3/12)$
Parallel axis theorem	$I_{\bar{x}} = I_x - Ad^2$
Section modulus	$S = I_{\bar{x}}/c$

We could use the formulas in the table to derive S as an explicit function of y, but that would involve a lot of error-prone algebra and result in an overly complicated expression. It makes more sense to let the computer do the work.

The program we used and its output are listed below. As we wish to maximize S with a minimization algorithm, the merit function is $-S$. There are no constraints in this problem.

```
#!/usr/bin/python
## example10_2
from goldSearch import *

def f(y):
    B = 48.0
    H = 60.0
    a = B*(H - y)/H
    b = (B - a)/2.0
```

```
        A = (B + a)*y/2.0
        Q = (a*y**2)/2.0 + (b*y**2)/3.0
        d = Q/A
        c = y - d
        I = (a*y**3)/3.0 + (b*y**3)/6.0
        Ibar = I - A*d**2
        return -Ibar/c

yStart = 60.0   # Starting value of y
h = 1.0         # Size of first step used in bracketing
a,b = bracket(f,yStart,h)
yOpt,fOpt = search(f,a,b)
print ''Optimal y ='',yOpt
print ''Optimal S ='',-fOpt
print ''S of triangle ='',-f(60.0)
raw_input(''Press return to exit'')

Optimal y = 52.1762738732
Optimal S = 7864.43094136
S of triangle = 7200.0
```

The printout includes the section modulus of the original triangle. The optimal section shows a 9.2% improvement over the triangle.

10.3 Conjugate Gradient Methods

Introduction

We now look at optimization in n-dimensional design space. The objective is to minimize $F(\mathbf{x})$, where the components of $\mathbf{x}$ are the n independent design variables. One way to tackle the problem is to use a succession of one-dimensional minimizations to close in on the optimal point. The basic strategy is

- Choose a point $\mathbf{x}_0$ in the design space.
- loop with $i = 1, 2, 3, \ldots$

    Choose a vector $\mathbf{v}_i$.

    Minimize $F(\mathbf{x})$ along the line through $\mathbf{x}_{i-1}$ in the direction of $\mathbf{v}_i$. Let the minimum point be $\mathbf{x}_i$.

    if $|\mathbf{x}_i - \mathbf{x}_{i-1}| < \varepsilon$ exit loop

- end loop

The minimization along a line can be accomplished with any one-dimensional optimization algorithm (such as the golden section search). The only question left open is how to choose the vectors $\mathbf{v}_i$.

Conjugate Directions

Consider the quadratic function

$$F(\mathbf{x}) = c - \sum_i b_i x_i + \frac{1}{2} \sum_i \sum_j A_{ij} x_i x_j$$

$$= c - \mathbf{b}^T \mathbf{x} + \frac{1}{2} \mathbf{x}^T \mathbf{A} \mathbf{x} \tag{10.5}$$

Differentiation with respect to x_i yields

$$\frac{\partial F}{\partial x_i} = -b_i + \sum_j A_{ij} x_j$$

which can be written in vector notation as

$$\nabla F = -\mathbf{b} + \mathbf{A}\mathbf{x} \tag{10.6}$$

where ∇F is the *gradient* of F.

Now consider the change in the gradient as we move from point $\mathbf{x}_0$ in the direction of a vector $\mathbf{u}$. The motion takes place along the line

$$\mathbf{x} = \mathbf{x}_0 + s\mathbf{u}$$

where s is the distance moved. Substitution into Eq. (10.6) yields the expression for the gradient at $\mathbf{x}$:

$$\nabla F|_{\mathbf{x}_0 + s\mathbf{u}} = -\mathbf{b} + \mathbf{A}(\mathbf{x}_0 + s\mathbf{u}) = \nabla F|_{\mathbf{x}_0} + s\mathbf{A}\mathbf{u}$$

Note that the change in the gradient is $s\mathbf{A}\mathbf{u}$. If this change is perpendicular to a vector $\mathbf{v}$; that is, if

$$\mathbf{v}^T \mathbf{A} \mathbf{u} = 0 \tag{10.7}$$

the directions of $\mathbf{u}$ and $\mathbf{v}$ are said to be mutually *conjugate* (noninterfering). The implication is that once we have minimized $F(\mathbf{x})$ in the direction of $\mathbf{v}$, we can move along $\mathbf{u}$ without ruining the previous minimization.

For a quadratic function of n independent variables it is possible to construct n mutually conjugate directions. Therefore, it would take precisely n line minimizations along these directions to reach the minimum point. If $F(\mathbf{x})$ is not a quadratic function, Eq. (10.5) can be treated as a local approximation of the merit function, obtained by

10.3 Conjugate Gradient Methods

truncating the Taylor series expansion of $F(\mathbf{x})$ about $\mathbf{x}_0$ (see Appendix A1):

$$F(\mathbf{x}) \approx F(\mathbf{x}_0) + \nabla F(\mathbf{x}_0)(\mathbf{x} - \mathbf{x}_0) + \frac{1}{2}(\mathbf{x} - \mathbf{x}_0)^T \mathbf{H}(\mathbf{x}_0)(\mathbf{x} - \mathbf{x}_0)$$

Now the conjugate directions based on the quadratic form are only approximations, valid in the close vicinity of $\mathbf{x}_0$. Consequently, it would take several cycles of n line minimizations to reach the optimal point.

The various conjugate gradient methods use different techniques for constructing conjugate directions. The so-called *zero-order methods* work with $F(\mathbf{x})$ only, whereas the *first-order methods* utilize both $F(\mathbf{x})$ and ∇F. The first-order methods are computationally more efficient, of course, but the input of ∇F, if it is available at all, can be very tedious.

Powell's Method

Powell's method is a zero-order method, requiring the evaluation of $F(\mathbf{x})$ only. If the problem involves n design variables, the basic algorithm is

- Choose a point $\mathbf{x}_0$ in the design space.
- Choose the starting vectors $\mathbf{v}_i$, $i = 1, 2, \ldots, n$ (the usual choice is $\mathbf{v}_i = \mathbf{e}_i$, where $\mathbf{e}_i$ is the unit vector in the x_i-coordinate direction).
- cycle

    do with $i = 1, 2, \ldots, n$

    Minimize $F(\mathbf{x})$ along the line through $\mathbf{x}_{i-1}$ in the direction of $\mathbf{v}_i$. Let the minimum point be $\mathbf{x}_i$.

    end do

    $\mathbf{v}_{n+1} \leftarrow \mathbf{x}_0 - \mathbf{x}_n$

    Minimize $F(\mathbf{x})$ along the line through $\mathbf{x}_0$ in the direction of $\mathbf{v}_{n+1}$. Let the minimum point be $\mathbf{x}_{n+1}$.

    if $|\mathbf{x}_{n+1} - \mathbf{x}_0| < \varepsilon$ exit loop

    do with $i = 1, 2, \ldots, n$

    $\mathbf{v}_i \leftarrow \mathbf{v}_{i+1}$ ($\mathbf{v}_1$ is discarded, the other vectors are reused)

    end do

- end cycle

Powell demonstrated that the vectors $\mathbf{v}_{n+1}$ produced in successive cycles are mutually conjugate, so that the minimum point of a quadratic surface is reached in precisely n cycles. In practice, the merit function is seldom quadratic, but as long as it can be approximated locally by Eq. (10.5), Powell's method will work. Of course, it usually takes more than n cycles to arrive at the minimum of a nonquadratic function. Note that it takes n line minimizations to construct each conjugate direction.

Figure 10.3(a) illustrates one typical cycle of the method in a two dimensional design space ($n = 2$). We start with point $\mathbf{x}_0$ and vectors $\mathbf{v}_1$ and $\mathbf{v}_2$. Then we find the distance s_1 that minimizes $F(\mathbf{x}_0 + s\mathbf{v}_1)$, finishing up at point $\mathbf{x}_1 = \mathbf{x}_0 + s_1\mathbf{v}_1$. Next, we determine s_2 that minimizes $F(\mathbf{x}_1 + s\mathbf{v}_2)$ which takes us to $\mathbf{x}_2 = \mathbf{x}_1 + s_2\mathbf{v}_2$. The last search direction is $\mathbf{v}_3 = \mathbf{x}_2 - \mathbf{x}_0$. After finding s_3 by minimizing $F(\mathbf{x}_0 + s\mathbf{v}_3)$ we get to $\mathbf{x}_3 = \mathbf{x}_0 + s_3\mathbf{v}_3$, completing the cycle.

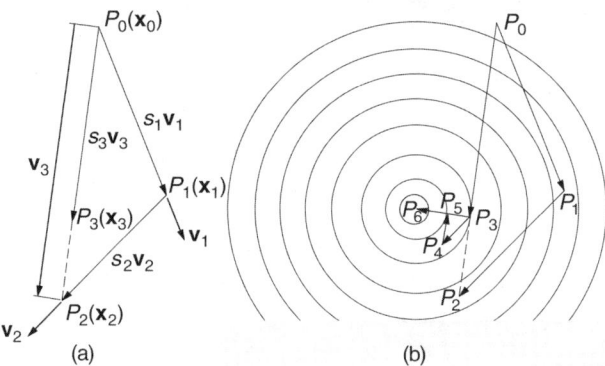

Figure 10.3. The method of Powell.

Figure 10.3(b) shows the moves carried out in two cycles superimposed on the contour map of a quadratic surface. As explained before, the first cycle starts at point P_0 and ends up at P_3. The second cycle takes us to P_6, which is the optimal point. The directions $P_0 P_3$ and $P_3 P_6$ are mutually conjugate.

Powell's method does have a major flaw that has to be remedied—if $F(\mathbf{x})$ is not a quadratic, the algorithm tends to produce search directions that gradually become linearly dependent, thereby ruining the progress towards the minimum. The source of the problem is the automatic discarding of $\mathbf{v}_1$ at the end of each cycle. It has been suggested that it is better to throw out the direction that resulted in the *largest decrease* of $F(\mathbf{x})$, a policy that we adopt. It seems counterintuitive to discard the best direction, but it is likely to be close to the direction added in the next cycle, thereby contributing to linear dependence. As a result of the change, the search directions cease to be mutually conjugate, so that a quadratic form is not minimized in n cycles any more. This is not a significant loss since in practice $F(\mathbf{x})$ is seldom a quadratic.

Powell suggested a few other refinements to speed up convergence. Since they complicate the bookkeeping considerably, we did not implement them.

10.3 Conjugate Gradient Methods

■ powell

The algorithm for Powell's method is listed below. It utilizes two arrays: df contains the decreases of the merit function in the first n moves of a cycle, and the matrix u stores the corresponding direction vectors $\mathbf{v}_i$ (one vector per row).

```
## module powell
''' xMin,nCyc = powell(F,x,h=0.1,tol=1.0e-6)
    Powell's method of minimizing user-supplied function F(x).
    x    = starting point
    h    = initial search increment used in 'bracket'
    xMin = mimimum point
    nCyc = number of cycles
'''
from numarray import identity,array,dot,zeros,Float64,argmax
from goldSearch import *
from math import sqrt

def powell(F,x,h=0.1,tol=1.0e-6):

    def f(s): return F(x + s*v)    # F in direction of v

    n = len(x)                          # Number of design variables
    df = zeros((n),type=Float64)        # Decreases of F stored here
    u = identity(n)*1.0                 # Vectors v stored here by rows
    for j in range(30):                 # Allow for 30 cycles:
        xOld = x.copy()                 # Save starting point
        fOld = F(xOld)
      # First n line searches record decreases of F
        for i in range(n):
            v = u[i]
            a,b = bracket(f,0.0,h)
            s,fMin = search(f,a,b)
            df[i] = fOld - fMin
            fOld = fMin
            x = x + s*v
      # Last line search in the cycle
        v = x - xOld
        a,b = bracket(f,0.0,h)
        s,fLast = search(f,a,b)
        x = x + s*v
```

```
    # Check for convergence
        if sqrt(dot(x-xOld,x-xOld)/n) < tol: return x,j+1
    # Identify biggest decrease & update search directions
        iMax = int(argmax(df))
        for i in range(iMax,n-1):
            u[i] = u[i+1]
        u[n-1] = v
    print ''Powell did not converge''
```

EXAMPLE 10.3

Find the minimum of the function[25]

$$F = 100(y - x^2)^2 + (1 - x)^2$$

with Powell's method starting at the point $(-1, 1)$. This function has an interesting topology. The minimum value of F occurs at the point $(1, 1)$. As seen in the figure, there is a hump between the starting and minimum points which the algorithm must negotiate.

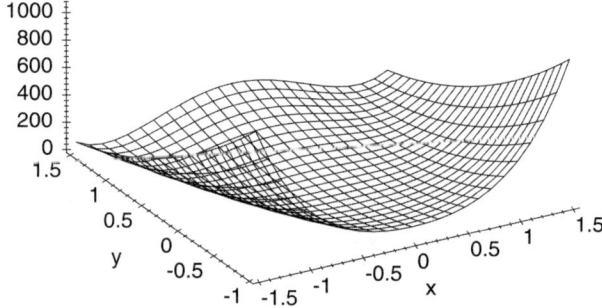

Solution The program that solves this unconstrained optimization problem is

```
#!/usr/bin/python
## example10_3
from powell import *
from numarray import array

def F(x): return 100.0*(x[1] - x[0]**2)**2 + (1 - x[0])**2

xStart = array([-1.0, 1.0])
xMin,nIter = powell(F,xStart)
print ''x ='',xMin
```

---

[25] From Shoup, T. E., and Mistree, F., *Optimization Methods with Applications for Personal Computers*, Prentice-Hall, 1987.

10.3 Conjugate Gradient Methods

```
print ''F(x) ='',F(xMin)
print ''Number of cycles ='',nIter
raw_input (''Press return to exit'')
```

As seen in the printout, the minimum point was obtained in 14 cycles.

```
x = [ 1.   1.]
F(x) = 4.93038065763e-030
Number of cycles = 14
```

EXAMPLE 10.4
Use `powell` to determine the smallest distance from the point $(5, 8)$ to the curve $xy = 5$.

Solution This is a constrained optimization problem: minimize $F(x, y) = (x - 5)^2 + (y - 8)^2$ (the square of the distance) subject to the equality constraint $xy - 5 = 0$. The following program uses Powell's method with penalty function:

```
#!/usr/bin/python
## example10_4
from powell import *
from numarray import array
from math import sqrt

def F(x):
    lam = 1.0                           # Penalty multiplier
    c = x[0]*x[1] - 5.0                 # Constraint equation
    return  distSq(x) + lam*c**2        # Penalized merit function

def distSq(x): return (x[0] - 5)**2 + (x[1] - 8)**2

xStart = array([1.0, 5.0])
x,numIter = powell(F,xStart,0.01)
print ''Intersection point ='',x
print ''Minimum distance ='', sqrt(distSq(x))
print ''xy ='', x[0]*x[1]
print ''Number of cycles ='',numIter
raw_input (''Press return to exit'')
```

As mentioned before, the value of the penalty function multiplier λ (called `lam` in the program) can have profound effects on the result. We chose $\lambda = 1$ (as in the program listing) with the following result:

```
Intersection point = [ 0.73306759   7.58776399]
Minimum distance = 4.28679959441
xy = 5.56234382324
Number of cycles = 6
```

The small value of λ favored speed of convergence over accuracy. Since the violation of the constraint $xy = 5$ is clearly unacceptable, we ran the program again with $\lambda = 10\,000$ and changed the starting point to $(0.733\,07, 7.587\,76)$, the end point of the first run. The results shown below are now acceptable:

```
Intersection point = [ 0.65561312   7.6265359 ]
Minimum distance = 4.36040970941
xy = 5.00005696388
Number of cycles = 5
```

Could we have used $\lambda = 10\,000$ in the first run? In this case we would be lucky and obtain the minimum in 19 cycles. Hence we save eight cycles by using two runs. However, a large λ often causes the algorithm to hang up, so that it generally wise to start with a small λ.

Fletcher–Reeves Method

Let us assume again that the merit function has the quadratic form in Eq. (10.5). Given a direction $\mathbf{v}$, it took Powell's method n line minimizations to construct a conjugate direction. We can reduce this to a single line minimization with a first-order method. Here is the procedure, known as the Fletcher–Reeves method:

- Choose a starting point $\mathbf{x}_0$.
- $\mathbf{g}_0 \leftarrow -\nabla F(\mathbf{x}_0)$
- $\mathbf{v}_0 \leftarrow \mathbf{g}_0$ (lacking a previous search direction, we choose the steepest descent).
- loop with $i = 0, 1, 2, \ldots$

  Minimize $F(\mathbf{x})$ along $\mathbf{v}_i$; let the minimum point be $\mathbf{x}_{i+1}$.

  $\mathbf{g}_{i+1} \leftarrow -\nabla F(\mathbf{x}_{i+1})$.

  if $|\mathbf{g}_{i+1}| < \varepsilon$ or $|F(\mathbf{x}_{i+1}) - F(\mathbf{x}_i)| < \varepsilon$ exit loop (convergence criterion).

  $\gamma \leftarrow (\mathbf{g}_{i+1} \cdot \mathbf{g}_{i+1})/(\mathbf{g}_i \cdot \mathbf{g}_i)$.

  $\mathbf{v}_{i+1} \leftarrow \mathbf{g}_{i+1} + \gamma \mathbf{v}_i$.

- end loop

It can be shown that $\mathbf{v}_i$ and $\mathbf{v}_{i+1}$ are mutually conjugate; that is, they satisfy the relationship $\mathbf{v}_i^T \mathbf{A} \mathbf{v}_{i+1} = 0$. Also $\mathbf{g}_i \cdot \mathbf{g}_{i+1} = 0$.

10.3 Conjugate Gradient Methods

The Fletcher–Reeves method will find the minimum of a quadratic function in n iterations. If $F(\mathbf{x})$ is not quadratic, it is necessary to restart the process after every n iterations. A variant of the Fletcher–Reeves method replaces the expression for γ by

$$\gamma = \frac{(\mathbf{g}_{i+1} - \mathbf{g}_i) \cdot \mathbf{g}_{i+1}}{\mathbf{g}_i \cdot \mathbf{g}_i} \qquad (10.6)$$

For a quadratic $F(\mathbf{x})$ this change makes no difference since $\mathbf{g}_i$ and $\mathbf{g}_{i+1}$ are orthogonal. However, for merit functions that are not quadratic, Eq. (10.6) is claimed to eliminate the need for a restart after n iterations.

■ fletcherReeves

```
## module fletcherReeves
''' xMin,nIter = optimize(F,gradF,x,h=0.01,tol=1.0e-6)
    Fletcher-Reeves method of minimizing a function.
    F(x)     = user-supplied function to be minimized.
    gradF(x) = user-supplied function for grad(F).
    x        = starting point.
    h        = initial search increment used in 'bracket'.
    xMin     = mimimum point.
    nIter    = number of iterations.
'''
from numarray import array,zeros,Float64,dot
from goldSearch import *
from math import sqrt

def optimize(F,gradF,x,h=0.1,tol=1.0e-6):

    def f(s): return F(x + s*v)  # Line function along v

    n = len(x)
    g0 = -gradF(x)
    v = g0.copy()
    F0 = F(x)
    for i in range(200):
        a,b = bracket(f,0.0,h)       # Minimization along
        s,fMin = search(f,a,b)       # a line
        x = x + s*v
        F1 = F(x)
        g1 = -gradF(x)
        if (sqrt(dot(g1,g1)) <= tol) or (abs(F0 - F1) < tol):
            return x,i+1
```

```
        gamma = dot((g1 - g0),g1)/dot(g0,g0)
        v = g1 + gamma*v
        g0 = g1.copy()
        F0 = F1
    print ''fletcherReeves did not converge''
```

EXAMPLE 10.5
Use the Fletcher–Reeves method to locate the minimum of

$$F(\mathbf{x}) = 10x_1^2 + 3x_2^2 - 10x_1x_2 + 2x_1$$

Start with $\mathbf{x}_0 = \begin{bmatrix} 0 & 0 \end{bmatrix}^T$.

Solution Since $F(\mathbf{x})$ is quadratic, we need only two iterations. The gradient of F is

$$\nabla F(\mathbf{x}) = \begin{bmatrix} \partial F/\partial x_1 \\ \partial F/\partial x_2 \end{bmatrix} = \begin{bmatrix} 20x_1 - 10x_2 + 2 \\ -10x_1 + 6x_2 \end{bmatrix}$$

First iteration:

$$\mathbf{g}_0 = -\nabla F(\mathbf{x}_0) = \begin{bmatrix} -2 \\ 0 \end{bmatrix} \quad \mathbf{v}_0 = \mathbf{g}_0 = \begin{bmatrix} -2 \\ 0 \end{bmatrix} \quad \mathbf{x}_0 + s\mathbf{v}_0 = \begin{bmatrix} -2s \\ 0 \end{bmatrix}$$

$$f(s) = F(\mathbf{x}_0 + s\mathbf{v}_0) = 10(2s)^2 + 3(0)^2 - 10(-2s)(0) + 2(-2s)$$
$$= 40s^2 - 4s$$

$$f'(s) = 80s - 4 = 0 \quad s = 0.05$$

$$\mathbf{x}_1 = \mathbf{x}_0 + s\mathbf{v}_0 = \begin{bmatrix} 0 \\ 0 \end{bmatrix} + 0.05 \begin{bmatrix} -2 \\ 0 \end{bmatrix} = \begin{bmatrix} -0.1 \\ 0 \end{bmatrix}$$

Second iteration:

$$\mathbf{g}_1 = -\nabla F(\mathbf{x}_1) = \begin{bmatrix} -20(-0.1) + 10(0) - 2 \\ 10(-0.1) - 6(0) \end{bmatrix} = \begin{bmatrix} 0 \\ -1.0 \end{bmatrix}$$

$$\gamma = \frac{\mathbf{g}_1 \cdot \mathbf{g}_1}{\mathbf{g}_0 \cdot \mathbf{g}_0} = \frac{1.0}{4} = 0.25$$

$$\mathbf{v}_1 = \mathbf{g}_1 + \gamma \mathbf{v}_0 = \begin{bmatrix} 0 \\ -1.0 \end{bmatrix} + 0.25 \begin{bmatrix} -2 \\ 0 \end{bmatrix} = \begin{bmatrix} -0.5 \\ -1.0 \end{bmatrix}$$

$$\mathbf{x}_1 + s\mathbf{v}_1 = \begin{bmatrix} -0.1 \\ 0 \end{bmatrix} + s \begin{bmatrix} -0.5 \\ -1.0 \end{bmatrix} = \begin{bmatrix} -0.1 - 0.5s \\ -s \end{bmatrix}$$

10.3 Conjugate Gradient Methods

$$f(s) = F(\mathbf{x}_1 + s\mathbf{v}_1)$$
$$= 10(-0.1 - 0.5s)^2 + 3(-s)^2 - 10(-0.1 - 0.5s)(-s) + 2(-0.1 - 0.5s)$$
$$= 0.5s^2 - s - 0.1$$

$$f'(s) = s - 1 = 0 \qquad s = 1.0$$

$$\mathbf{x}_2 = \mathbf{x}_1 + s\mathbf{v}_1 = \begin{bmatrix} -0.1 \\ 0 \end{bmatrix} + 1.0 \begin{bmatrix} -0.5 \\ -1.0 \end{bmatrix} = \begin{bmatrix} -0.6 \\ -1.0 \end{bmatrix}$$

We have now reached the minimum point.

EXAMPLE 10.6

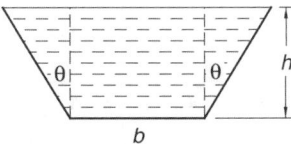

The figure shows the cross section of a channel carrying water. Determine h, b and θ that minimize the length of the wetted perimeter while maintaining a cross-sectional area of 8 m². (Minimizing the wetted perimeter results in least resistance to the flow.) Use the Fletcher–Reeves method.

Solution The cross-sectional area of the channel is

$$A = \frac{1}{2}[b + (b + 2h\tan\theta)]h = (b + h\tan\theta)h$$

and the length of the wetted perimeter is

$$S = b + 2(h\sec\theta)$$

The optimization problem can be cast as

$$\text{minimize } b + 2h\sec\theta$$
$$\text{subject to } (b + h\tan\theta)h = 8$$

Equality constraints can often be used to eliminate some of the design variables. In this case we can solve the area constraint for b, obtaining

$$b = \frac{8}{h} - h\tan\theta$$

Substituting the result into the expression for S, we get

$$S = \frac{8}{h} - h\tan\theta + 2h\sec\theta$$

Introduction to Optimization

We have now arrived at an unconstrained optimization problem of finding h and θ that minimize S. The gradient of the merit function is

$$\nabla S = \begin{bmatrix} \partial S/\partial h \\ \partial S/\partial \theta \end{bmatrix} = \begin{bmatrix} -8/h^2 - \tan\theta + 2\sec\theta \\ -h\sec^2\theta + 2h\sec\theta\tan\theta \end{bmatrix}$$

Letting $\mathbf{x} = \begin{bmatrix} h & \theta \end{bmatrix}^T$ and starting with $\mathbf{x}_0 = \begin{bmatrix} 2 & 0 \end{bmatrix}^T$, we arrive at the following program:

```
#!/usr/bin/python
## example10_6
from fletcherReeves import *
from numarray import array,zeros,Float64
from math import cos,tan,pi

def F(x):
    return  8.0/x[0] - x[0]*(tan(x[1]) - 2.0/cos(x[1]))

def gradF(x):
    g = zeros((2),type=Float64)
    g[0] = -8.0/(x[0]**2) - tan(x[1]) + 2.0/cos(x[1])
    g[1] = x[0]*(-1.0/cos(x[1]) + 2.0*tan(x[1]))/cos(x[1])
    return g

x = array([2.0, 0.0])
x,nIter = optimize(F,gradF,x)
b = 8.0/x[0] - x[0]*tan(x[1])
print ''h ='',x[0],''m''
print ''b ='',b,''m''
print ''theta ='',x[1]*180.0/pi,''deg''
print ''perimeter ='', F(x),''m''
print ''Number of iterations ='',nIter
raw_input(''Press return to exit'')
```

The results are:

```
h = 2.14914172295 m
b = 2.48162366149 m
theta = 29.9997001208 deg
perimeter = 7.44483887289 m
Number of iterations = 4
```

10.3 Conjugate Gradient Methods

PROBLEM SET 10.1

1. ■ The Lennard–Jones potential between two molecules is

$$V = 4\varepsilon \left[\left(\frac{\sigma}{r}\right)^{12} - \left(\frac{\sigma}{r}\right)^{6}\right]$$

where ε and σ are constants, and r is the distance between the molecules. Use the module `goldSearch` to find σ/r that minimizes the potential and verify the result analytically.

2. ■ One wave function of the hydrogen atom is

$$\psi = C\left(27 - 18\sigma + 2\sigma^2\right) e^{-\sigma/3}$$

where

$$\sigma = zr/a_0$$

$$C = \frac{1}{81\sqrt{3\pi}} \left(\frac{z}{a_0}\right)^{2/3}$$

$$z = \text{nuclear charge}$$

$$a_0 = \text{Bohr radius}$$

$$r = \text{radial distance}$$

Find σ where ψ is at a minimum. Verify the result analytically.

3. ■ Determine the parameter p that minimizes the integral

$$\int_0^\pi \sin x \cos px \, dx$$

Hint: use numerical quadrature to evaluate the integral.

4. ■

Kirchoff's equations for the two loops of the electrical circuit are

$$R_1 i_1 + R_3 i_1 + R(i_1 - i_2) = E$$

$$R_2 i_2 + R_4 i_2 + R_5 i_2 + R(i_2 - i_1) = 0$$

Find the resistance R that maximizes the power dissipated by R. *Hint*: solve Kirchoff's equations numerically with one of the functions in Chapter 2.

5. ■

A wire carrying an electric current is surrounded by rubber insulation of outer radius r. The resistance of the wire generates heat, which is conducted through the insulation and convected into the surrounding air. The temperature of the wire can be shown to be

$$T = \frac{q}{2\pi}\left(\frac{\ln(r/a)}{k} + \frac{1}{hr}\right) + T_\infty$$

where

q = rate of heat generation in wire = 50 W/m
a = radius of wire = 5 mm
k = thermal conductivity of rubber = 0.16 W/m · K
h = convective heat-transfer coefficient = 20 W/m² · K
T_∞ = ambient temperature = 280 K

Find r that minimizes T.

6. ■ Minimize the function

$$F(x, y) = (x - 1)^2 + (y - 1)^2$$

subject to the constraints $x + y \leq 1$ and $x \geq 0.6$.

7. ■ Find the minimum of the function

$$F(x, y) = 6x^2 + y^3 + xy$$

in $y \geq 0$. Verify the result analytically.

8. ■ Solve Prob. 7 if the constraint is changed to $y \geq -2$.

9. ■ Determine the smallest distance from the point (1, 2) to the parabola $y = x^2$.

10.3 Conjugate Gradient Methods

10. ■

Determine x that minimizes the distance d between the base of the area shown and its centroid C.

11. ■

The cylindrical vessel of mass M has its center of gravity at C. The water in the vessel has a depth x. Determine x so that the center of gravity of the vessel–water combination is as low as possible. Use $M = 115$ kg, $H = 0.8$ m and $r = 0.25$ m.

12. ■

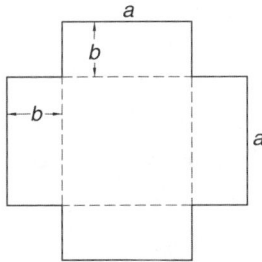

The sheet of cardboard is folded along the dashed lines to form a box with an open top. If the volume of the box is to be 1.0 m³, determine the dimensions a and b that would use the least amount of cardboard. Verify the result analytically.

13.

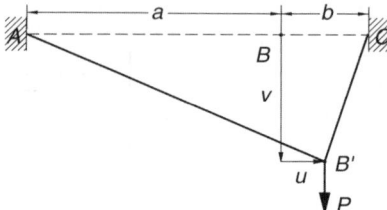

The elastic cord ABC has an extensional stiffness k. When the vertical force P is applied at B, the cord deforms to the shape $AB'C$. The potential energy of the system in the deformed position is

$$V = -Pv + \frac{k(a+b)}{2a}\delta_{AB}^2 + \frac{k(a+b)}{2b}\delta_{BC}^2$$

where

$$\delta_{AB} = \sqrt{(a+u)^2 + v^2} - a$$
$$\delta_{BC} = \sqrt{(b-u)^2 + v^2} - b$$

are the elongations of AB and BC. Determine the displacements u and v by minimizing V (this is an application of the principle of minimum potential energy: a system is in stable equilibrium if its potential energy is at a minimum). Use $a = 150$ mm, $b = 50$ mm, $k = 0.6$ N/mm and $P = 5$ N.

14.

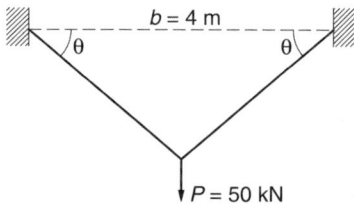

Each member of the truss has a cross-sectional area A. Find A and the angle θ that minimize the volume

$$V = \frac{bA}{\cos\theta}$$

of the material in the truss without violating the constraints

$$\sigma \leq 150 \text{ MPa} \qquad \delta \leq 5 \text{ mm}$$

10.3 Conjugate Gradient Methods

where

$$\sigma = \frac{P}{2A\sin\theta} = \text{stress in each member}$$

$$\delta = \frac{Pb}{2EA\sin 2\theta \sin\theta} = \text{displacement at the load } P$$

and $E = 200 \times 10^9$ P$_a$.

15. ■ Solve Prob. 14 if the allowable displacement is changed to 2.5 mm.

16. ■

The cantilever beam of circular cross section is to have the smallest volume possible subject to constraints

$$\sigma_1 \leq 180 \text{ MPa} \qquad \sigma_2 \leq 180 \text{ MPa} \qquad \delta \leq 25 \text{ mm}$$

where

$$\sigma_1 = \frac{8PL}{\pi r_1^3} = \text{maximum stress in left half}$$

$$\sigma_2 = \frac{4PL}{\pi r_2^3} = \text{maximum stress in right half}$$

$$\delta = \frac{4PL^3}{3\pi E}\left(\frac{7}{r_1^4} + \frac{1}{r_2^4}\right) = \text{displacement at free end}$$

and $E = 200$ GPa. Determine r_1 and r_2.

17. ■ Find the minimum of the function

$$F(x, y, z) = 2x^2 + 3y^2 + z^2 + xy + xz - 2y$$

and confirm the result analytically.

18. ■

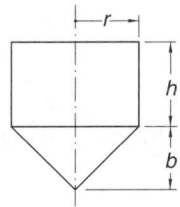

The cylindrical container has a conical bottom and an open top. If the volume V of the container is to be 1.0 m^3, find the dimensions r, h and b that minimize the surface area S. Note that

$$V = \pi r^2 \left(\frac{b}{3} + h\right)$$

$$S = \pi r \left(2h + \sqrt{b^2 + r^2}\right)$$

19. ■

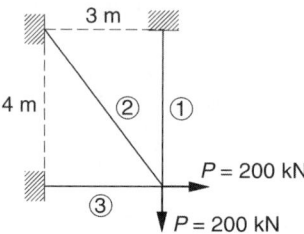

The equilibrium equations of the truss shown are

$$\sigma_1 A_1 + \frac{4}{5}\sigma_2 A_2 = P \qquad \frac{3}{5}\sigma_2 A_2 + \sigma_3 A_3 = P$$

where σ_i is the axial stress in member i and A_i are the cross-sectional areas. The third equation is supplied by compatibility (geometrical constraints on the elongations of the members):

$$\frac{16}{5}\sigma_1 - 5\sigma_2 + \frac{9}{5}\sigma_3 = 0$$

Find the cross-sectional areas of the members that minimize the weight of the truss without the stresses exceeding 150 MPa.

20. ■

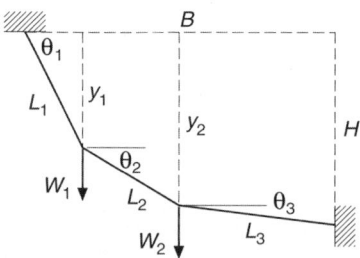

A cable supported at the ends carries the weights W_1 and W_2. The potential energy of the system is

$$V = -W_1 y_1 - W_2 y_2$$

$$= -W_1 L_1 \sin\theta_1 - W_2(L_1 \sin\theta_1 + L_2 \sin\theta_2)$$

and the geometric constraints are

$$L_1 \cos\theta_1 + L_2 \cos\theta_2 + L_3 \cos\theta_3 = B$$

$$L_1 \sin\theta_1 + L_2 \sin\theta_2 + L_3 \sin\theta_3 = H$$

The principle of minimum potential energy states that the equilibrium configuration of the system is the one that satisfies geometric constraints and minimizes the potential energy. Determine the equilibrium values of θ_1, θ_2 and θ_3 given that $L_1 = 1.2$ m, $L_2 = 1.5$ m, $L_3 = 1.0$ m, $B = 3.5$ m, $H = 0$, $W_1 = 20$ kN and $W_2 = 30$ kN.

10.4 Other Methods

The *Nelder–Mead method*, also know as the *downhill simplex* algorithm, is a popular and robust method of optimization. Its main attraction is a common sense geometrical approach that requires no mathematical sophistication. In speed of execution of the downhill simplex is not competitive with Powell's method.

Simulated annealing methods have been successfully employed for complex problems involving many design variables. These methods are based on an analogy with the annealing as a slowly cooled liquid metal solidifies into a crystalline, minimum energy structure. One distinguishing feature of simulated annealing is its ability to pass over local minima in its search for the global minimum.

A topic that we reluctantly omitted is the *simplex method* of linear programming. Linear programming deals with optimization problems where the merit function and the constraints are linear expressions of the independent variables. The general linear programming problem is to minimize the objective function

$$F = \sum_{i=1}^{n} a_i x_i$$

subject to the constraints

$$\sum_{j=1}^{n} B_{ij} x_j \leq b_i, \quad i = 1, 2, \ldots, m_1$$

$$\sum_{j=1}^{n} C_{ij} x_j \geq c_i, \quad i = 1, 2, \ldots, m_2$$

$$\sum_{j=1}^{n} D_{ij} x_j = d_i, \quad i = 1, 2, \ldots, m_3$$

$$x_i \geq 0, \quad i = 1, 2, \ldots n$$

where the constants b_i, c_i and d_i are nonnegative. The roots of linear programming lie in cost analysis, operations research and related fields. We skip this topic because there are very few engineering applications that can be formulated as linear programming problems. In addition, a fail-safe implementation of the simplex method results in a rather complicated algorithm. This not to say that the simplex method has no place nonlinear optimization. There are several effective methods that rely in part on the simplex method. For example, problems with nonlinear constraints can often be solved by a piecewise application of linear programming. The simplex method is also used to compute search directions in the so-called *method of feasible directions*.

Appendices

A1 Taylor Series

Function of a Single Variable

The Taylor series expansion of a function $f(x)$ about the point $x = a$ is the infinite series

$$f(x) = f(a) + f'(a)(x-a) + f''(a)\frac{(x-a)^2}{2!} + f'''(a)\frac{(x-a)^3}{3!} + \cdots \tag{A1}$$

In the special case $a = 0$ the series is also known as the *MacLaurin series*. It can be shown that the Taylor series expansion is unique in the sense that no two functions have identical Taylor series.

A Taylor series is meaningful only if all the derivatives of $f(x)$ exist at $x = a$ and the series converges. In general, convergence occurs only if x is sufficiently close to a; i.e., if $|x - a| \leq \varepsilon$, where ε is called the *radius of convergence*. In many cases ε is infinite.

Another useful form of the Taylor series is the expansion about an arbitrary value of x:

$$f(x+h) = f(x) + f'(x)h + f''(x)\frac{h^2}{2!} + f'''(x)\frac{h^3}{3!} + \cdots \tag{A2}$$

Since it is not possible to evaluate all the terms of an infinite series, the effect of truncating the series in Eq. (A2) is of great practical importance. Keeping the first $n+1$ terms, we have

$$f(x+h) = f(x) + f'(x)h + f''(x)\frac{h^2}{2!} + \cdots + f^{(n)}(x)\frac{h^n}{n!} + E_n \tag{A3}$$

where E_n is the *truncation error* (sum of the truncated terms). The bounds on the truncation error are given by *Taylor's theorem*:

$$E_n = f^{(n+1)}(\xi)\frac{h^{n+1}}{(n+1)!} \tag{A4}$$

where ξ is some point in the interval $(x, x + h)$. Note that the expression for E_n is identical to the first discarded term of the series, but with x replaced by ξ. Since the value of ξ is undetermined (only its limits are known), the most we can get out of Eq. (A4) are the upper and lower bounds on the truncation error.

If the expression for $f^{(n+1)}(\xi)$ is not available, the information conveyed by Eq. (A4) is reduced to

$$E_n = \mathcal{O}(h^{n+1}) \tag{A5}$$

which is a concise way of saying that the truncation error is *of the order of* h^{n+1}, or behaves as h^{n+1}. If h is within the radius of convergence, then

$$\mathcal{O}(h^n) > \mathcal{O}(h^{n+1})$$

i.e., the error is always reduced if a term is added to the truncated series (this may not be true for the first few terms).

In the special case $n = 1$, Taylor's theorem is known as the *mean value theorem*:

$$f(x + h) = f(x) + f'(\xi)h, \quad x \leq \xi \leq x + h \tag{A6}$$

Function of Several Variables

If f is a function of the m variables $x_1, x_2, \ldots, x_m$, then its Taylor series expansion about the point $\mathbf{x} = [x_1, x_2, \ldots, x_m]^T$ is

$$f(\mathbf{x} + \mathbf{h}) = f(\mathbf{x}) + \sum_{i=1}^{m} \left.\frac{\partial f}{\partial x_i}\right|_{\mathbf{x}} h_i + \frac{1}{2!} \sum_{i=1}^{m} \sum_{j=1}^{m} \left.\frac{\partial^2 f}{\partial x_i \partial x_j}\right|_{\mathbf{x}} h_i h_j + \cdots \tag{A7}$$

This is sometimes written as

$$f(\mathbf{x} + \mathbf{h}) = f(\mathbf{x}) + \nabla f(\mathbf{x}) \cdot \mathbf{h} + \frac{1}{2}\mathbf{h}^T \mathbf{H}(\mathbf{x})\mathbf{h} + \cdots \tag{A8}$$

The vector ∇f is known as the *gradient* of f and the matrix $\mathbf{H}$ is called the *Hessian matrix* of f.

EXAMPLE A1
Derive the Taylor series expansion of $f(x) = \ln(x)$ about $x = 1$.

Solution The derivatives of f are

$$f'(x) = \frac{1}{x} \quad f''(x) = -\frac{1}{x^2} \quad f'''(x) = \frac{2!}{x^3} \quad f^{(4)} = -\frac{3!}{x^4} \text{ etc.}$$

Evaluating the derivatives at $x = 1$, we get

$$f'(1) = 1 \quad f''(1) = -1 \quad f'''(1) = 2! \quad f^{(4)}(1) = -3! \text{ etc.}$$

A1 Taylor Series

which upon substitution into Eq. (A1) together with $a = 1$ yields

$$\ln(x) = 0 + (x-1) - \frac{(x-1)^2}{2!} + 2!\frac{(x-1)^3}{3!} - 3!\frac{(x-1)^4}{4!} + \cdots$$

$$= (x-1) - \frac{1}{2}(x-1)^2 + \frac{1}{3}(x-1)^3 - \frac{1}{4}(x-1)^4 + \cdots$$

EXAMPLE A2

Use the first five terms of the Taylor series expansion of e^x about $x = 0$:

$$e^x = 1 + x + \frac{x^2}{2!} + \frac{x^3}{3!} + \frac{x^4}{4!} + \cdots$$

together with the error estimate to find the bounds of e.

Solution

$$e = 1 + 1 + \frac{1}{2} + \frac{1}{6} + \frac{1}{24} + E_4 = \frac{65}{24} + E_4$$

$$E_4 = f^{(4)}(\xi)\frac{h^5}{5!} = \frac{e^\xi}{5!}, \quad 0 \leq \xi \leq 1$$

The bounds on the truncation error are

$$(E_4)_{\min} = \frac{e^0}{5!} = \frac{1}{120} \qquad (E_4)_{\max} = \frac{e^1}{5!} = \frac{e}{120}$$

Thus the lower bound on e is

$$e_{\min} = \frac{65}{24} + \frac{1}{120} = \frac{163}{60}$$

and the upper bound is given by

$$e_{\max} = \frac{65}{24} + \frac{e_{\max}}{120}$$

which yields

$$\frac{119}{120}e_{\max} = \frac{65}{24} \qquad e_{\max} = \frac{325}{119}$$

Therefore,

$$\frac{163}{60} \leq e \leq \frac{325}{119}$$

EXAMPLE A3

Compute the gradient and the Hessian matrix of

$$f(x, y) = \ln \sqrt{x^2 + y^2}$$

at the point $x = -2$, $y = 1$.

Solution

$$\frac{\partial f}{\partial x} = \frac{1}{\sqrt{x^2+y^2}}\left(\frac{1}{2}\frac{2x}{\sqrt{x^2+y^2}}\right) = \frac{x}{x^2+y^2} \qquad \frac{\partial f}{\partial y} = \frac{y}{x^2+y^2}$$

$$\nabla f(x, y) = \begin{bmatrix} x/(x^2+y^2) & y/(x^2+y^2) \end{bmatrix}^T$$

$$\nabla f(-2, 1) = \begin{bmatrix} -0.4 & 0.2 \end{bmatrix}^T$$

$$\frac{\partial^2 f}{\partial x^2} = \frac{(x^2+y^2) - x(2x)}{(x^2+y^2)^2} = \frac{-x^2+y^2}{(x^2+y^2)^2}$$

$$\frac{\partial^2 f}{\partial y^2} = \frac{x^2-y^2}{(x^2+y^2)^2}$$

$$\frac{\partial^2 f}{\partial x \partial y} = \frac{\partial^2 f}{\partial y \partial x} = \frac{-2xy}{(x^2+y^2)^2}$$

$$\mathbf{H}(x, y) = \begin{bmatrix} -x^2+y^2 & -2xy \\ -2xy & x^2-y^2 \end{bmatrix}\frac{1}{(x^2+y^2)^2}$$

$$\mathbf{H}(-2, 1) = \begin{bmatrix} -0.12 & 0.16 \\ 0.16 & 0.12 \end{bmatrix}$$

A2 Matrix Algebra

A matrix is a rectangular array of numbers. The *size* of a matrix is determined by the number of rows and columns, also called the *dimensions* of the matrix. Thus a matrix of m rows and n columns is said to have the size $m \times n$ (the number of rows is always listed first). A particularly important matrix is the square matrix, which has the same number of rows and columns.

An array of numbers arranged in a single column is called a *column vector*, or simply a vector. If the numbers are set out in a row, the term *row vector* is used. Thus a column vector is a matrix of dimensions $n \times 1$ and a row vector can be viewed as a matrix of dimensions $1 \times n$.

We denote matrices by boldface, upper case letters. For vectors we use boldface, lower case letters. Here are examples of the notation:

$$\mathbf{A} = \begin{bmatrix} A_{11} & A_{12} & A_{13} \\ A_{21} & A_{22} & A_{23} \\ A_{31} & A_{32} & A_{33} \end{bmatrix} \qquad \mathbf{b} = \begin{bmatrix} b_1 \\ b_2 \\ b_3 \end{bmatrix} \qquad (A9)$$

A2 Matrix Algebra

Indices of the elements of a matrix are displayed in the same order as its dimensions: the row number comes first, followed by the column number. Only one index is needed for the elements of a vector.

Transpose

The transpose of a matrix $\mathbf{A}$ is denoted by $\mathbf{A}^T$ and defined as

$$A^T_{ij} = A_{ji}$$

The transpose operation thus interchanges the rows and columns of the matrix. If applied to vectors, it turns a column vector into a row vector and *vice versa*. For example, transposing $\mathbf{A}$ and $\mathbf{b}$ in Eq. (A9), we get

$$\mathbf{A}^T = \begin{bmatrix} A_{11} & A_{21} & A_{31} \\ A_{12} & A_{22} & A_{32} \\ A_{13} & A_{23} & A_{33} \end{bmatrix} \qquad \mathbf{b}^T = \begin{bmatrix} b_1 & b_2 & b_3 \end{bmatrix}$$

An $n \times n$ matrix is said to be *symmetric* if $\mathbf{A}^T = \mathbf{A}$. This means that the elements in the upper triangular portion (above the diagonal connecting A_{11} and A_{nn}) of a symmetric matrix are mirrored in the lower triangular portion.

Addition

The sum $\mathbf{C} = \mathbf{A} + \mathbf{B}$ of two $m \times n$ matrices $\mathbf{A}$ and $\mathbf{B}$ is defined as

$$C_{ij} = A_{ij} + B_{ij}, \quad i = 1, 2, \ldots, m; \quad j = 1, 2, \ldots, n \tag{A10}$$

Thus the elements of $\mathbf{C}$ are obtained by adding elements of $\mathbf{A}$ to the elements of $\mathbf{B}$. Note that addition is defined only for matrices that have the same dimensions.

Multiplication

The scalar or *dot product* $c = \mathbf{a} \cdot \mathbf{b}$ of the vectors $\mathbf{a}$ and $\mathbf{b}$, each of size m, is defined as

$$c = \sum_{k=1}^{m} a_k b_k \tag{A11}$$

It can also be written in the form $c = \mathbf{a}^T \mathbf{b}$.

The matrix product $\mathbf{C} = \mathbf{AB}$ of an $l \times m$ matrix $\mathbf{A}$ and an $m \times n$ matrix $\mathbf{B}$ is defined by

$$C_{ij} = \sum_{k=1}^{m} A_{ik} B_{kj}, \quad i = 1, 2, \ldots, l; \quad j = 1, 2, \ldots, n \tag{A12}$$

The definition requires the number of columns in **A** (the dimension m) to be equal to the number of rows in **B**. The matrix product can also be defined in terms of the dot product. Representing the ith row of **A** as the vector $\mathbf{a}_i$ and the jth column of **B** as the vector $\mathbf{b}_j$, we have

$$C_{ij} = \mathbf{a}_i \cdot \mathbf{b}_j \tag{A13}$$

A square matrix of special importance is the identity or *unit matrix*

$$\mathbf{I} = \begin{bmatrix} 1 & 0 & 0 & \cdots & 0 \\ 0 & 1 & 0 & \cdots & 0 \\ 0 & 0 & 1 & \cdots & 0 \\ \vdots & \vdots & \vdots & \ddots & \vdots \\ 0 & 0 & 0 & 0 & 1 \end{bmatrix} \tag{A14}$$

It has the property $\mathbf{AI} = \mathbf{IA} = \mathbf{A}$.

Inverse

The inverse of an $n \times n$ matrix **A**, denoted by $\mathbf{A}^{-1}$, is defined to be an $n \times n$ matrix that has the property

$$\mathbf{A}^{-1}\mathbf{A} = \mathbf{AA}^{-1} = \mathbf{I} \tag{A15}$$

Determinant

The determinant of a square matrix **A** is a scalar denoted by $|\mathbf{A}|$ or $\det(\mathbf{A})$. There is no concise definition of the determinant for a matrix of arbitrary size. We start with the determinant of a 2×2 matrix, which is defined as

$$\begin{vmatrix} A_{11} & A_{12} \\ A_{21} & A_{22} \end{vmatrix} = A_{11}A_{22} - A_{12}A_{21} \tag{A16}$$

The determinant of a 3×3 matrix is then defined as

$$\begin{vmatrix} A_{11} & A_{12} & A_{13} \\ A_{21} & A_{22} & A_{23} \\ A_{31} & A_{32} & A_{33} \end{vmatrix} = A_{11}\begin{vmatrix} A_{22} & A_{23} \\ A_{32} & A_{33} \end{vmatrix} - A_{12}\begin{vmatrix} A_{21} & A_{23} \\ A_{31} & A_{33} \end{vmatrix} + A_{13}\begin{vmatrix} A_{21} & A_{22} \\ A_{31} & A_{32} \end{vmatrix}$$

Having established the pattern, we can now define the determinant of an $n \times n$ matrix in terms of the determinant of an $(n-1) \times (n-1)$ matrix:

$$|\mathbf{A}| = \sum_{k=1}^{n} (-1)^{k+1} A_{1k} M_{1k} \tag{A17}$$

where M_{ik} is the determinant of the $(n-1) \times (n-1)$ matrix obtained by deleting the ith row and kth column of **A**. The term $(-1)^{k+i} M_{ik}$ is called a *cofactor* of A_{ik}.

Equation (A17) is known as *Laplace's development* of the determinant on the first row of **A**. Actually Laplace's development can take place on any convenient row. Choosing the ith row, we have

$$|\mathbf{A}| = \sum_{k=1}^{n} (-1)^{k+i} A_{ik} M_{ik} \qquad (A18)$$

The matrix **A** is said to be *singular* if $|\mathbf{A}| = 0$.

Positive Definiteness

An $n \times n$ matrix A is said to be positive definite if

$$\mathbf{x}^T \mathbf{A} \mathbf{x} > 0 \qquad (A19)$$

for all nonvanishing vectors **x**. It can be shown that a matrix is positive definite if the determinants of all its leading minors are positive. The leading minors of **A** are the n square matrices

$$\begin{bmatrix} A_{11} & A_{12} & \cdots & A_{1k} \\ A_{12} & A_{22} & \cdots & A_{2k} \\ \vdots & \vdots & \ddots & \vdots \\ A_{k1} & A_{k2} & \cdots & A_{kk} \end{bmatrix}, \quad k = 1, 2, \ldots, n$$

Therefore, positive definiteness requires that

$$A_{11} > 0, \quad \begin{vmatrix} A_{11} & A_{12} \\ A_{21} & A_{22} \end{vmatrix} > 0, \quad \begin{vmatrix} A_{11} & A_{12} & A_{13} \\ A_{21} & A_{22} & A_{23} \\ A_{31} & A_{32} & A_{33} \end{vmatrix} > 0, \ldots, |A| > 0 \qquad (A20)$$

Useful Theorems

We list without proof a few theorems that are utilized in the main body of the text. Most proofs are easy and could be attempted as exercises in matrix algebra.

$$(\mathbf{AB})^T = \mathbf{B}^T \mathbf{A}^T \qquad (A21a)$$

$$(\mathbf{AB})^{-1} = \mathbf{B}^{-1} \mathbf{A}^{-1} \qquad (A21b)$$

$$|\mathbf{A}^T| = |\mathbf{A}| \qquad (A21c)$$

$$|\mathbf{AB}| = |\mathbf{A}| |\mathbf{B}| \qquad (A21d)$$

$$\text{if } \mathbf{C} = \mathbf{A}^T \mathbf{B} \mathbf{A} \text{ where } \mathbf{B} = \mathbf{B}^T, \text{ then } \mathbf{C} = \mathbf{C}^T \qquad (A21e)$$

EXAMPLE A4
Letting

$$A = \begin{bmatrix} 1 & 2 & 3 \\ 1 & 2 & 1 \\ 0 & 1 & 2 \end{bmatrix} \quad \mathbf{u} = \begin{bmatrix} 1 \\ 6 \\ -2 \end{bmatrix} \quad \mathbf{v} = \begin{bmatrix} 8 \\ 0 \\ -3 \end{bmatrix}$$

compute $\mathbf{u} + \mathbf{v}$, $\mathbf{u} \cdot \mathbf{v}$, $A\mathbf{v}$ and $\mathbf{u}^T A \mathbf{v}$.

Solution

$$\mathbf{u} + \mathbf{v} = \begin{bmatrix} 1+8 \\ 6+0 \\ -2-3 \end{bmatrix} = \begin{bmatrix} 9 \\ 6 \\ -5 \end{bmatrix}$$

$$\mathbf{u} \cdot \mathbf{v} = 1(8) + 6(0) + (-2)(-3) = 14$$

$$A\mathbf{v} = \begin{bmatrix} \mathbf{a}_1 \cdot \mathbf{v} \\ \mathbf{a}_2 \cdot \mathbf{v} \\ \mathbf{a}_3 \cdot \mathbf{v} \end{bmatrix} = \begin{bmatrix} 1(8) + 2(0) + 3(-3) \\ 1(8) + 2(0) + 1(-3) \\ 0(8) + 1(0) + 2(-3) \end{bmatrix} = \begin{bmatrix} -1 \\ 5 \\ -6 \end{bmatrix}$$

$$\mathbf{u}^T A \mathbf{v} = \mathbf{u} \cdot (A\mathbf{v}) = 1(-1) + 6(5) + (-2)(-6) = 41$$

EXAMPLE A5
Compute $|A|$, where A is given in Example A4. Is A positive definite?

Solution Laplace's development of the determinant on the first row yields

$$|A| = 1\begin{vmatrix} 2 & 1 \\ 1 & 2 \end{vmatrix} - 2\begin{vmatrix} 1 & 1 \\ 0 & 2 \end{vmatrix} + 3\begin{vmatrix} 1 & 2 \\ 0 & 1 \end{vmatrix}$$

$$= 1(3) - 2(2) + 3(1) = 2$$

Development on the third row is somewhat easier due to the presence of the zero element:

$$|A| = 0\begin{vmatrix} 2 & 3 \\ 2 & 1 \end{vmatrix} - 1\begin{vmatrix} 1 & 3 \\ 1 & 1 \end{vmatrix} + 2\begin{vmatrix} 1 & 2 \\ 1 & 2 \end{vmatrix}$$

$$= 0(-4) - 1(-2) + 2(0) = 2$$

To verify positive definiteness, we evaluate the determinants of the leading minors:

$$A_{11} = 1 > 0 \quad \text{O.K.}$$

$$\begin{vmatrix} A_{11} & A_{12} \\ A_{21} & A_{22} \end{vmatrix} = \begin{vmatrix} 1 & 2 \\ 1 & 2 \end{vmatrix} = 0 \quad \text{Not O.K.}$$

A is not positive definite.

A2 Matrix Algebra

EXAMPLE A6

Evaluate the matrix product **AB**, where **A** is given in Example A4 and

$$\mathbf{B} = \begin{bmatrix} -4 & 1 \\ 1 & -4 \\ 2 & -2 \end{bmatrix}$$

Solution

$$\mathbf{AB} = \begin{bmatrix} \mathbf{a}_1 \cdot \mathbf{b}_1 & \mathbf{a}_1 \cdot \mathbf{b}_2 \\ \mathbf{a}_2 \cdot \mathbf{b}_1 & \mathbf{a}_2 \cdot \mathbf{b}_2 \\ \mathbf{a}_3 \cdot \mathbf{b}_1 & \mathbf{a}_3 \cdot \mathbf{b}_2 \end{bmatrix}$$

$$= \begin{bmatrix} 1(-4) + 2(1) + 3(2) & 1(1) + 2(-4) + 3(-2) \\ 1(-4) + 2(1) + 1(2) & 1(1) + 2(-4) + 1(-2) \\ 0(-4) + 1(1) + 2(2) & 0(1) + 1(-4) + 2(-2) \end{bmatrix} = \begin{bmatrix} 4 & -13 \\ 0 & -9 \\ 5 & -8 \end{bmatrix}$$

Index

adaptive Runge–Kutta method, 275–283
algebra. *See* linear algebraic equations systems; matrix algebra
appendices, 409–412
arithmetic operators, in Python, 7
arrays
   accessing/changing, 21
   copying, 23
   creating, 19–21
   functions, 22–23
   operations on, 21–22
augmented assignment operators, 7
augmented coefficient matrix, 28

backward finite difference approximations, 183
banded matrix, 56–66
`bisect`, 145–146
bisection method, for equation root, 145–148
`brent`, 150–151
Brent's method, 148–153
Bulirsch–Stoer algorithm, 286–290
Bulirsch–Stoer method, 278–279, 283
   algorithm, 286–290
   midpoint method, 283–284
   Richardson extrapolation, 284–286
`bulStoer`, 287
byte code, 1

`choleski(a)`, 47–48
Choleski's decomposition, 45–52
`cmath` module, 18–19
coefficient matrices, symmetric/banded, 56–66
   symmetric, 59–60
   symmetric/pentadiagonal, 61–66
   tridiagonal, 57–59
comparison operators, in Python, 8

composite Simpson's 1/3 rule, 204
composite trapezoidal rule, 200–201
conditionals, in Python, 9
`conjGrad`, 86, 90
conjugate gradient methods, 86, 88–96, 389–400
   conjugate directions, 390–391
   Fletcher–Reeves method, 396–400
   Powell's method, 391–396
continuation character, 6
`cubicSpline`, 117–119
cubic splines, 115–121, 195
curve fitting. *See* interpolation/curve fitting
cyclic tridiagonal equation, 94

deflation of polynomials, 172–173
diagonal dominance, 68
Doolittle's decomposition, 42–45
`dot`, 3

`eigenvals3`, 371–372
eigenvalue problems. *See* symmetric matrix eigenvalue problems
elementary operations, linear algebra, 31
embedded integration formula, 275
equivalent linear equation, 31
error
   control in Python, 15
   in finite difference approximations, 186–187
Euhler's method, stability of, 272
Euler–Maclaurin summation formula, 201
`evalPoly`, 172
evaluation of polynomials, 170–172
exponential functions, fitting, 130–131

false position method, roots of equations, 179
finite difference approximations, 181–187
   errors in, 186–187

finite difference approximations (*cont.*)
  first central difference approximations, 182–183
  first noncentral, 183–184
  second noncentral, 185
finite elements, 233
first central difference approximations, 182–183
first noncentral finite difference approximations, 183–184
`fletcherReeves`, 397–398
Fletcher–Reeves method, 396–400
forward finite difference approximations, 183
fourth-order differential equation, 316–320
fourth-order Runge–Kutta method, 257–265
functions, in Python, 15–16

`gaussElimin`, 37–38
Gauss elimination method, 33–41
  algorithm for, 35–38
    back substitution phase, 37
    elimination phase, 35–36
  back substitution phase, 35
  elimination phase, 34–35
  multiple sets of equations, 38–41
Gauss elimination with scaled row pivoting, 68–76
Gaussian integration, 215–230
  abscissas/weights for Guaussian quadratures, 221–230
    Gauss–Chebyshev quadrature, 222
    Gauss–Hermite quadrature, 223
    Gauss–Laguerre quadrature, 222–223
    Gauss–Legendre quadrature, 221–222
    Gauss quadrature with logarithmic singularity, 224
    determination of nodal abscissas/weights, 219–221
  formulas for, 215
  orthogonal polynomials, 217–219
Gauss–Jordan elimination, 32
Gauss–Legendre quadrature over quadrilateral element, 233–240
`gaussNodes`, 224–225
`gaussPivot`, 71–72
`gaussQuad`, 225–226
`gaussQuad2`, 235–237
`gaussSeidel`, 87–88
Gauss–Seidel method, 85–88
`gerschgorin`, 368–369
Gerschgorin's theorem, 368–369
golden section search, 383–389
`goldSearch`, 385–386

Higher-order equations, shooting method, 301–307
`householder`, 362–363
  householder reduction to tridiagonal form, 357–365
    accumulated transformation matrix, 361–365
    householder matrix, 358–359
    householder reduction of symmetric matrix, 359–361

Idle (code editor), 3
ill-conditioning, 29–30
incremental search method, roots of equations, 143–145
indirect methods. *See* iterative methods
initial value problems
  adaptive Runge–Kutta method, 275–283
  Bulirsch–Stoer method, 278–279, 283
    algorithm, 286–290
    midpoint method, 283–284
    Richardson extrapolation, 284–286
  introduction, 248–249
  multistep methods, 294
  problem set, 266–271, 290–294
  Runge–Kutta methods, 255–265
    fourth-order, 257–265
    second-order, 255–257
  stability/stiffness, 271–275
    stability of Euhler's method, 272
    stiffness, 272–275
  Taylor series method, 249–255
Input/output
  printing, 14–15
  reading, 13
integration order, 234
interpolation, derivatives by, 190–195
  cubic spline interpolant, 195
  polynomial interpolant, 190–191
interpolation/curve fitting
  interpolation with cubic spline, 115–121
  introduction, 103
  least-squares fit, 124–137
    fitting a straight line, 126
    fitting linear forms, 126–127
    polynomial fit, 127–129
    weighting of data, 129–137
      fitting exponential functions, 130–131
      weighted linear regression, 130
  polynomial interpolation, 104–115
    Lagrange's method, 104–105
    limits of, 110–115

Index

Neville's method, 108–110
Newton's method, 105–108
problem set, 121–124, 137–140
rational function interpolation, 140–141
interval halving method. *See* bisection method
inversePower, 345–346
inversePower3, 373–374
iterative methods, 85–96
    conjugate gradient method, 86, 88–96
    Gauss–Seidel method, 85–88

jacobi, 331
Jacobian matrix, 235
Jacobi method, 326–342
    Jacobi diagonalization, 328–333
    Jacobi rotation, 327–328
    similarity transformation/diagonalization, 326–327
    transformation to standard form, 333–342
Jenkins–Traub algorithm, 179–180

knots of spline, 115
Kronecker delta, 104

Lagrange's method, 104–105
Laguerre's method, 173–178
lamRange, 369–370
LAPACK (Linear Algebra PACKage), 27
least-squares fit, 124–137
    fitting linear forms, 126–127
    fitting straight line, 126
    polynomial fit, 127–129
    weighting data, 129–137
        fitting exponential functions, 130–131
        weighted linear regression, 130
linear algebraic equations systems. *See also* matrix algebra
    back substitution, 32
    direct methods overview, 31–33
    elementary operations, 31
    equivalent equations, 31
    forward substitution, 32
    Gauss elimination method, 33–41
        algorithm for, 35–38
            back substitution phase, 37
            elimination phase, 35–36
        back substitution phase, 35
        elimination phase, 34–35
        multiple sets of equations, 38–41
    ill-conditioning, 29–30
    introduction, 27–33

iterative methods, 85–96
    conjugate gradient method, 86, 88–96
    Gauss–Seidel method, 85–88
linear systems, 30
LU decomposition methods, 41–52
    Choleski's decomposition, 45–52
    Doolittle's decomposition, 42–45
matrix inversion, 82–84
methods of solution, 30–31
notation in, 27–28
pivoting, 66–76
    diagonal dominance, 68
    Gauss elimination with scaled row pivoting, 68–76
    when to pivot, 74–76
problem set, 53–56, 77, 86, 96–101
QR decomposition, 101
singular value decomposition, 101
symmetric/banded coefficient matrices, 56–66
    symmetric coefficient, 59–60
    symmetric/pentadiagonal coefficient, 61–66
    tridiagonal coefficient, 57–59
uniqueness of solution, 28–29
linear forms, fitting, 126–127
linear systems, 30
linInterp, 297
lists, 5–7
loops, 10–11
LR algorithm, 380
LUdecomp, 44
LUdecomp3, 59
LUdecomp5, 66
LU decomposition methods, 41–52
    Choleski's decomposition, 45–52
    Doolittle's decomposition, 42–45
LUpivot, 72–73

mathematical functions, 12–13
math module, 17–18
MATLAB, 2–3
matrix algebra, 412–417
    addition, 413
    determinant, 414–415
    example, 416–417
    inverse, 414
    multiplication, 413–414
    positive definiteness, 415
    transpose, 413
    useful theorems, 415
matrix inversion, 82–84
methods of feasible directions, 408

methods of solution, 30–31
`midpoint`, 285–286
minimization along line, 383–389
   bracketing, 383
   golden section search, 383–389
modules, in Python, 17
multiple integrals, 233
   Gauss–Legendre quadrature over quadrilateral element, 233–240
   quadrature over triangular element, 240–245
multistep methods, for initial value problems, 294

`NameError`, 23
Namespace, 23
natural cubic spline, 115
Nelder–Mead method, 407
`neville`, 109–110
Neville's method, 108–110
Newton–Cotes formulas, 199–207
   composite trapezoidal rule, 200–201
   recursive trapezoidal rule, 202–203
   Simpson's rules, 203–207
   trapezoidal rule, 200
`newtonPoly`, 107–108
`newtonRaphson`, 155–156
`newtonRaphson2`, 160–161
Newton–Raphson method, 153–158
norm of matrix, 29
notation, 27–28
Numarray module, 3
`numarry` module, 19–23, 7
   accessing/changing array, 21
   array functions, 22–23
   copying arrays, 23
   creating an array, 19–21
   operations on arrays, 21–22
numerical differentiation
   derivatives by interpolation, 190–195
     cubic spline interpolant, 195
     polynomial interpolant, 190–191
   finite difference approximations, 181–187
     errors in, 186–187
     first central difference approximations, 182–183
     first noncentral, 183–184
     second noncentral, 185
   introduction, 181
   problem set, 195–197
   Richardson extrapolation, 187–190
numerical instability, 243, 262

numerical integration
   Gaussian integration, 215–230
     abscissas/weights for Guaussian quadratures, 221–230
       Gauss–Chebyshev quadrature, 222
       Gauss–Hermite quadrature, 223
       Gauss–Laguerre quadrature, 222–223
       Gauss–Legendre quadrature, 221–222
       Gauss quadrature with logarithmic singularity, 224
     determination of nodal abscissas/weights, 219–221
     formulas for, 215
     orthogonal polynomials, 217–219
   introduction, 198–199, 217
   multiple integrals, 233
     Gauss–Legendre quadrature over quadrilateral element, 233–240
     quadrature over triangular element, 240–245
   Newton–Cotes formulas, 199–207
     composite trapezoidal rule, 200–201
     recursive trapezoidal rule, 202–203
     Simpson's rules, 203–207
     trapezoidal rule, 200
   problem set, 212–215, 230–233, 245–247
   Romberg integration, 207–212

operators
   arithmetic, 7
   comparison, 8
optimization
   conjugate gradient methods, 389–400
     conjugate directions, 390, 391
     Fletcher–Reeves method, 396–400
     Powell's method, 391–396
   introduction, 381–383
   minimization along line, 383–389
     bracketing, 383
     golden section search, 383–389
   Nelder–Mead method, 407
   problem set, 401–407
   simplex method, 407–408
   simulated annealing method, 407
orthogonal polynomials, 217–219
overrelaxation factor, 86, 96–101

piecewise cubic curve, 116
pivoting, 66–76
   diagonal dominance, 68
   Gauss elimination with scaled row pivoting, 68–76
   when to pivot, 74–76

Index

polyFit, 128–129
polynomial fit, 127–129
polynomial interpolant, 190–191
polynomial interpolation, 104–115
   Lagrange's method, 104–105
   limits of, 110–115
   Neville's method, 108–110
   Newton's method, 105–108
polynomials, zeroes of, 170–178
   deflation of polynomials, 172–173
   evaluation of polynomials, 170–172
   Laguerre's method, 173–178
polyRoots, 174–176
powell, 393–394
Powell's method, 391–396
Prandtl stress function, 235
printing input, 14–15
printSoln, 251–252
Python
   arithmetic operators, 7
   cmath module, 18–19
   comparison operators, 8
   conditionals, 9
   error control, 15
   functions, 15–16
   general information, 1–4
      obtaining Python, 3–4
      overview, 1–3
   lists, 5–7
   loops, 10–11
   mathematical functions, 12–13
   math module, 17–18
   modules, 17
   numarry module, 19–23
      accessing/changing array, 21
      array functions, 22–23
      copying arrays, 23
      creating an array, 19–21
      operations on arrays, 21–22
   printing input, 14–15
   reading input, 13
   scoping of variables, 23–24
   strings, 5
   tuples, 5
   type conversion, 11–12
   variables, 4
   writing/running programs, 24–26
Python interpreter, 1, 1

QR algorithm, 380
quadrature. *See* numerical integration
quadrature over triangular element, 240–245

rational function interpolation, 140–141
reading input, 13
recursive trapezoidal rule, 202–203
relaxation, 86
relaxation factor, 86, 88–96
Richardson extrapolation, 187–190, 284–286
romberg, 209–210
Romberg integration, 207–212
rootsearch, 144
roots of equations
   Brent's method, 148–153
   false position method, 179
   incremental search method, 143–145
   introduction, 142–143
   Jenkins–Traub algorithm, 179–180
   method of bisection, 145–148
   Newton–Raphson method, 153–158
   problem set, 164–169, 178–179
   secant method, 179
   systems of equations, 158–163
      Newton–Raphson method, 159–163
   zeroes of polynomials, 170–178
      deflation of polynomials, 172–173
      evaluation of polynomials, 170–172
      Laguerre's method, 173–178
row pivoting, 56, 67
Runge–Kutta–Fehlberg formulas, 275–276
Runge–Kutta methods, 255–265
   fourth-order, 257–265
   second-order, 255–257
run_kut4, 258–259
run_kut5, 278–279, 283

scaled row pivoting, 68–76
secent formula, 179
second forward finite difference approximations, 185
second noncentral finite difference approximations, 185
second-order differential equation, 296–301, 311–316
second-order Runge–Kutta method, 255–257
shape functions, 234
shooting method, 296–307
   higher-order equations, 301–307
   second-order differential equation, 296–301
Shur's factorization, 380
similarity transformation, 104
Simpson's 3/8 rule, 204
Simpson's rules, 203–207
slicing operator, 3
sortJacobi, 333
sparsely populated matrix, 56

stability/stiffness, 271–275
    stability of Euhler's method, 272
    stiffness, 272–275
`stdForm`, 335–336
stiffness, 272–275
straight line, fitting, 126
strings, 5
Strum sequence, 365–375
`sturmSeq`, 366–367, 375
`swapCols`, 70
`swapRows`, 70
symmetric/banded coefficient matrices, 56–66
    symmetric coefficient, 59–60
    symmetric/pentadiagonal coefficient, 61–66
    tridiagonal coefficient, 57–59
symmetric coefficient matrix, 59–60
symmetric matrix eigenvalue problems
    eigenvalues of symmetric tridiagonal matrices, 365–375
        bracketing eigenvalues, 369–371
        computation of eigenvalues, 371–372
        computation of eigenvectors, 373–375
        Gerschgorin's theorem, 368–369
        Strum sequence, 365–375
    householder reduction to tridiagonal form, 357–365
        accumulated transformation matrix, 361–365
        householder matrix, 358–359
        householder reduction of symmetric matrix, 359–361
    introduction, 324–326
    inverse power/power methods, 342–351
        eigenvalue shifting, 344
        inverse power method, 342–344
        power method, 345–351
    Jacobi method, 326–342
        Jacobi diagonalization, 328–333
        Jacobi rotation, 327–328
        similarity transformation/diagonalization, 326–327
        transformation to standard form, 333–342
    LR algorithm, 380
    problem set, 351–357, 375–380
    QR algorithm, 380
    Shur's factorization, 380

symmetric/pentadiagonal coefficient matrix, 61–66
synthetic division, 172–173
systems of equations
    Newton–Raphson method, 159–163
    roots of equations, 158–163

`taylor`, 250–251
Taylor series, 249–255, 409–412
    function of several variables, 410–412
    function of single variable, 409–410
transpose operator, 413
`trapezoid`, 202–203
trapezoidal rule, 200
`triangleQuad`, 242
tridiagonal coefficient matrix, 57–59
tuples, 5
two-point boundary value problems
    finite difference method, 310–320
        fourth-order differential equation, 316–320
        second-order differential equation, 311–316
    introduction, 295–296
    problem set, 307–310, 320–323
    shooting method, 296–307
        higher-order equations, 301–307
        second-order differential equation, 296–301
`type(a)`, 13
type conversion, 11–12

underrelaxation factor, 86, 90

variables
    Python, 4
    scoping, 23–24
    Taylor series, 409–410, 410–412

weighted linear regression, 130
writing/running programs, in Python, 24–26

`ZeroDivisionError`, 15
zeroes of polynomials, 170–178
    deflation of polynomials, 172–173
    evaluation of polynomials, 170–172
    Laguerre's method, 173–178
zero offset, 3